Advances in Industrial Control

Other titles published in this Series:

Nonlinear Identification and Control
Guoping Liu

Digital Controller Implementation and Fragility
Robert S.H. Istepanian and James F. Whidborne (Eds.)

Optimisation of Industrial Processes at Supervisory Level
Doris Sáez, Aldo Cipriano and Andrzej W. Ordys

Applied Predictive Control
Huang Sunan, Tan Kok Kiong and Lee Tong Heng

Hard Disk Drive Servo Systems
Ben M. Chen, Tong H. Lee and Venkatakrishnan Venkataramanan

Robust Control of Diesel Ship Propulsion
Nikolaos Xiros

Hydraulic Servo-systems
Moheiddine Jelali and Andreas Kroll

Model-based Fault Diagnosis in Dynamic Systems Using Identification Techniques
Silvio Simani, Cesare Fantuzzi and Ron J. Patton

Strategies for Feedback Linearisation: A Dynamic Neural Network Approach
Freddy Garces, Victor M. Becerra, Chandrasekhar Kambhampati and Kevin Warwick

Robust Autonomous Guidance
Alberto Isidori, Lorenzo Marconi and Andrea Serrani

Dynamic Modelling of Gas Turbines
Gennady G. Kulikov and Haydn A. Thompson (Eds.)

Control of Fuel Cell Power Systems
Jay T. Pukrushpan, Anna G. Stefanopoulou and Huei Peng

Fuzzy Logic, Identification and Predictive Control
Jairo Espinosa, Joos Vanderwalle and Vincent Wertz

Optimal Real-time Control of Sewer Networks
Magdalene Marinaki and Markos Papageorgiou

Computational Intelligence in Time Series Forecasting
Ajoy K. Palit and Dobrivoje Popovic

Modelling and Control of mini-Flying Machines
Pedro Castillo, Rogelio Lozano and Alejandro Dzul

Rudder and Fin Ship Roll Stabilization
Tristan Perez

Control of Passenger Traffic Systems in Buildings
Sandor Markon
Publication due January 2006

Nonlinear H_2/H_∞ Feedback control
Murad Abu-Khalaf, Frank L. Lewis and Jie Juang
Publication due January 2006

Benoît Codrons

Process Modelling for Control

A Unified Framework Using Standard Black-box Techniques

With 74 Figures

Springer

Benoît Codrons, Dr.Eng.
Laborelec, 125 Rue de Rhode, B-1630 Linkebeek, Belgium

British Library Cataloguing in Publication Data
Codrons, Benoît
Process modelling for control : a unified framework using standard black-box techniques. — (Advances in industrial control)
1. Process control — Mathematical models
I. Title
629.8′312

Advances in Industrial Control series ISSN 1430-9491
ISBN 978-1-84996-960-4 e-ISBN 978-1-84628-247-8
Springer Science+Business Media
springeronline.com

Printed in the United States of America

69/3830-543210 Printed on acid-free paper

Advances in Industrial Control

Professor Emeritus O.P. Malik
Department of Electrical and Computer Engineering
University of Calgary
2500, University Drive, NW
Calgary
Alberta
T2N 1N4
Canada

Professor K.-F. Man
Electronic Engineering Department
City University of Hong Kong
Tat Chee Avenue
Kowloon
Hong Kong

Professor G. Olsson
Department of Industrial Electrical Engineering and Automation
Lund Institute of Technology
Box 118
S-221 00 Lund
Sweden

Professor A. Ray
Pennsylvania State University
Department of Mechanical Engineering
0329 Reber Building
University Park
PA 16802
USA

Professor D.E. Seborg
Chemical Engineering
3335 Engineering II
University of California Santa Barbara
Santa Barbara
CA 93106
USA

Doctor I. Yamamoto
Technical Headquarters
Nagasaki Research & Development Center
Mitsubishi Heavy Industries Ltd
5-717-1, Fukahori-Machi
Nagasaki 851-0392
Japan

Series Editors' Foreword

The series *Advances in Industrial Control* aims to report and encourage technology transfer in control engineering. The rapid development of control technology has an impact on all areas of the control discipline. New theory, new controllers, actuators, sensors, new industrial processes, computer methods, new applications, new philosophies . . . , new challenges. Much of this development work resides in industrial reports, feasibility study papers and the reports of advanced collaborative projects. The series offers an opportunity for researchers to present an extended exposition of such new work in all aspects of industrial control for wider and rapid dissemination.

The standard framework for process modelling and control design has been: Determine nominal model – Validate model – Progress control design – Implement controller. Benoît Codrons brings new depth and insight to these standard steps in this new *Advances in Industrial Control* monograph *Process Modelling for Control*. His approach is to pose a searching set of questions for each component of the modelling – control cycle. The first two chapters set the scene for model identification including a short survey of transfer function system representation, prediction-error identification and balanced model truncation methods. This is followed by four investigative chapters each constructed around answering some questions very pertinent to the use of model identification for control design. As an example, look at Chapter 3 where the motivating question is simply: "How should we identify a model $(\hat{G}, \hat{H})$ in such a way that it be good for control design?" Dr. Codrons then presents a careful theoretical analysis and supporting examples to demonstrate that "it is generally preferable to identify a (possible low-order) model in *closed loop* when the purpose is to use the model for control design". This type of presentation is used to examine similar questions for the other stages in the model identification – control design cycle. The mix of theory and illustrative examples is used by Dr. Codrons to lead on to recommendations and sets of guidelines that the industrial control engineer will find invaluable.

This new volume in the *Advances in Industrial Control* series has thorough theoretical material that will appeal to the academic control community lecturers and postgraduate students alike. The rhetorical question-and-answer style along with the provision of recommendations and guidelines will give the industrial engineer and practitioner the motivation to consider and try new approaches in the model identification field in industrial control. Hence the volume is a very welcome addition to the *Advances in Industrial Control* monograph series which we hope will encourage the industrial control community to look anew at the current model identification paradigm.

M.J. Grimble and M.A. Johnson
Industrial Control Centre
Glasgow, Scotland, U.K.

Preface

The contents of this book are, for the most part, the result of the experience I gained during the five years I spent as a researcher at the CENTRE FOR SYSTEMS ENGINEERING AND APPLIED MECHANICS (CESAME), UNIVERSITÉ CATHOLIQUE DE LOUVAIN, Belgium, from September 1995 to June 2000. I am very grateful to Michel Gevers, Brian D.O. Anderson, Xavier Bombois and Franky De Bruyne for the good times we spent, the great ideas we had and the good work we did together. I also want to give credit to Pascale Bendotti and Clément-Marc Falinower of ÉLECTRICITÉ DE FRANCE for their support and collaboration in the elaboration of the case study in Chapter 6.

After completion of my Ph.D. thesis (Codrons, 2000), I landed in an industrial R&D laboratory, fully loaded with scientific knowledge but with almost no practical experience. This was the real beginning of the genesis of this book. As I have spent the last four years discovering the industrial reality and trying to put my theoretical knowledge into practice, I have also discovered the gap that often separates both worlds. For instance, when the issue is the realisation of identification tests on an industrial system in order to design or tune some controller, in a rather amusing way, both the scientist and the plant operator will recommend making the tests in closed loop. The latter, because it is often more comfortable for him and less risky for plant operation than open-loop tests; the former, for the reasons exposed in this book. In spite of this, however, the usual practice consists of opening the loop, applying a step on the input on the process, and measuring the output response from which essential values like the dominant time constant and the gain of the process can be determined. What is the problem, then? Who is right? Is it the scientist and his closed-loop... whim? Or is it the control or process engineer, with his rules of good practice or her pragmatic approach? To be fair, both are probably right. The open-loop approach is satisfactory most of the time when the objective is to tune PID controllers. *grosso modo*, those only require knowledge of the process time constant and gain for their tuning. However, this approach deprives the engineer of the necessary knowledge of the process dynamics that would help him to design more sophisticated control structures, especially in the case of coupled subsystems that would require a global handling with multivariable control solutions in lieu of monovariable but interacting (and mutually disturbing) PID loops.

The problem is even worse when the objective is to design a controller that has to be simultaneously *optimal* (with respect to some performance indicator, *e.g.*, the variance of the output or an LQG criterion), *robust* (with respect to modelling errors, disturbances, changes in operating conditions, *etc.*) and *implementable* in an industrial control system (which will generally put a constraint on its order and/or complexity). Most of the time, suboptimal solutions are worked out on the basis of, again, monovariable control loops and sometimes indecipherable logic programming aimed at managing the working status of the various PID controllers in function of the process conditions. This results, generally, from a lack of knowledge of the existing modelling and control design tools, or of the way they can be used in industrial practice.

Actually, one of the most critical and badly tuned loops I have been faced with is probably the learning loop between industry and the academic world. Without pretentiousness, my intention when starting to write of this book was to make some of the most recent results in process modelling for control available to the industrial community. The objective is twofold: firstly, it is to provide the control engineer with the necessary theory on modelling for control using some chosen specific black-box techniques in a linear, time-invariant framework (it sounds a reasonable first step before nonlinear techniques can be addressed); secondly, and this is perhaps the most important issue, it is to initiate a change is the way modelling and control problems are perceived and tackled in industrial practice. This is just a small part of what a feed-forward action from University to Industry might be. Also very important is the feedback from Industry to Academia that happens essentially through collaborations.

Benoît Codrons
Brussels, 1st September, 2004

Contents

List of Figures

List of Tables

Symbols and Abbreviations

Abbreviations

ARMAX	Auto-Regressive Moving-Average with eXogenous inputs
ARX	Auto-Regressive with eXogenous inputs
BJ	Box-Jenkins
EDF	Électricité de France
FIR	Finite Impulse Response
GPC	Generalised Predictive Control
LFT	Linear Fractional Transformation
LMI	Linear Matrix Inequality
LQG	Linear Quadratic Gaussian
LTI	Linear Time Invariant
MIMO	Multi-Input Multi-Output
MISO	Multi-Input Single-Output
MPC	Model-based Predictive Control
OE	Output Error
PI	Proportional Integral
PID	Proportional Integral Derivative
PRBS	Pseudo-Random Binary Signal
PWR	Pressurised Water Reactor
QFT	Quantitative Feedback Theory
SISO	Single-Input Single-Output
SNR	Signal-to-Noise Ratio
w.p.	with probability
w.r.t.	with respect to

Scalar and Vector Signals

$d(t)$	measured and/or unmeasured disturbances vector
$e(t)$	white noise signal
$f(t)$	controller output signal in the general closed-loop representation
$g(t)$	controller input signal in the general closed-loop representation
$h(t)$	controller input signal in an LFT
$l(t)$	controller output signal in an LFT
$p(t)$	step disturbance signal
$r(t)$	reference signal
$r_1(t)$	reference signal
$r_2(t)$	reference or feed-forward signal
$u(t)$	input signal
$v(t)$	stochastic disturbance signal
$w(t)$	exogenous input signal of an LFT
$x(t)$	state vector of a system
$y(t)$	output signal
$\hat{y}(t)$	simulated output using a model $\hat{G}$ of the plant
$\hat{y}(t \mid t-1, \theta)$	predicted output at time t using a model $\mathcal{M}(\theta)$ and based on past data Z^{t-1}
$y_p(t \mid \mathcal{M}(\hat{\theta}))$	predicted output using a model $\mathcal{M}(\hat{\theta})$
$y_s(t \mid \mathcal{M}(\hat{\theta}))$	simulated output using a model $\mathcal{M}(\hat{\theta})$
$z(t)$	output signal of an LFT
$\varepsilon(t)$, $\varepsilon(t, \theta)$, $\varepsilon(t \mid \mathcal{M}(\theta))$	prediction error, simulation error or residuals
$\varepsilon_F(t, \theta)$	filtered prediction error
$\psi(t, \theta)$	negative gradient of the prediction error: $\psi(t, \theta) = -\frac{d}{d\theta}\varepsilon(t, \theta)$

General Symbols

$A > 0$	the matrix A is positive definite
$A \geq 0$	the matrix A is positive semi-definite
a_{ij}	i-th row and j-th column entry of the matrix A
A_{ij}	i-th row and j-th column block entry (or submatrix) of a block (or partitioned) matrix A
$A_{m \times n}$	indicates that the matrix A has dimension $m \times n$
(A, B, C, D)	state-space realisation of a transfer function

$(\breve{A}, \breve{B}, \breve{C}, \breve{D})$	balanced or frequency-weighted balanced state-space realisation of a transfer function
ARMAX$[n_a, n_b, n_c, n_k]$	ARMAX model with numerator of order $n_b - 1$, denominator of order n_a, noise numerator of order n_c and delay n_k
ARX$[n_a, n_b, n_k]$	ARX model with numerator of order $n_b - 1$, denominator of order n_a and delay n_k
$AsN(m, P)$	$x \in AsN(m, P)$ means that the sequence of random variables x converges in distribution to the normal distribution with mean m and covariance matrix P.
$As\chi^2(n)$	$x \in As\chi^2(n)$ means that the sequence of random variables x converges in distribution to the χ^2 distribution with mean n degrees of freedom.
$b_{G,K}$	generalised stability margin
$\mathbf{C}$	controllability matrix
C, $C(z)$, $C(s)$	two-degree-of-freedom controller, $C = \begin{bmatrix} F & K \end{bmatrix}$ or $C = \begin{bmatrix} K & F \end{bmatrix}$
$\breve{C}$	two-degree-of-freedom controller, $\breve{C} = \begin{bmatrix} \breve{F} & \breve{K} \end{bmatrix}$
$\mathbb{C}$	space of complex numbers or complex plane
$\mathbb{C}_0$	$\mathbb{C} \setminus \{0\}$
$\mathbb{C}^+$	closed right half plane: $\mathbb{C}^+ = \{s \in \mathbb{C} \mid \mathrm{Re}(s) \geq 0\}$
$\mathbb{C}_0^+$	open right half plane: $\mathbb{C}_0^+ = \{s \in \mathbb{C} \mid \mathrm{Re}(s) > 0\}$
$\mathcal{C}_\gamma^\kappa$	controller set of chordal-distance size $\gamma(\omega)$
$\mathcal{C}_\beta^\nu$	controller set of ν-gap size β
$\complement_Y X$	complement of X in $Y \supseteq X$: $\complement_Y X = Y \setminus X$
$\mathcal{D}$	uncertainty region in transfer function space
$\mathbb{D}^-$	closed unit disc (enlarged domain of stability): $\mathbb{D}^- = \{z \in \mathbb{C} \mid \lvert z \rvert \leq 1\}$
$\mathbb{D}_1^-$	open unit disc (restricted domain of stability): $\mathbb{D}_1^- = \{z \in \mathbb{C} \mid \lvert z \rvert < 1\}$
$\mathbb{D}_1^+$	complement in $\mathbb{C}$ of the closed unit disc (restricted domain of instability): $\mathbb{D}_1^+ = \{z \in \mathbb{C} \mid \lvert z \rvert > 1\}$
$\partial\mathbb{D}$	unit circle: $\partial\mathbb{D} = \{z \in \mathbb{C} \mid \lvert z \rvert = 1\}$
$\vec{e}_k$	$\vec{e}_k \in \mathbb{R}^n$, $1 \leq k \leq n$, is the k-th orthonormal basis vector of $\mathbb{R}^n$. Its entries are all 0 except the k-th one which is 1
$\mathcal{E}$	uncertainty region around an estimated closed-loop transfer function
f_s	sampling frequency
F, $F(z)$, $F(s)$	either feed-forward part of a two-degree-of-freedom controller C or filter
$\breve{F}$	feed-forward part of a two-degree-of-freedom controller $\breve{C}$

F_{id}	feed-forward controller present in the loop during identification or model validation
G, $G(z)$, $G(s)$	system or model (input-output dynamics)
$G(z, \theta)$	parametrised plant model
G_n, $G_n(z)$, $G_n(s)$	either system of order n or system number n
$\breve{G}_n$	balanced or frequency-weighted balanced realisation of a system G_n
$\hat{G}$, $\breve{G}$	plant model (in case of a parametrised model, $\hat{G} \triangleq G(\hat{\theta})$)
$\tilde{G}$	augmented plant model
G_0, $G_0(z)$, $G_0(s)$	true system (input-output transfer function)
G_{nom}	nominal model
$\hat{G}_r$	approximation of order r of a system or model G_n of order $n > r$
$\mathcal{G}$	model set for the input-output dynamics
$\mathcal{G}_{K_{id}}$	set of all LTI systems that are stabilised by K_{id}
$\mathcal{G}^d_\beta$	directed-gap uncertainty set of size β
$\mathcal{G}^\nu_\beta$	ν-gap uncertainty set of size β
$\mathcal{G}^\kappa_\gamma$	chordal-distance uncertainty set of size $\gamma(\omega)$
$H(z, \theta)$	parametrised noise model
$\hat{H}$	noise model (in case of a parametrised model, $\hat{H} \triangleq H(\hat{\theta})$)
H_0, $H_0(z)$, $H_0(s)$	true noise process
$\mathcal{H}_2$	Hardy space: closed subspace of $\mathcal{L}_2$ with transfer functions (or matrices) $P(s)$ analytic in $\mathbb{C}_0^+$ (continuous case) or with transfer functions (or matrices) $P(z)$ analytic in $\mathbb{D}_1^+$ (discrete case)
$\mathcal{H}_\infty$	Hardy space: closed subspace of $\mathcal{L}_\infty$ with transfer functions (or matrices) $P(s)$ analytic in $\mathbb{C}_0^+$ (continuous case) or with transfer functions (or matrices) $P(z)$ analytic in $\mathbb{D}_1^+$ (discrete case)
I	identity matrix (of appropriate dimension)
j	$\sqrt{-1}$
$j\mathbb{R}$	set of imaginary numbers, imaginary axis
K, $K(z)$, $K(s)$	feedback controller or feedback part of a two-degree-of-freedom controller C
$\breve{K}$	feedback part of a two-degree-of-freedom controller $\breve{C}$
$\tilde{K}$	controller for an augmented plant model
K_{id}	feedback controller present in the loop during identification or model validation

$\mathcal{L}_2$	Hilbert space: set of transfer functions (or matrices) $P(s)$ (continuous case) or $P(z)$ (discrete case) such that the following integral is bounded: $\int_{-\infty}^{\infty} \text{trace}[P^\star(j\omega)P(j\omega)]d\omega < \infty$ (continuous case) or $\int_{-\pi}^{\pi} \text{trace}[P^\star(e^{j\omega})P(e^{j\omega})]d\omega < \infty$ (discrete case)
$\mathcal{L}_\infty$	Banach space: set of transfer functions (or matrices) $P(s)$ bounded on $j\mathbb{R}$ (continuous case), or set of transfer functions (or matrices) $P(z)$ bounded on $\partial\mathbb{D}$ (discrete case)
$\mathcal{M}$	model set for the input-output and noise dynamics
M, $M(z)$, $M(s)$	denominator of a right coprime factorisation of a system $G = NM^{-1}$; $M \in \mathcal{RH}_\infty$
$\tilde{M}$, $\tilde{M}(z)$, $\tilde{M}(s)$	denominator of a left coprime factorisation of a system $G = \tilde{M}^{-1}\tilde{N}$; $\tilde{M} \in \mathcal{RH}_\infty$
$\mathcal{M}(\theta)$	parametrised model $(G(\theta), H(\theta))$ in a set $\mathcal{M}$
N, $N(z)$, $N(s)$	numerator of a right coprime factorisation of a system $G = NM^{-1}$; $N \in \mathcal{RH}_\infty$
$\tilde{N}$, $\tilde{N}(z)$, $\tilde{N}(s)$	numerator of a left coprime factorisation of a system $G = \tilde{M}^{-1}\tilde{N}$; $\tilde{N} \in \mathcal{RH}_\infty$
$N(G, K)$	closed-loop transfer matrix (noise dynamics)
$N_i(G, K)$	closed-loop transfer matrix (noise dynamics)
$N_o(G, K)$	closed-loop transfer matrix (noise dynamics)
$N(m, P)$	$x \in N(m, P)$ means that the random variable x has a normal distribution with mean m and covariance matrix P.
$0_{m\times n}$	zero matrix of size $m \times n$
$\mathbf{O}$	observability matrix
OE$[n_b,n_f,n_k]$	OE model with numerator of order $n_b - 1$, denominator of order n_f and delay n_k
$\mathbf{P}$, $\mathbf{P}(t)$	controllability Gramian
P, $P(z)$, $P(s)$	generic notation for a transfer function (or transfer matrix) of any system or controller
P_θ	covariance matrix of the parameter vector θ
$\hat{P}_\theta$	estimated covariance matrix of the parameter vector θ
$\Pr(x \leq \alpha)$	probability that the random variable x is less than α
Q	generalised controller in an LFT
$\mathbf{Q}$, $\mathbf{Q}(t)$	observability Gramian
$\mathbb{R}$	space of real numbers
$\mathbb{R}_0$	$\mathbb{R} \setminus \{0\}$
$\mathbb{R}^+$	set of positive real numbers
$\mathbb{R}^n$	Euclidean space of dimension n
$\mathbb{R}^{n\times m}$	real matrix space of dimension $n \times m$

$\mathcal{RH}_\infty$	real rational subspace of $\mathcal{H}_\infty$: ring of proper and real rational stable transfer functions or matrices
$\mathcal{S}$	the true system (G_0, H_0)
$S(G, K)$	closed-loop sensitivity function: $S(G, K) = \frac{1}{1+GK}$
t	time (in seconds in the continuous case, or in multiples of t_s in the sampled case)
t_s	sampling period
$T(G, K)$	generalised closed-loop transfer matrix
$T_i(G, K)$	generalised closed-loop transfer matrix
$T_o(G, K)$	generalised closed-loop transfer matrix
$T_{zw}(\Gamma, Q)$	transfer function or matrix of an LFT
$\mathcal{U}$	uncertainty region in parameter space
$U, U(z), U(s)$	numerator of a right coprime factorisation of a controller $K = UV^{-1}$; $U \in \mathcal{RH}_\infty$
$\tilde{U}, \tilde{U}(z), \tilde{U}(s)$	numerator of a left coprime factorisation of a controller $K = \tilde{V}^{-1}\tilde{U}$; $\tilde{U} \in \mathcal{RH}_\infty$
$V, V(z), V(s)$	denominator of a right coprime factorisation of a controller $K = UV^{-1}$; $V \in \mathcal{RH}_\infty$
$\tilde{V}, \tilde{V}(z), \tilde{V}(s)$	denominator of a left coprime factorisation of a controller $K = \tilde{V}^{-1}\tilde{U}$; $\tilde{V} \in \mathcal{RH}_\infty$
W_{in}	input frequency-weighting filter
W_l	left-hand side frequency-weighting filter
W_{out}	output frequency-weighting filter
W_r	right-hand side frequency-weighting filter
X^{cont}	controllable state subspace of a given system G_n: $X^{cont} \subseteq \mathbb{R}^n$
X^{obs}	observable state subspace of a given system G_n: $X^{obs} \subseteq \mathbb{R}^n_0$
Z^N	data set $\{u(1), y(1), \ldots, u(N), y(N)\}$
$\chi^2(n)$	χ^2-distributed random variable with n degrees of freedom
$\chi^2_p(n)$	p-level of the χ^2-distribution with n degrees of freedom
ΔG	model error $\Delta G = G_0 - G_{nom}$
$\Delta(z)$	discrete-time differentiator
η	parameter vector $\eta \in \mathbb{R}^r$ of a reduced-order model obtained by $\mathcal{L}_2$ reduction
$\hat{\eta}$	estimate of η
η^*	asymptotic estimate of η
η_0	true value of η
Γ, Γ_0	generalised plant in an LFT
λ_0	variance of $e(t)$ (scalar case)

Λ_0	covariance matrix of $e(t)$ (vectorial case)
ω	frequency or normalised frequency $[rad/s]$
$\varphi(t)$	regression vector
$\phi_y(\omega)$	power spectral density of the signal $y(t)$
$\phi_y^x(\omega)$	power spectral density of that part of $y(t)$ originating from $x(t)$
$\phi_{yx}(\omega)$	cross-spectrum of $y(t)$ and $x(t)$
σ_x	standard deviation of the random variable x
ς_i	i-th Hankel singular value
Σ	diagonal observability and controllability Gramian of a balanced system $\breve{G}_n$: $\Sigma = \mathrm{diag}(\varsigma_1, \ldots, \varsigma_n) = \mathbf{P} = \mathbf{Q}$
θ	parameter vector $\theta \in D_\theta \subseteq \mathbb{R}^n$
$\hat{\theta}$	estimate of θ
θ^*	asymptotic estimate of θ
θ_0	true value of θ
ξ	parameter vector
$\hat{\xi}$	estimate of ξ
ξ_0	true value of ξ

Operators and Functions

A^T	transpose of the matrix A
$A^\star$	complex conjugate transpose of the matrix A
A^{-1}	inverse of the matrix A
$\arg\min_x f(x)$	minimising argument of $f(x)$
$\mathrm{bt}\,(P, r)$	r-th order system obtained by unweighted balanced truncation of a higher-order system P
$\mathrm{col}(A)$	column vector of length mn obtained by stacking the columns of the matrix $A_{m\times n}$
$\mathrm{col}(x, y, \ldots)$	column vector of length $n_x + n_y + \ldots$ obtained by stacking x, y, ..., which are column vectors of respective lengths n_x, n_y, ...
$\mathrm{cov}\, x$	covariance matrix of the random variable x
$\det A$	determinant of the matrix A
$\mathrm{diag}(a_1, \ldots, a_n)$	$n \times n$ diagonal matrix with entries $a_1 \ldots a_n$
$E\, x$	mathematical expectation of the random variable x
$\bar{E}\, f(t)$	$\lim_{N\to\infty} \frac{1}{N}\sum_{t=1}^{N} E\, f(t)$
$f'(x_0)$, $f''(x_0)$,...	$\frac{df(x)}{dx}\,\vert_{x=x_0}$, $\frac{d^2 f(x)}{dx^2}\,\vert_{x=x_0}$, ...
$\mathrm{Im}(x)$	imaginary part of $x \in \mathbb{C}$

$\operatorname{im} A$	range space (image) of the real matrix $A_{m \times n}$: $\operatorname{im} A = \{y \in \mathbb{R}^m \mid y = Ax,\, x \in \mathbb{R}^n\}$
$\ker A$	null space (kernel) of the real matrix $A_{m \times n}$: $\ker A = \{x \in \mathbb{R}^n \mid Ax = 0\}$
$\mathcal{F}_l\,(\Gamma,\, Q)$	lower LFT of Γ and Q
$\operatorname{fwbt}(P,\, W_l,\, W_r,\, r)$	r-th order system obtained by frequency-weighted balanced truncation of a higher-order system P with left and right-hand side filters W_l and W_r
s	differential operator: $s\, f(t) = \frac{\partial f(t)}{\partial t}$, $s^{-1} f(t) = \int f(t)\, dt$, or Laplace variable
$\operatorname{Re}(x)$	real part of $x \in \mathbb{C}$
$R_y(\tau)$	autocorrelation function of the signal $y(t)$
$R_{yx}(\tau)$	cross-correlation function of the signals $y(t)$ and $x(t)$
$\hat{R}_y^N(\tau)$	autocorrelation function of the signal $y(t)$ estimated from N samples
$\hat{R}_{yx}^N(\tau)$	cross-correlation function of the signals $y(t)$ and $x(t)$ estimated from N samples
$\operatorname{trace} A$	sum of the diagonal entries of the matrix A
$\operatorname{var} x$	variance of the random variable x
$\dot{x}(t)$	$\frac{\partial x(t)}{\partial t}$
$\operatorname{wno}(P(s))$	winding number of $P(s)$
z	time-shift operator: $z\, f(t) = f(t+1)$, $z^{-1} f(t) = f(t-1)$, or Z-transform variable
$\delta_\nu(G_1,\, G_2)$	ν-gap between two systems G_1 and G_2
$\vec{\delta}_g(G_1,\, G_2)$	directed gap between two systems G_1 and G_2
$\delta_{WC}(G,\, \mathcal{D})$	worst-case ν-gap between a system G and a set of systems $\mathcal{D}$
$\eta(P(s))$	number of poles of $P(s)$ in $\mathbb{C}_0^+$
$\eta_0(P(s))$	number of imaginary-axis poles of $P(s)$
$\kappa(G_1,\, G_2)$	chordal distance between two systems G_1 and G_2
$\kappa_{WC}(G,\, \mathcal{D})$	worst-case chordal distance between a system G and a set of systems $\mathcal{D}$
$\lambda_i(A)$	i-th eigenvalue of the matrix A
$\lambda_{max}(A)$	largest eigenvalue of the matrix A
$\mu(M_{\mathcal{D}}^K(e^{j\omega}))$	stability radius associated to the uncertainty region $\mathcal{D}$ and the controller K
$\bar{\sigma}(A)$	largest singular value of the (transfer) matrix A
$\underline{\sigma}(A)$	smallest singular value of the (transfer) matrix A
$\varsigma_i(P)$	i-th Hankel singular value of the transfer function (or matrix) P
$\bar{\varsigma}(P)$	largest Hankel singular value of the transfer function (or matrix) P: $\bar{\varsigma}(P) = \varsigma_1(P) = \\|P\\|_H$

$\|x\|$	absolute value of $x \in \mathbb{C}$
$\|P\|_H$	Hankel norm of P: $\|P\|_H = \bar{\varsigma}(P)$
$\|P\|_2$	2-norm of $P \in \mathcal{L}_2$: $\|P\|_2^2 = \frac{1}{2\pi}\int_{-\infty}^{\infty} \text{trace}[P^\star(j\omega)P(j\omega)]d\omega$ (continuous case) or $\|P\|_2^2 = \frac{1}{2\pi}\int_{-\pi}^{\pi} \text{trace}[P^\star(e^{j\omega})P(e^{j\omega})]d\omega$ (discrete case)
$\|P\|_\infty$	∞-norm of $P \in \mathcal{L}_\infty$: $\|P\|_\infty = \sup_{s\in j\mathbb{R}} \bar{\sigma}(P(s))$ (continuous case) or $\|P\|_\infty = \sup_{z\in\partial\mathbb{D}} \bar{\sigma}(P(z))$ (discrete case)
$\|x(t)\|_Q^2$	$x^T(t)Qx(t)$
$A \otimes B$	Kronecker product of the matrices A and B: $A_{m\times n} \otimes B_{p\times q} = C_{mp\times nq} = \begin{bmatrix} a_{11}B & \dots & a_{1n}B \\ \vdots & \ddots & \vdots \\ a_{m1}B & \dots & a_{mn}B \end{bmatrix}$

Criteria

$J_{clbt}(\hat{K}_r)$	closed-loop balanced truncation criterion
$J_{GPC}(u)$	GPC design criterion
$J_{LQG}(u)$	LQG design criterion
$J_{perf}(\hat{G}_r)$	closed-loop performance oriented model reduction criterion
$J_{perf}^{(0)}(\hat{G}_r)$	zeroth-order approximation of $J_{perf}(\hat{G}_r)$
$J_{perf}^{(1)}(\hat{G}_r)$	first-order approximation of $J_{perf}(\hat{G}_r)$
$J_{perf}(\hat{K}_r)$	closed-loop performance oriented controller reduction criterion
$J_{perf}^{(0)}(\hat{K}_r)$	zeroth-order approximation of $J_{perf}(\hat{K}_r)$
$J_{perf}^{(1)}(\hat{K}_r)$	first-order approximation of $J_{perf}(\hat{K}_r)$
$J_s(\mathcal{M}(\hat{\theta}))$, $J_p(\mathcal{M}(\hat{\theta}))$	model fit indicators
$J_{stab}(\hat{G}_r)$	closed-loop stability oriented model reduction criterion
$J_{WC}(\mathcal{D}, K, \omega)$	worst-case performance of the controller K over all systems in the uncertainty set $\mathcal{D}$
$R_s(\mathcal{M}(\hat{\theta}))$, $R_p(\mathcal{M}(\hat{\theta}))$	normalised model fit indicators
$\bar{V}(\theta)$	asymptotic identification criterion
$V_N(\theta, Z^N)$	identification criterion

CHAPTER 1

Introduction

1.1 An Overview of the Recent History of Process Control in Industry

Leading industry has always required tools for increasing production rate and/or product quality while keeping the costs as low as possible. Doubtless, one of these tools is automatic process control.

In the early days, the control engineer essentially used their knowledge of the process and their understanding of the underlying physics to design very simple controllers, such as PID controllers. Analysis methods were combined with an empirical approach of the problem to design controllers with an acceptable level of performance.

During the last decades, with the increasing competition on the markets, more powerful control techniques have been developed and used in various industrial sectors in order to increase their productivity. A typical example is that of Model-based Predictive Control (MPC) in the petrochemy, where improving the pureness of some products by a few percent can yield very important profits. An interesting aspect of MPC is that it was born more than 30 years ago and developed in the industrial world, based on a very pragmatic approach of what optimal multivariable industrial control could be. It only began to arouse the interest of the scientific community much later.

More recently, advances in numerical computation and in digital electronics have allowed process control to permeate everybody's life: it is present in your kitchen, in your car, in your CD player, *etc.*

Industrial sectors that have until now been very traditional in their approach of process control, like the electricity industry, are also beginning to pay interest to optimal control techniques in order to remain competitive in the newly deregulated energy market.

Nearly all modern optimal control design techniques rely on the use of a model of the system to be controlled. As the performance specifications became more stringent with the advent of new technologies, the need for precise models from which complex controllers could be designed became a major issue, resulting in the theory of system identification. The theorists of system identification quickly oriented their research towards the computation of the 'best' estimate of a system, from which an optimal controller could be designed using the so-called *certainty equivalence principle*: the controller was designed as if the model was an exact representation of the system.

However, as good as it can be, a model is never perfect, with the result that a controller designed to achieve some performance with this model may fail to meet the minimum requirements when applied to the true system. Robust control design is an answer to this problem. It uses bounds on the modelling error to design controllers with guaranteed stability and/or performance. Such bounds are generally defined using some prior knowledge of the physical system (*e.g.*, an *a priori* description of the noise process acting on the system such as a hard bound on its magnitude, or confidence intervals around physical parameters of the system). As the identification theorists did not spend much effort trying to produce such bounds in complement to their models, a huge gap appeared, at the end of the 1980s, between robust control and system identification.

During the last twenty years, many efforts have been accomplished in order to bridge this gap and a new sub-discipline emerged, which has been called *identification for control*. However, the earlier works on control-oriented identification only aimed at producing uncertainty descriptions that were compatible with those required for doing robust control design, without paying much attention to the interaction between identification and model-based control design and to the production of uncertainty descriptions that would be ideally tuned towards control design (*i.e.*, allowing the computation of high-performance controllers). It later became clear that a major issue is the interplay between identification and control and that identification for control should be seen as a design problem. This was highlighted in (Gevers, 1991, 1993).

1.2 Where are We Today?

During the 1990s, identification for control became a field of tremendous research activity. Some notable results were produced, all concerning the tuning of the bias and variance errors occurring in prediction-error identification in a way that would be suited to control design.

The fact that closed-loop identification could be helpful to tune the variance distribution of a model aimed at designing a controller for the true system was shown for the first time in (Gevers and Ljung, 1986) for the case of an unbiased model and a minimum-variance control law. On the other hand, the publication of (Bitmead *et al.*, 1990) has initiated a line of research consisting of producing models with a bias error that is small at frequencies that are critical with respect to closed-loop stability or performance. A major result was to show that closed-loop identification with appropriate data filters is generally required to deliver such models.

On the basis of the ideas expressed in these two publications, much research has been done in order to combine the identification and control design in a mutually supportive way. These efforts have resulted in iterative schemes where steps of closed-loop identification (with the latest designed controller operating in the loop during data collection) alternate with steps of control design (using the latest identified model), the identification criterion matching the control design criterion; see, *e.g.*, (Åström, 1993), (Liu and Skelton, 1990), (Schrama, 1992*a*, 1992*b*) and (Zang *et al.*, 1991, 1995). Other very interesting recent results on the interplay between identification and control can be found in (Hjalmarsson *et al.*, 1994*a*, 1996), (Van den Hof and Schrama, 1995), (De Bruyne, 1996) and references therein.

The very end of the previous century saw the emergence of another line of research in the field of *model validation.* Until recently, model validation (in the case of prediction-error identification) essentially consisted in computing some statistics about the residuals (*i.e.*, the simulation or prediction errors) of the model. However, as stressed in (Ljung and Guo, 1997) and (Ljung, 1997, 1998), there is more information in these residuals than used by classical validation tests. In particular, the residuals can be used to estimate the modelling error and to build a frequency-domain confidence region guaranteed to contain the true system with some specified probability. The claimed advantage of this approach is that such a confidence region can be used for robust control design. Another important issue revealed later in (Gevers *et al.*, 1999) is that the shape of this confidence region depends highly on the way the identification is carried out and that an appropriate choice of the experimental conditions leads to uncertainty regions that are better tuned to design a robust controller. We call this the interplay between identification and control and it will be the thread of this book.

1.3 The Contribution of this Book

Simply said, the contribution of this book to the literature is essentially to give an answer to the following question:

> *As a practitioner, how should I carry out identification experiments on my process in order to obtain a model (and possibly an uncertainty region attached to it) that be ideal for control design?*

Clearly, our intention when writing this book was not to provide the reader with yet another book on optimal or robust control design techniques. In fact, the way a controller can be designed from a model will receive little attention in this book. It is assumed that the reader has already got a significant background in this field. Many very good reference works exist; see, *e.g.*, (Åström and Hägglund, 1995) for an overview of PID control, (Goodwin *et al.*, 2001) for the fundamentals of SISO and MIMO control, (Åström and Wittenmark, 1995; Bitmead *et al.*, 1990) for adaptive control, (Skogestad and Postlethwaite, 1996; Glad and Ljung, 2000) for multivariable and nonlinear control, (Anderson and Moore, 1990) for optimal linear quadratic control, (Zhou and Doyle, 1998) for robust control and (Bitmead *et al.*, 1990; Maciejowski, 2002) for predictive control. We strongly encourage the reader to consult these references to extend their knowledge in this matter.

Our decision to write the present monograph was motivated by the lack of information contained in most books on process control about the modelling step that precedes the design of a controller. Most of the time, they simply assume that a model is already available (and possibly some uncertainty region attached to it as well), which could be explained by the fact that, as said before, the control and modelling communities of researchers have for a long time been living quite apart. This book aims to bridge this gap and to present the most recent results in *modelling for control* in a comprehensive way for the practitioner. Rather than on the modelling tools themselves, the focus is on the way they should be used in (industrial) practice. Our ultimate objective is to offer the control community a unified set of rules of good practice for the modelling step inherent in the design of most controllers. It will turn out that the principles of these rules are common to most techniques used for system modelling (prediction-error identification, reduction of high-order models, *etc.*) or control design (LQG, $\mathcal{H}_\infty$, *etc.*), even though we restricted ourselves in this book essentially to two well-known black-box techniques, namely prediction-error system identification and model order reduction *via* balanced truncation. As a result, techniques like frequency-domain identification, modelling of systems with hard uncertainty bounds or prior knowledge on the disturbances, *etc.*, are not addressed.

For the sake of simplicity, the mathematical developments concerning prediction-error identification and validation are presented for the SISO case. Most of them remain true and can be extended in a straightforward way to the multivariable case. Yet, some controller robustness verification (or quantification) tools proposed in this book cannot be extended as such to the MIMO case because they rely on some algebraic properties of system transfer functions that do not hold in this case. The underlying heuristics and rules of good practice for the design of the modelling experiment remain true however, and we can expect that MIMO robustness verification tools, based on prediction-error methods, will be found in the future.

1.4 Synopsis

This monograph is organised as follows.

Chapter 2: Preliminary Material. This chapter reviews the modelling and analysis tools that are used in this book, namely prediction-error identification, balanced truncation, coprime factorisations and the ν-gap metric. It also contains a description of the closed-loop set-up that is considered throughout the book, as well as the definitions of closed-loop stability and of generalised stability margin.

Chapter 3: Identification in Closed Loop for Better Control Design. This chapter shows how performing the identification of a system in closed loop usually leads to a better model, *i.e.*, to better tuned modelling errors, when the objective is to use the model for control design.

Chapter 4: Dealing with Controller Singularities in Closed-loop Identification. It often happens that the controller used during identification is unstable (*e.g.*, a PID controller). Sometimes it may also have a nonminimum phase (for instance, this can happen with state-feedback controllers containing an observer). This chapter shows that, in this case, closed-loop stability of the resulting model (when connected to the same controller) cannot always be guaranteed (sometimes instability is even guaranteed) if precautions are not taken regarding the identification method and the excitation source. Furthermore, it is shown that a model that is destabilised by the controller used during identification is not good for control design. Guidelines are given to avoid this problem.

Chapter 5: Model Validation for Control and Controller Validation. This chapter explains how closed-loop data can be used to compute a parametrised set of transfer functions that is guaranteed to contain the true system with some probability and that is better tuned towards robust control design than a similar set that would be computed from open-loop data. Such a set can be used for the validation of a given controller, *i.e.*, to assess closed-loop robust stability and performance margins before the controller is implemented on the real system. Two realistic case studies are presented at the end of the chapter: a flexible transmission system (resonant mechanical system subject to step disturbances) and a ferrosilicon production process (chemical process subject to stochastic disturbances).

Chapter 6: Control-oriented Model Reduction and Controller Reduction. The same reasoning that lead to the choice of closed-loop identification and validation methods in the previous chapters, when the goal is to obtain a model and an uncertainty region around it that be ideally shaped for control design, can be extended to the case of model reduction, when the goal is to preserve the states of an initial high-order model that are the most important for control design while discarding the others. Because controller order reduction is just the dual problem, it turns out that it should also be carried out in closed loop in order to preserve the stability and the performance of the initial closed-loop system. An original criterion based on this reasoning is proposed for both model and controller reduction. Other criteria and methods are also reviewed. The chapter ends with a case study illustrating the design of a low-order controller for a complex Pressurised Water Reactor nuclear power plant model.

Chapter 7: Some Final Words. This chapter shows how the various tools proposed in this book can be used in a common framework for *control-oriented modelling*, in such a way as to maximise the chance of obtaining a model-based controller that will stabilise the real process and achieve a pre-specified level of closed-loop performance. Perspectives of improvement of the whole scheme are also presented. The book ends with a few words about Iterative Feedback Tuning, a model-free control design method that has already been used successfully in industry.

CHAPTER 2

Preliminary Material

2.1 Introduction

This chapter describes the modelling and analysis tools that will be used throughout the book. They all concern Linear Time Invariant systems or transfer functions and can be classified in three categories.

Closed-loop system representation and analysis: We briefly define two possible representations of a system G_0 in closed loop with some stabilising controller K. The first one is a general representation based on a standard block-diagram description of the closed loop (Section 2.2) and on which we shall base the notions of closed-loop stability and generalised stability margin. The second one is based on a linear fractional transformation (LFT) (Section 2.3), which is sometimes more convenient for the manipulation of multivariable closed-loop systems or for robust control design and analysis. We show, in Section 2.3, that any closed-loop system represented by means of an LFT can be put in the general form defined in Section 2.2, allowing use of standard stability analysis tools.

Linear systems analysis: Two important tools are presented. The first one is coprime factorisation (Section 2.4). A coprime factorisation is a way of representing a possibly unstable transfer function or matrix by two stable ones (a numerator and a denominator) with particular properties related, *e.g.*, to closed-loop stability. Procedures are given to build such (possibly normalised) factorisations. It is also shown how the generalised closed-loop transfer matrix of a system can be expressed in function of the plant and controller coprime factors. The second tool is the ν-gap metric between two transfer functions (Section 2.5). It is a control-oriented measure of the distance between two transfer functions or matrices, of great importance for robustness analysis.

System modelling: The two classical black-box modelling tools that we consider in this book are prediction-error identification and balanced truncation.

The first one, described in Section 2.6, uses data measured on the actual plant to compute the best model of this plant in some set of parametrised transfer functions, with respect to a criterion that penalises the prediction errors attached to this model. The second one, described in Section 2.7, is a tool aimed to reduce the order of a given linear high-order model of the plant (obtained, for instance, by identification, first-principles-based physical modelling, finite-element modelling, or linearisation of a nonlinear simulator) to derive a lower order model, or of any high-order transfer function, by discarding the least controllable and observable modes. It will be used to design low-order controllers for high-order processes.

2.2 General Representation of a Closed-loop System and Closed-loop Stability

2.2.1 General Closed-loop Set-up

Let us consider a (possibly multi-input multi-output) linear time-invariant system described by

$$y(t) = G_0(z)u(t) + v(t) \quad \text{or} \quad y(t) = G_0(s)u(t) + v(t) \tag{2.1}$$

where $G_0(z)$ (discrete-time case) or $G_0(s)$ (continuous-time case) is a rational transfer function or matrix. Here, z^{-1} is the backward shift operator[1] ($z^{-1}x(t) = x(t-1)$ with t expressed in multiples of the sampling period) and s is the time differentiation operator ($sx(t) = \dot{x}(t)$). $u(t)$ is the input of the system, $y(t)$ its output, $v(t)$ an output disturbance.

We shall often consider the representation of Figure 2.1 when the plant G_0 operates in closed loop with a controller K. In this representation, $r_1(t)$ and $r_2(t)$ are two possible sources of exogenous signals (typically, $r_1(t)$ will be a reference or set-point signal for $y(t)$, while $r_2(t)$ will be either a feed-forward control signal or an input disturbance). $g(t)$ and $f(t)$ denote respectively the input and the output of the controller.

[1]The time-domain backward shift operator used in discrete-time systems is often represented by q^{-1} in the literature, while the notation z is generally used for the corresponding frequency-domain Z-transform variable. Here, for the sake of simplicity and although mathematical rigour would require such distinction, we shall use the same notation for both the operator and the variable. Similarly, the time differentiation operator used in continuous-time systems is often represented by p, but we shall make no distinction between it and the frequency-domain Laplace-transform variable s.

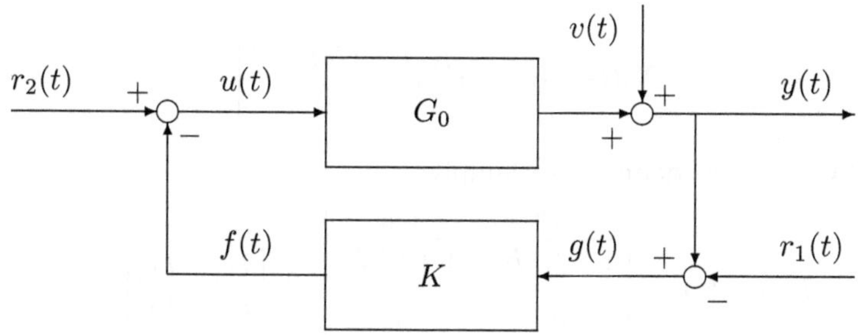

Figure 2.1. General representation of a system in closed loop

2.2.2 Closed-loop Transfer Functions and Stability

In mainstream robust control, the following generalised closed-loop transfer matrices are often considered[2]:

$$T_i(G_0, K) = \begin{bmatrix} -I & G_0 \\ K & I \end{bmatrix}^{-1} + \begin{bmatrix} I & 0 \\ 0 & 0 \end{bmatrix}$$
$$= \begin{bmatrix} G_0(I + KG_0)^{-1}K & G_0(I + KG_0)^{-1} \\ (I + KG_0)^{-1}K & (I + KG_0)^{-1} \end{bmatrix} \quad (2.2)$$

and

$$T_o(G_0, K) = \begin{bmatrix} -I & K \\ G_0 & I \end{bmatrix}^{-1} + \begin{bmatrix} I & 0 \\ 0 & 0 \end{bmatrix}$$
$$= \begin{bmatrix} K(I + G_0K)^{-1}G_0 & K(I + G_0K)^{-1} \\ (I + G_0K)^{-1}G_0 & (I + G_0K)^{-1} \end{bmatrix} \quad (2.3)$$

The entries of $T_i(G_0, K)$ are the transfer functions between the exogenous reference signals and the input and output signals of the plant defined in Figure 2.1:

$$\begin{bmatrix} y(t) \\ u(t) \end{bmatrix} = T_i(G_0, K) \begin{bmatrix} r_1(t) \\ r_2(t) \end{bmatrix} + N_i(G_0, K)v(t) \quad (2.4)$$

where

$$N_i(G_0, K) = \begin{bmatrix} (I + G_0K)^{-1} \\ -(I + KG_0)^{-1}K \end{bmatrix} \quad (2.5)$$

(the entry $(I + G_0K)^{-1}$ is called the *closed-loop sensitivity function* of the system), while those of $T_o(G_0, K)$ are the transfer functions between the exogenous reference signals and the output and input signals of the controller:

$$\begin{bmatrix} f(t) \\ g(t) \end{bmatrix} = T_o(G_0, K) \begin{bmatrix} r_2(t) \\ -r_1(t) \end{bmatrix} + N_o(G_0, K)v(t) \quad (2.6)$$

[2] All transfer functions and matrices must be understood as rational functions of z (discrete-time case) or s (continuous-time case). However, when no confusion is possible, we shall often omit these symbols to ease the notations.

where

$$N_o(G_0, K) = \begin{bmatrix} K(I+G_0K)^{-1} \\ (I+G_0K)^{-1} \end{bmatrix} \tag{2.7}$$

In the SISO case, we define more simply

$$T(G_0, K) = \begin{bmatrix} \dfrac{G_0K}{1+G_0K} & \dfrac{G_0}{1+G_0K} \\ \dfrac{K}{1+G_0K} & \dfrac{1}{1+G_0K} \end{bmatrix} \triangleq \begin{bmatrix} T_{11} & T_{12} \\ T_{21} & T_{22} \end{bmatrix} \tag{2.8}$$

and

$$N(G_0, K) = \begin{bmatrix} \dfrac{1}{1+G_0K} \\ \dfrac{-K}{1+G_0K} \end{bmatrix} = \begin{bmatrix} T_{22} \\ -T_{21} \end{bmatrix} \triangleq \begin{bmatrix} S \\ -KS \end{bmatrix} \triangleq \begin{bmatrix} N_1 \\ N_2 \end{bmatrix} \tag{2.9}$$

so that

$$\begin{bmatrix} y(t) \\ u(t) \end{bmatrix} = T(G_0, K) \begin{bmatrix} r_1(t) \\ r_2(t) \end{bmatrix} + N(G_0, K)v(t) \tag{2.10}$$

Definition 2.1. (INTERNAL STABILITY) *The closed loop (G_0, K) of Figure 2.1 is called 'internally stable' if all four entries of $T_i(G_0, K)$ or, equivalently, all four entries of $T_o(G_0, K)$, are stable, i.e., if they belong to $\mathcal{H}_\infty$.*

The generalised stability margin $b_{G_0,K}$ is an important measure of the internal stability of the closed loop. It is defined as

$$b_{G_0,K} \triangleq \begin{cases} \|T_i(G_0, K)\|_\infty^{-1} & \text{if } (G_0, K) \text{ is stable} \\ 0 & \text{otherwise} \end{cases} \tag{2.11}$$

Note that $\|T_i(G_0, K)\|_\infty = \|T_o(G_0, K)\|_\infty$, as shown in (Georgiou and Smith, 1990). An alternative definition, in the SISO case, is the following:

$$b_{G_0,K} = \min_\omega \kappa \left(G_0(e^{j\omega}), \frac{-1}{K(e^{j\omega})} \right) \tag{2.12}$$

where $\kappa\left(G_0(e^{j\omega}), \frac{-1}{K(e^{j\omega})}\right)$ is the chordal distance at frequency ω between G_0 and $\frac{-1}{K(e^{j\omega})}$, as defined in (Vinnicombe, 1993*a*): see Section 2.5.

The margin $b_{G_0,K}$ plays an important role in robust optimal control design. The following results hold in the SISO case (Vinnicombe, 1993*b*):

$$\text{gain margin} \geq \frac{1+b_{G_0,K}}{1-b_{G_0,K}} \tag{2.13}$$

and

$$\text{phase margin} \geq 2 \arcsin\left(b_{G_0,K}\right) \tag{2.14}$$

Note that there is a maximum attainable value of the generalised stability margin $b_{G_0,K}$ over all controllers stabilising G_0 (Vinnicombe, 1993*b*):

$$\sup_K b_{G_0,K} = \sqrt{1 - \lambda_{max}(\mathbf{P}_{rcf}\mathbf{Q}_{rcf})} = \sqrt{1 - \left\| \begin{bmatrix} N_0 \\ M_0 \end{bmatrix} \right\|_H^2} \tag{2.15}$$

and (Georgiou and Smith, 1990)

$$\sup_K b_{G_0,K} \leq \inf_{\substack{z \in \mathbb{D}_1^+ \\ \text{or} \\ s \in \mathbb{C}_0^+}} \underline{\sigma}\left(\begin{bmatrix} N_0 \\ M_0 \end{bmatrix} \right) \tag{2.16}$$

Here, $\begin{bmatrix} N_0 \\ M_0 \end{bmatrix}$ is a normalised right coprime factorisation of G_0 (see Section 2.4). $\mathbf{P}_{rcf}$ and $\mathbf{Q}_{rcf}$ are, respectively, the controllability and observability Gramians (see Section 2.7) of the right coprime factors $\begin{bmatrix} N_0 \\ M_0 \end{bmatrix}$. $\|\cdot\|_H$ denotes the Hankel norm. It should be clear that it is easier to design a stabilising controller for a system with a large $\sup_K b_{G_0,K}$ than for a system with a small one.

We refer the reader to (Zhou and Doyle, 1998) and references therein for more detail about the links between the generalised stability margin and controller performance.

The following proposition will be used several times throughout this book.

Proposition 2.1. (Anderson *et al.*, 1998) *Let* (G_0, K) *and* (G, K) *be two stable closed-loop systems such that*

$$\|T_i(G, K) - T_i(G_0, K)\|_\infty < \varepsilon \tag{2.17a}$$

or, equivalently,

$$\|T_o(G, K) - T_o(G_0, K)\|_\infty < \varepsilon \tag{2.17b}$$

for some $\varepsilon > 0$. *Then,*

$$|b_{G,K} - b_{G_0,K}| < \varepsilon \tag{2.18}$$

It tells us that the stability margin achieved by a controller K connected to a system G will be close to that achieved by the same K with G_0 if the closed-loop transfer matrices $T_i(G, K)$ and $T_i(G_0, K)$ are close in the $\mathcal{H}_\infty$ norm.

2.2.3 Some Useful Algebra for the Manipulation of Transfer Matrices

The following relations are very useful when it comes to manipulating closed-loop transfer matrices.

A. Block-matrix inversion.

$$\begin{bmatrix} A & B \\ C & D \end{bmatrix}^{-1}$$
$$= \begin{bmatrix} A^{-1} + A^{-1}B(D-CA^{-1}B)^{-1}CA^{-1} & -A^{-1}B(D-CA^{-1}B)^{-1} \\ -(D-CA^{-1}B)^{-1}CA^{-1} & (D-CA^{-1}B)^{-1} \end{bmatrix} \quad (2.19)$$

provided A and D are square, and A and $(D - CA^{-1}B)$ are invertible.

B. Other formulae.

$$A(I+BA)^{-1} = (I+AB)^{-1}A \quad (2.20a)$$
$$BA(I+BA)^{-1} = (I+BA)^{-1}BA$$
$$= B(I+AB)^{-1}A$$
$$= I-(I+BA)^{-1} \quad (2.20b)$$

2.3 LFT-based Representation of a Closed-loop System

In the MIMO case, it is often easier to represent a closed-loop system using Linear Fractional Transformations, as depicted in Figure 2.2, where Γ_0 is called the generalised plant and Q the generalised controller.

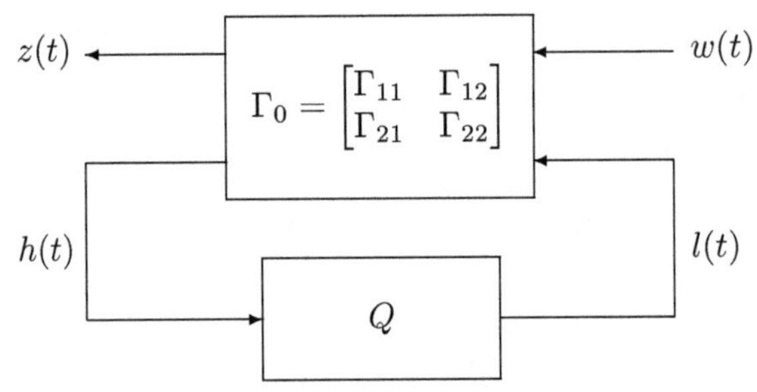

Figure 2.2. LFT representation of a system in closed loop

In this representation, all exogenous signals are contained in $w(t)$. $l(t)$ is the control signal and $h(t)$ is the input of the controller Q, *e.g.*, $l(t) = f(t)$ and $h(t) = g(t)$ if $Q = K$. $z(t)$ contains all inner signals that are useful for the considered application. For instance, if the objective is to design a control law for the tracking of the reference signal $r_1(t)$, $z(t)$ will typically contain the tracking error signal $g(t) = y(t) - r_1(t)$ and, possibly, the control signal $f(t)$ if the control energy is penalised by the control law.

The closed-loop transfer function between $w(t)$ and $z(t)$ is given by

$$T_{zw} = \mathcal{F}_l(\Gamma_0, Q) \triangleq \Gamma_{11} + \Gamma_{12}Q(I - \Gamma_{22}Q)^{-1}\Gamma_{21} \quad (2.21)$$

This representation can easily be transposed into the standard one of Figure 2.1 as we now show by means of three examples.

▶ **Example 2.1.** A single-degree-of-freedom controller K is used, and the signals of interest are the tracking error $g(t)$ and the control signal $f(t)$. Let us set

$$
\begin{aligned}
&Q = K \\
&z(t) = \mathrm{col}\big(f(t),\, g(t)\big) && w(t) = \mathrm{col}\big(r_2(t),\, -r_1(t),\, v(t)\big) \\
&h(t) = g(t) && l(t) = f(t) \\
&\Gamma_{11} = \begin{bmatrix} 0 & 0 & 0 \\ G_0 & I & I \end{bmatrix} && \Gamma_{12} = \begin{bmatrix} I \\ -G_0 \end{bmatrix} \\
&\Gamma_{21} = \begin{bmatrix} G_0 & I & I \end{bmatrix} && \Gamma_{22} = \begin{bmatrix} -G_0 \end{bmatrix}
\end{aligned}
$$

in Figure 2.2. Then,

$$T_{zw}(\Gamma_0,\, Q) = \begin{bmatrix} T_o(G_0,\, K) & N_o(G_0,\, K) \end{bmatrix}$$

◀

▶ **Example 2.2.** A single-degree-of-freedom controller K is used, and the signals of interest are the output $y(t)$ and the control signal $u(t)$. Let us consider Figure 2.2 and set

$$
\begin{aligned}
&Q = K \\
&z(t) = \mathrm{col}\big(y(t),\, u(t)\big) && w(t) = \mathrm{col}\big(r_1(t),\, r_2(t),\, v(t)\big) \\
&h(t) = g(t) && l(t) = f(t) \\
&\Gamma_{11} = \begin{bmatrix} 0 & G_0 & I \\ 0 & I & 0 \end{bmatrix} && \Gamma_{12} = \begin{bmatrix} -G_0 \\ -I \end{bmatrix} \\
&\Gamma_{21} = \begin{bmatrix} -I & G_0 & I \end{bmatrix} && \Gamma_{22} = \begin{bmatrix} -G_0 \end{bmatrix}
\end{aligned}
$$

In this case,

$$T_{zw}(\Gamma_0,\, Q) = \begin{bmatrix} T_i(G_0,\, K) & N_i(G_0,\, K) \end{bmatrix}$$

◀

▶ **Example 2.3.** A two-degree-of-freedom controller $C = \begin{bmatrix} K & F \end{bmatrix}$ is used and the signals of interest are the output $y(t)$ and the control signal $u(t)$. The control law is

$$u(t) = Fr_2(t) - Kg(t)$$

If we set

$$
\begin{aligned}
Q &= C = \begin{bmatrix} K & F \end{bmatrix} \\
z(t) &= \operatorname{col}\big(y(t),\, u(t)\big) & w(t) &= \operatorname{col}\big(v(t),\, r_1(t),\, r_2(t)\big) \\
h(t) &= \operatorname{col}\big(-g(t),\, r_2(t)\big) & l(t) &= u(t) \\
\Gamma_{11} &= \begin{bmatrix} I & 0 & 0 \\ 0 & 0 & 0 \end{bmatrix} & \Gamma_{12} &= \begin{bmatrix} G_0 \\ I \end{bmatrix} \\
\Gamma_{21} &= \begin{bmatrix} -I & I & 0 \\ 0 & 0 & I \end{bmatrix} & \Gamma_{22} &= \begin{bmatrix} -G_0 \\ 0 \end{bmatrix}
\end{aligned}
$$

we find that

$$
T_{zw}(\Gamma_0,\, Q) = \left[N_i(G_0,\, K) \quad T_i(G_0,\, K) \begin{bmatrix} I & 0 \\ 0 & F \end{bmatrix} \right]
$$

Observe that it is possible to rewrite

$$
T_i(G_0,\, K) \begin{bmatrix} I & 0 \\ 0 & F \end{bmatrix} = \begin{bmatrix} I & 0 & 0 \\ 0 & 0 & I \end{bmatrix} T_i \left(\begin{bmatrix} G_0 \\ 0 \end{bmatrix},\, \begin{bmatrix} K & F \end{bmatrix} \right) \begin{bmatrix} I & 0 \\ 0 & I \\ 0 & 0 \end{bmatrix}
$$

with 0 matrices of appropriate dimensions, which can be convenient if one desires to treat $\begin{bmatrix} K & F \end{bmatrix}$ as a single object C, for instance if F and K share a common state-space representation. ◄

2.4 Coprime Factorisations

2.4.1 Coprime Factorisations of Transfer Functions or Matrices

We shall often use left and right coprime factorisations of systems and controllers in the sequel. Here, we define formally the notion of coprimeness and we explain how to compute the coprime factors of a dynamic system. Details can be found in, *e.g.*, (Francis, 1987), (Vidyasagar, 1985, 1988) and (Varga, 1998).

Definition 2.2. (COPRIMENESS) *Two matrices $\tilde{M}$ and $\tilde{N}$ in $\mathcal{RH}_\infty$ are left coprime over $\mathcal{RH}_\infty$ if they have the same number of rows and if there exist matrices X_l and Y_l in $\mathcal{RH}_\infty$ such that*

$$
\begin{bmatrix} \tilde{M} & \tilde{N} \end{bmatrix} \begin{bmatrix} X_l \\ Y_l \end{bmatrix} = \tilde{M} X_l + \tilde{N} Y_l = I \tag{2.25}
$$

Similarly, two matrices M and N in $\mathcal{RH}_\infty$ are right coprime over $\mathcal{RH}_\infty$ if they have the same number of columns and if there exist matrices X_r and Y_r in $\mathcal{RH}_\infty$ such that

$$
\begin{bmatrix} Y_r & X_r \end{bmatrix} \begin{bmatrix} N \\ M \end{bmatrix} = Y_r N + X_r M = I \tag{2.26}
$$

Definition 2.3. (LEFT COPRIME FACTORISATION) *Let P be a proper real rational transfer matrix. A left coprime factorisation of P is a factorisation $P = \tilde{M}^{-1}\tilde{N}$ where $\tilde{M}$ and $\tilde{N}$ are left coprime over $\mathcal{RH}_\infty$.*

If P has the following state-space realisation:

$$P = \left[\begin{array}{c|c} A & B \\ \hline C & D \end{array}\right] \tag{2.27a}$$

i.e.,

$$P(s) = C(sI - A)^{-1}B + D \quad \text{(continuous-time case)} \tag{2.27b}$$

or

$$P(z) = C(zI - A)^{-1}B + D \quad \text{(discrete-time case)} \tag{2.27c}$$

with $(A,\, C)$ detectable[3], a left coprime factorisation $\begin{bmatrix}\tilde{N} & \tilde{M}\end{bmatrix}$ can be constructed according to the following proposition.

Proposition 2.2. *Let $(A,\, B,\, C,\, D)$ be a detectable realisation of a transfer matrix P, let L be any constant injection matrix stabilising the output of P and let Z be any nonsingular matrix. Define*

$$\begin{bmatrix}\tilde{N} & \tilde{M}\end{bmatrix} = \left[\begin{array}{c|cc} A+LC & B+LD & L \\ \hline ZC & ZD & Z \end{array}\right] \tag{2.28}$$

i.e., (continuous-time case)

$$\begin{bmatrix}\tilde{N}(s) & \tilde{M}(s)\end{bmatrix} = ZC\big(sI - (A+LC)\big)^{-1}\begin{bmatrix}B+LD & L\end{bmatrix} + \begin{bmatrix}ZD & Z\end{bmatrix} \tag{2.29a}$$

or (discrete-time case)

$$\begin{bmatrix}\tilde{N}(z) & \tilde{M}(z)\end{bmatrix} = ZC\big(zI - (A+LC)\big)^{-1}\begin{bmatrix}B+LD & L\end{bmatrix} + \begin{bmatrix}ZD & Z\end{bmatrix} \tag{2.29b}$$

Then, $\begin{bmatrix}\tilde{N} & \tilde{M}\end{bmatrix} \in \mathcal{RH}_\infty$ and

$$P = \tilde{M}^{-1}\tilde{N} \tag{2.30}$$

A normalised left coprime factorisation, *i.e.*, a left coprime factorisation such that $\tilde{N}\tilde{N}^\star + \tilde{M}\tilde{M}^\star = I$, can be obtained as follows.

Proposition 2.3. *Let $(A,\, B,\, C,\, D)$ be a detectable realisation of a continuous-time transfer matrix $P(s)$. Define*

$$\begin{bmatrix}\tilde{N} & \tilde{M}\end{bmatrix} = \left[\begin{array}{c|cc} A+LC & B+LD & L \\ \hline JC & JD & J \end{array}\right] \tag{2.31}$$

where

$$L = -\big(XC^T + BD^T\big)\tilde{R}^{-1} \tag{2.32}$$

$$J = \tilde{R}^{-1/2} \tag{2.33}$$

[3]The pair $(A,\, C)$ is called *detectable* if there exists a real matrix L of appropriate dimensions such that $A + LC$ is Hurwitz, *i.e.*, if all eigenvalues of $A + LC$ have a strictly negative real part: $\mathrm{Re}(\lambda_i(A + LC)) < 0$.

X is the stabilising solution of the following Riccati equation:

$$(A - BD^T\tilde{R}^{-1}C)X + X(A - BD^T\tilde{R}^{-1}C)^T - X(C^T\tilde{R}^{-1}C)X + BR^{-1}B^T = 0 \quad (2.34)$$

and

$$R = I + D^T D \quad (2.35)$$

$$\tilde{R} = I + DD^T \quad (2.36)$$

Then, $\begin{bmatrix}\tilde{N}(s) & \tilde{M}(s)\end{bmatrix} \in \mathcal{RH}_\infty$*,*

$$\forall \omega \quad \tilde{N}(j\omega)\tilde{N}^\star(j\omega) + \tilde{M}(j\omega)\tilde{M}^\star(j\omega) = I \quad (2.37)$$

and

$$P(s) = \tilde{M}^{-1}(s)\tilde{N}(s) \quad (2.38)$$

▶ **Remark.** The same result holds in the discrete-time case with X the stabilising solution of the following discrete algebraic Riccati equation:

$$X = AXA^T + BB^T - (AXC^T + BD^T)(\tilde{R} + CXC^T)^{-1}(CXA^T + DB^T) \quad (2.39)$$

and L and J respectively given by

$$L = -(AXC^T + BD^T)(\tilde{R} + CXC^T)^{-1} \quad (2.40)$$

and

$$J = (\tilde{R} + CXC^T)^{-1/2} \quad (2.41)$$

◀

Definition 2.4. (RIGHT COPRIME FACTORISATION) *Let P be a proper real rational transfer matrix. A right coprime factorisation of P is a factorisation $P = NM^{-1}$ where N and M are right coprime over $\mathcal{RH}_\infty$.*

If P is stabilisable[4], such a right coprime factorisation can be constructed as follows.

Proposition 2.4. *Let $(A,\ B,\ C,\ D)$ be a stabilisable realisation of a transfer matrix P, let F be any constant feedback matrix stabilising P and let Z be any nonsingular matrix. Define*

$$\begin{bmatrix} N \\ M \end{bmatrix} = \left[\begin{array}{c|c} A + BF & BZ \\ \hline C + DF & DZ \\ F & Z \end{array}\right] \quad (2.42)$$

[4] A transfer matrix P with realisation $(A,\ B,\ C,\ D)$ is stabilisable if the pair $(A,\ B)$ is stabilisable, *i.e.*, if there exists a real matrix F of appropriate dimensions such that $A + BF$ is Hurwitz, meaning that all eigenvalues of $A + BF$ have a strictly negative real part: $\mathrm{Re}(\lambda_i(A + BF)) < 0$.

i.e., (continuous-time case)

$$\begin{bmatrix} N(s) \\ M(s) \end{bmatrix} = \begin{bmatrix} C+DF \\ F \end{bmatrix} \left(sI-(A+BF)\right)^{-1} BZ + \begin{bmatrix} DZ \\ Z \end{bmatrix} \tag{2.43a}$$

or (discrete-time case)

$$\begin{bmatrix} N(z) \\ M(z) \end{bmatrix} = \begin{bmatrix} C+DF \\ F \end{bmatrix} \left(zI-(A+BF)\right)^{-1} BZ + \begin{bmatrix} DZ \\ Z \end{bmatrix} \tag{2.43b}$$

Then, $\begin{bmatrix} N \\ M \end{bmatrix} \in \mathcal{RH}_\infty$ *and*

$$P = NM^{-1} \tag{2.44}$$

The following proposition gives the procedure to build a normalised right coprime factorisation, *i.e.*, a right coprime factorisation such that $N^\star N + M^\star M = I$.

Proposition 2.5. *Let (A, B, C, D) be a stabilisable realisation of a continuous-time transfer matrix $P(s)$. Define*

$$\begin{bmatrix} N \\ M \end{bmatrix} = \left[\begin{array}{c|c} A+BF & BH \\ \hline C+DF & DH \\ F & H \end{array}\right] \tag{2.45}$$

where

$$F = -R^{-1}(B^T X + D^T C) \tag{2.46}$$

$$H = R^{-1/2} \tag{2.47}$$

X is the stabilising solution of the following Riccati equation:

$$(A - BR^{-1}D^T C)^T X + X(A - BR^{-1}D^T C) + X(-BR^{-1}B^T)X + C^T \tilde{R}^{-1} C = 0 \tag{2.48}$$

and

$$R = I + D^T D \tag{2.49}$$

$$\tilde{R} = I + DD^T \tag{2.50}$$

Then, $\begin{bmatrix} N(s) \\ M(s) \end{bmatrix} \in \mathcal{RH}_\infty$,

$$\forall \omega \quad N^\star(j\omega)N(j\omega) + M^\star(j\omega)M(j\omega) = I \tag{2.51}$$

and

$$P(s) = N(s)M^{-1}(s) \tag{2.52}$$

▶ **Remark.** The same result holds in the discrete-time case with X the stabilising solution of the following discrete algebraic Riccati equation:

$$X = A^T XA + C^T C - (A^T XB + C^T D)(R + B^T XB)^{-1}(B^T XA + D^T C) \tag{2.53}$$

and F and H respectively given by

$$F = -(R + B^T XB)^{-1}(B^T XA + D^T C) \tag{2.54}$$

and

$$H = (R + B^T XB)^{-1/2} \tag{2.55}$$

◀

In the sequel, we shall generally use the notations $[\tilde{N} \quad \tilde{M}]$ (respectively $\begin{bmatrix} N \\ M \end{bmatrix}$) for the left (respectively right) coprime factorisations of a system $G = \tilde{M}^{-1}\tilde{N} = NM^{-1}$ and $[\tilde{U} \quad \tilde{V}]$ (respectively $\begin{bmatrix} U \\ V \end{bmatrix}$) for the left (respectively right) coprime factorisations of a controller $K = \tilde{V}^{-1}\tilde{U} = UV^{-1}$.

2.4.2 The Bezout Identity and Closed-loop Stability

Consider a closed-loop system as in Figure 2.1. Its closed-loop transfer matrix $T_o(G_0, K)$, given by (2.3), can be expressed in function of any pair of left coprime factors $[\tilde{N} \quad \tilde{M}]$ of the plant $G_0 = \tilde{M}^{-1}\tilde{N}$ and of any pair of right coprime factors $\begin{bmatrix} U \\ V \end{bmatrix}$ of the controller $K = UV^{-1}$ as

$$\begin{aligned} T_o(G_0, K) &= \begin{bmatrix} K \\ I \end{bmatrix} (I + G_0 K)^{-1} \begin{bmatrix} G_0 & I \end{bmatrix} \\ &= \begin{bmatrix} UV^{-1} \\ I \end{bmatrix} (I + \tilde{M}^{-1}\tilde{N}UV^{-1})^{-1} \begin{bmatrix} \tilde{M}^{-1}\tilde{N} & I \end{bmatrix} \\ &= \begin{bmatrix} U \\ V \end{bmatrix} \Phi^{-1} \begin{bmatrix} \tilde{N} & \tilde{M} \end{bmatrix} \end{aligned} \tag{2.56}$$

where

$$\Phi = \begin{bmatrix} \tilde{N} & \tilde{M} \end{bmatrix} \begin{bmatrix} U \\ V \end{bmatrix} \tag{2.57}$$

Lemma 2.1. *The closed-loop transfer matrix $T_o(G_0, K)$ of (2.56) is stable if and only if Φ is a unit (i.e., $\Phi, \Phi^{-1} \in \mathcal{RH}_\infty$).*

Proof. This follows from the fact that $[\tilde{N} \quad \tilde{M}]$, $\begin{bmatrix} U \\ V \end{bmatrix} \in \mathcal{RH}_\infty$, by definition of the coprime factors, hence the only remaining necessary and sufficient condition of closed-loop stability is that Φ be inversely stable. □

Lemma 2.2. (Bezout identity) *The closed-loop transfer matrix $T_o(G_0, K)$ of (2.56) is stable if and only if there exist plant and controller coprime factors $[\tilde{N} \quad \tilde{M}]$ and $\begin{bmatrix} U \\ V \end{bmatrix}$ such that*

$$\begin{bmatrix} \tilde{N} & \tilde{M} \end{bmatrix} \begin{bmatrix} U \\ V \end{bmatrix} = I \tag{2.58}$$

This equation is called a Bezout identity.

Proof. The sufficient condition is a direct consequence of Lemma 2.1, since I is a unit. The proof of the necessary condition is constructive: let $\begin{bmatrix}\tilde{N} & \tilde{M}\end{bmatrix}$ and $\begin{bmatrix}U \\ V\end{bmatrix}$ be any pairs of plant and controller coprime factors. By closed-loop stability hypothesis, $\Phi = \begin{bmatrix}\tilde{N} & \tilde{M}\end{bmatrix}\begin{bmatrix}U \\ V\end{bmatrix}$ is a unit, hence $\Phi^{-1} \in \mathcal{RH}_\infty$. Let us define $\begin{bmatrix}U' \\ V'\end{bmatrix} \triangleq \begin{bmatrix}U \\ V\end{bmatrix}\Phi^{-1}$. Then, $\begin{bmatrix}U' \\ V'\end{bmatrix} \in \mathcal{RH}_\infty$ and $U'V'^{-1} = U\Phi^{-1}(V\Phi^{-1})^{-1} = UV^{-1} = K$, which means that $\begin{bmatrix}U' \\ V'\end{bmatrix}$ is a coprime factorisation of K, and it follows from its definition that $\begin{bmatrix}\tilde{N} & \tilde{M}\end{bmatrix}\begin{bmatrix}U' \\ V'\end{bmatrix} = I$. Hence, $\begin{bmatrix}\tilde{N} & \tilde{M}\end{bmatrix}$ and $\begin{bmatrix}U' \\ V'\end{bmatrix}$ are plant and controller coprime factors satisfying the Bezout identity. □

In a similar way, one could define $\begin{bmatrix}\tilde{N}' & \tilde{M}'\end{bmatrix} \triangleq \Phi^{-1}\begin{bmatrix}\tilde{N} & \tilde{M}\end{bmatrix}$ and verify that it is a pair of plant coprime factors satisfying the Bezout identity with $\begin{bmatrix}U \\ V\end{bmatrix}$. This means that, starting from any two pairs of plant and controller coprime factors $\begin{bmatrix}\tilde{N} & \tilde{M}\end{bmatrix}$ and $\begin{bmatrix}U \\ V\end{bmatrix}$ and if the closed-loop system is stable, it is always possible to satisfy the Bezout identity by altering only one pair of coprime factors (either those of the plants or those of the controller), which leaves all freedom for the other pair which could be, for instance, normalised. The following proposition is a direct consequence of this observation.

Proposition 2.6. *Consider a stable closed-loop system with transfer matrix $T_o(G_0, K)$ given by (2.56). Let $G_0 = \tilde{M}^{-1}\tilde{N}$ and $K = UV^{-1}$ define, respectively, a left coprime factorisation of the plant and a right coprime factorisation of the controller. Then, two of the following three equalities can always be satisfied simultaneously:*

- *normalisation of the left coprime factors of G_0:*

$$\tilde{N}\tilde{N}^\star + \tilde{M}\tilde{M}^\star = I \tag{2.59}$$

- *normalisation of the right coprime factors of K:*

$$U^\star U + V^\star V = I \tag{2.60}$$

- *Bezout identity:*

$$\Phi = \tilde{N}U + \tilde{M}V = I \tag{2.61}$$

All these derivations could also be made for the closed-loop transfer matrix $T_i(G_0, K)$ of (2.2), which can be recast as

$$\begin{aligned} T_i(G_0, K) &= \begin{bmatrix}G_0 \\ I\end{bmatrix}(I + KG_0)^{-1}\begin{bmatrix}K & I\end{bmatrix} \\ &= \begin{bmatrix}NM^{-1} \\ I\end{bmatrix}(I + \tilde{V}^{-1}\tilde{U}NM^{-1})^{-1}\begin{bmatrix}\tilde{V}^{-1}\tilde{U} & I\end{bmatrix} \\ &= \begin{bmatrix}N \\ M\end{bmatrix}\tilde{\Phi}^{-1}\begin{bmatrix}\tilde{U} & \tilde{V}\end{bmatrix} \end{aligned} \tag{2.62}$$

where

$$\tilde{\Phi} = \begin{bmatrix} \tilde{U} & \tilde{V} \end{bmatrix} \begin{bmatrix} N \\ M \end{bmatrix} \tag{2.63}$$

using plant right coprime factors $G_0 = NM^{-1}$ and controller left coprime factors $K = \tilde{V}^{-1}\tilde{U}$. The existence of plant and controller coprime factors yielding the Bezout identity $\tilde{U}N + \tilde{V}M = I$ is then a necessary and sufficient condition for closed-loop stability. The following proposition summarises these results.

Proposition 2.7. (DOUBLE BEZOUT IDENTITY) *Consider the closed-loop system of Figure 2.1. This system is internally stable if and only if there exist plant and controller left and right coprime factorisations $G_0 = \tilde{M}^{-1}\tilde{N} = NM^{-1}$ and $K = \tilde{V}^{-1}\tilde{U} = UV^{-1}$ satisfying the double Bezout identity*

$$\begin{bmatrix} \tilde{M} & -\tilde{N} \\ \tilde{U} & \tilde{V} \end{bmatrix} \begin{bmatrix} V & N \\ -U & M \end{bmatrix} = I \tag{2.64}$$

2.5 The ν-gap Metric

2.5.1 Definition

The ν-gap metric between two continuous-time transfer matrices $G_1(s)$ and $G_2(s)$ is a measure of distance between these two systems. It was first introduced by G. Vinnicombe in (Vinnicombe, 1993*a*, 1993*b*).

Definition 2.5. (ν-GAP METRIC) *The ν-gap metric between two transfer matrices G_1 and G_2 is defined as*

$$\delta_\nu(G_1, G_2) = \begin{cases} \left\| \kappa\big(G_1(j\omega), G_2(j\omega)\big) \right\|_\infty & \text{if } \det \Xi(j\omega) \neq 0 \quad \forall \omega \\ & \text{and } \operatorname{wno}\big(\det \Xi(s)\big) = 0 \\ 1 & \text{otherwise} \end{cases} \tag{2.65}$$

where

- $\Xi(s) \triangleq N_2^\star(s)N_1(s) + M_2^\star(s)M_1(s)$;
- $\kappa(G_1(j\omega), G_2(j\omega)) \triangleq -\tilde{N}_2(j\omega)M_1(j\omega) + \tilde{M}_2(j\omega)N_1(j\omega)$ *is called the* chordal distance *between G_1 and G_2 at frequency ω;*
- $G_1(s) = N_1(s)M_1^{-1}(s)$ *and* $G_2(s) = N_2(s)M_2^{-1}(s) = \tilde{M}_2^{-1}(s)\tilde{N}_2(s)$ *are normalised coprime factorisations of G_1 and G_2;*
- $\operatorname{wno}(P(s)) = \eta(P^{-1}(s)) - \eta(P(s))$ *is called the* winding number *of the transfer function $P(s)$ and is defined as the number of counterclockwise encirclements (a clockwise encirclement counts as a negative encirclement) around the origin of the Nyquist contour of $P(s)$ indented around the right of any imaginary axis pole of $P(s)$;*
- $\eta(P(s))$ *denotes the number of poles of $P(s)$ in $\mathbb{C}_0^+$.*

The definition also holds in the discrete-time case by means of the use of the bilinear transformation $s = \frac{z-1}{z+1}$.

It has been shown in (Vinnicombe, 1993*a*) that

$$\delta_\nu(G_1, G_2) = \delta_\nu(G_2, G_1) = \delta_\nu(G_1^T, G_2^T) \tag{2.66}$$

An alternative definition of the ν-gap is the following:

$$\delta_\nu(G_1, G_2) = \begin{cases} \left\|\kappa\big(G_1(j\omega), G_2(j\omega)\big)\right\|_\infty \\ \quad \text{if } \det\big(I + G_2^\star(j\omega)G_1(j\omega)\big) \neq 0 \ \forall\omega \text{ and} \\ \quad \text{wno}\Big(\det\big(I + G_2^\star(s)G_1(s)\big)\Big) + \eta\big(G_1(s)\big) \\ \qquad\qquad -\eta\big(G_2(s)\big) - \eta_0\big(G_2(s)\big) = 0 \\ 1 \quad \text{otherwise} \end{cases} \tag{2.67}$$

where $\eta_0(P(s))$ is the number of imaginary axis poles of $P(s)$ and where $\kappa(G_1(j\omega), G_2(j\omega))$ can be written as

$$\kappa\big(G_1(j\omega), G_2(j\omega)\big) = \big(I + G_2(j\omega)G_2^\star(j\omega)\big)^{-1/2} \times \big(G_1(j\omega) - G_2(j\omega)\big) \times \big(I + G_1^\star(j\omega)G_1(j\omega)\big)^{-1/2} \tag{2.68}$$

In the SISO case, the ν-gap metric has a nice geometric interpretation. Indeed, the chordal distance at frequency ω, $\kappa(G_1(j\omega), G_2(j\omega))$, is the distance between the projections onto the Riemann sphere of the points of the Nyquist plots of G_1 and G_2 corresponding to that frequency (hence the appellation 'chordal distance'). The Riemann sphere is a unit-diameter sphere tangent at its south pole to the complex plane at its origin and the points of the Nyquist plots are projected onto the sphere using its north pole as centre of projection. Due to this particular projection, the chordal distance has a maximum resolution at frequencies where $|G_1| \approx 1$ and/or $|G_2| \approx 1$, *i.e.* around the cross-over frequencies of G_1 and G_2, since the corresponding points are projected onto the equator of the Riemann sphere. This property makes the ν-gap a control-oriented measure of distance between two systems.

▶ **Example 2.4.** Consider the systems $G_1(s) = \frac{1}{s+1}$ and $G_2(s) = (\frac{1}{s+1})^3$. Their Nyquist diagrams and their projections onto the Riemann sphere are depicted in Figure 2.3. Consider, for instance, the points $G_1(j\omega_1)$ and $G_1(j\omega_1)$ with $\omega_1 = 0.938$ rad/s. They are respectively located at the coordinates $(0.532, -0.499, 0)$ and $(-0.247, -0.299, 0)$. Their projections on the Riemann sphere are respectively located at the coordinates $(0.347, -0.326, 0.347)$ and $(-0.214, -0.260, 0.131)$. The distance between these two points, represented by a line segment inside the sphere in Figure 2.3, is 0.606, which is precisely $|\kappa(G_1(j\omega_1), G_2(j\omega_1))|$. ◀

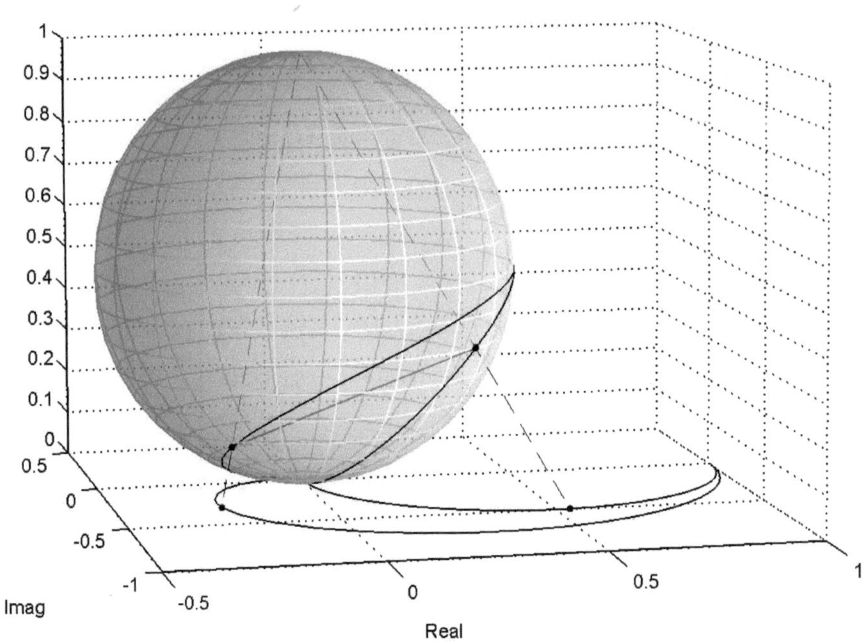

Figure 2.3. Projection onto the Riemann sphere of the Nyquist plots of G_1 and G_2, and chordal distance between $G_1(j\omega_1)$ and $G_2(j\omega_1)$

2.5.2 Stabilisation of a Set of Systems by a Given Controller and Comparison with the Directed Gap Metric

The definition of the ν-gap metric and in particular the winding number condition it involves, is based on a robust stability argument. In the theory of robust control, coprime-factor uncertainties are often considered. Let $G_1(s) = N_1(s)M_1^{-1}(s)$ define the normalised right coprime factorisation of a nominal system G_1 and let $\Delta = \begin{bmatrix} \Delta_N \\ \Delta_M \end{bmatrix}$ be a coprime-factor perturbation. If K is a feedback controller that stabilises G_1, then K also stabilises all G_2 in the set

$$\mathcal{G}_\beta^d = \left\{ G_2 = (N_1 + \Delta_N)(M_1 + \Delta_M)^{-1} \mid \Delta \in \mathcal{H}_\infty,\ \|\Delta\|_\infty \leq \beta \right\} \tag{2.69}$$

if and only if (Georgiou and Smith, 1990)

$$b_{G_1,K} > \beta \tag{2.70}$$

An alternative definition of this set is the following:

$$\mathcal{G}_\beta^d = \left\{ G_2 \mid \vec{\delta}_g(G_1, G_2) < \beta \right\} \tag{2.71}$$

where $\vec{\delta}_g(G_1, G_2)$ is the directed gap defined as

$$\vec{\delta}_g(G_1, G_2) = \inf_{Q(s)\in\mathcal{H}_\infty} \left\| \begin{bmatrix} N_1(s) \\ M_1(s) \end{bmatrix} - \begin{bmatrix} N_2(s) \\ M_2(s) \end{bmatrix} Q(s) \right\|_\infty \tag{2.72}$$

and where $G_2(s) = N_2(s)M_2^{-1}(s)$ defines the normalised right coprime factorisation of G_2. However, it can be shown (Vinnicombe, 1993*b*) that the largest class of systems that can be guaranteed to be stabilised *a priori* by K consists of those G_2 satisfying

$$\inf_{Q(s)\in\mathcal{L}_\infty} \left\| \begin{bmatrix} N_1(s) \\ M_1(s) \end{bmatrix} - \begin{bmatrix} N_2(s) \\ M_2(s) \end{bmatrix} Q(s) \right\|_\infty < \beta, \quad \text{wno}\big(\det \Xi(s)\big) = 0 \tag{2.73}$$

(where $\Xi(s)$ is defined in Definition 2.5), which is precisely the set

$$\mathcal{G}_\beta^\nu = \Big\{ G_2 = (N_1 + \Delta_N)(M_1 + \Delta_M)^{-1} \mid \Delta \in \mathcal{L}_\infty,\ \|\Delta\|_\infty \leq \beta \\ \text{and } \eta(G_2) = \text{wno}(M_1 + \Delta_M) \Big\} = \Big\{ G_2 \mid \delta_\nu(G_1, G_2) \leq \beta \Big\} \tag{2.74}$$

Hence, one can define a larger set of plants that are guaranteed to be stabilised by a given controller K with the ν-gap metric than with directed gap, since the ν-gap allows coprime-factor perturbations in $\mathcal{L}_\infty$ rather than in $\mathcal{H}_\infty$. Another serious advantage of the ν-gap over the directed gap is the fact that the former is much easier to compute than the latter. However, in order to be valid, the robust stability theory with the ν-gap metric requires the verification of the winding number condition (which can take various forms depending on the chosen definition of the ν-gap). The demonstration of the necessity of the winding number condition is very complicated and is outside the scope of this book. The interested reader is referred to (Vinnicombe, 1993*a*, 1993*b*, 2000).

2.5.3 The ν-gap Metric and Robust Stability

The main interest of the ν-gap metric is its use in a range of robust stability results. One of these results relates the size of the set of robustly stabilising controllers of a ν-gap uncertainty set (*i.e.*, a set of the form (2.74) defined with the ν-gap) to the size of this uncertainty set, as summarised in the following two propositions.

Proposition 2.8. (Vinnicombe, 2000) *Let us consider the uncertainty set $\mathcal{G}_\gamma^\kappa$, centred at a model G_1, defined by*

$$\mathcal{G}_\gamma^\kappa = \Big\{ G_2 \mid \kappa\big(G_1(e^{j\omega}), G_2(e^{j\omega})\big) \leq \gamma(\omega)\ \forall\omega \text{ and } \delta_\nu(G_1, G_2) < 1 \Big\} \tag{2.75}$$

with $0 \leq \gamma(\omega) < 1\ \forall\omega$. Then, a controller K stabilising G_1 stabilises all plants in the uncertainty set $\mathcal{G}_\gamma^\kappa$ if and only if it lies in the controller set

$$\mathcal{C}_\gamma^\kappa = \Big\{ K(z) \mid \kappa\Big(G_1(e^{j\omega}), \tfrac{-1}{K(e^{j\omega})}\Big) > \gamma(\omega)\quad \forall\omega \Big\} \tag{2.76}$$

The second proposition is a Min-Max version of the first one:

Proposition 2.9. (Vinnicombe, 2000) *Let us consider the ν-gap uncertainty set $\mathcal{G}_\beta^\nu$ of size $\beta < 1$ centred at a model G_1:*

$$\mathcal{G}_\beta^\nu = \Big\{ G_2 \mid \delta_\nu(G_1, G_2) \leq \beta \Big\} \tag{2.77}$$

Then, a controller K stabilising G_1 stabilises all plants in the uncertainty set $\mathcal{G}_\beta^\nu$ if and only if it lies in the controller set

$$\mathcal{C}_\beta^\nu = \Big\{ K(z) \mid b_{G_1,K} > \beta \Big\} \tag{2.78}$$

The size β of a ν-gap uncertainty set $\mathcal{G}_\beta^\nu$ is thus directly connected to the size of the set of all controllers that robustly stabilise $\mathcal{G}_\beta^\nu$. Moreover, the smaller this size β, the larger the set of controllers that robustly stabilises $\mathcal{G}_\beta^\nu$. This result will be of the highest importance in Chapter 5.

2.6 Prediction-error Identification

Prediction-error identification is the only modelling tool, considered in this book, that uses data collected on the process to obtain a mathematical representation of it under the form of a transfer function or matrix. Contrary to, *e.g.*, first-principles modelling, the objective here is not to build a good *knowledge* model of the process, but to obtain a (generally black-box) *representation* model that exhibits a good qualitative and quantitative matching of the process behaviour. Such a model is generally delivered with an uncertainty region around it, which can be used for robust control design.

Our intention here is not to give a thorough theoretical description of the method. The interested reader is kindly referred to the existing literature on the subject and more particularly to (Ljung, 1999) for the detailed theory of prediction-error identification and (Zhu, 2001) for its application to multivariable systems in a process control framework.

We make the assumption that the true system is the possibly multi-input multi-output, linear time-invariant[5] system described by

$$\mathcal{S} : \begin{cases} y(t) = G_0(z)u(t) + v(t) \\ v(t) = H_0(z)e(t) \end{cases} \tag{2.79}$$

where $G_0(z)$ and $H_0(z)$ are rational transfer functions or matrices. $G_0(z)$ is strictly proper and has p outputs and m inputs. $H_0(z)$ is a stable and inversely

[5] This may seem a restrictive hypothesis as, in practice, all industrial systems exhibit at least a little amount of nonlinearities and have a tendency to alter with the time. Conceptually, however, the idea of an LTI system is generally perfectly acceptable if the plant is regarded around a given operating point.

stable, minimum-phase, proper and monic[6] $p \times p$ transfer matrix. $u(t) \in \mathbb{R}^m$ is the control input signal, $y(t) \in \mathbb{R}^p$ is the observed output signal and $e(t) \in \mathbb{R}^p$ is a white noise process with zero mean and covariance matrix Λ_0 (or variance λ_0 in the single-input single-output case).

For the sake of simplicity, however, we shall restrict the derivations of this section to the SISO case except where explicitely indicated.

2.6.1 Signals Properties

The assumption is made that all signals are quasi-stationary (Ljung, 1999). A quasi-stationary signal $y(t)$ is a signal for which the following auto-correlation function exists:

$$R_y(\tau) = \lim_{N\to\infty} \frac{1}{N} \sum_{t=1}^{N} Ey(t)y(t-\tau) \triangleq \bar{E}y(t)y(t-\tau) \tag{2.80}$$

where the expectation is taken with respect to the stochastic components of the signal. *e.g.*, if $y(t)$ is the output of the system (2.79) with $u(t)$ deterministic and $v(t)$ zero-mean stochastic, then $Ey(t) = G_0(z)u(t)$. This quasi-stationarity assumption is useful to treat deterministic, stochastic or mixed signals in a common framework with theoretical exactness and it allows defining spectra and cross-spectra of signals as follows.

The spectrum (or power spectral density) of a quasi-stationary signal $y(t)$ is defined by

$$\phi_y(\omega) = \sum_{\tau=-\infty}^{\infty} R_y(\tau)e^{-j\omega\tau} \tag{2.81}$$

The cross-correlation function and the cross-spectrum of two quasi-stationary signals are respectively defined by

$$R_{yu}(\tau) = \bar{E}y(t)u(t-\tau) \tag{2.82}$$

and

$$\phi_{yu}(\omega) = \sum_{\tau=-\infty}^{\infty} R_{yu}(\tau)e^{-j\omega\tau} \tag{2.83}$$

When (2.79) is satisfied, the following relations hold:

$$\phi_v(\omega) = \left|H_0(e^{j\omega})\right|^2 \phi_e(\omega) = \left|H_0(e^{j\omega})\right|^2 \lambda_0 \tag{2.84}$$

$$\phi_y(\omega) = \phi_y^u(\omega) + \phi_y^e(\omega) \triangleq \left|G_0(e^{j\omega})\right|^2 \phi_u(\omega) + \left|H_0(e^{j\omega})\right|^2 \lambda_0 \tag{2.85}$$

$$\phi_{yu}(\omega) = G_0(e^{j\omega})\phi_u(\omega) \tag{2.86}$$

[6] A monic filter is a filter whose impulse-response zeroth coefficient is 1 or the unit matrix.

etc. Spectra are real-valued functions of the frequency ω, while cross-spectra are complex-valued functions of ω. $\phi_y^u(\omega)$ denotes the spectrum of that part of $y(t)$ that originates from $u(t)$. It is not the same as the cross-spectrum $\phi_{yu}(\omega)$.

In the MIMO case, these expressions become a little more complicated:

$$\phi_y^u(\omega) = G_0(e^{j\omega})\phi_u(\omega)\left[G_0(e^{j\omega})\right]^\star \tag{2.87}$$

etc.

The following equality, called Parseval's relationship, holds:

$$E\,|y(t)|^2 = \frac{1}{2\pi}\int_{-\pi}^{\pi} \phi_y(\omega)\,d\omega \tag{2.88}$$

It says that the power of the signal $y(t)$ is equal to the power contained in its spectrum. This relationship will have important consequences regarding the distribution of the modelling error.

In practice, the following formulae can be used to estimate auto-correlation or cross-correlation functions:

$$\hat{R}_y^N(\tau) = \frac{1}{N}\sum_{t=1}^{N} y(t)y(t-\tau) \tag{2.89}$$

$$\hat{R}_{yu}^N(\tau) = \frac{1}{N}\sum_{t=1}^{N} y(t)u(t-\tau) \tag{2.90}$$

and these estimates can be used to calculate spectra and cross-spectra as

$$\hat{\phi}_y(\omega) = \sum_{\tau=-\tau_m}^{\tau_m} \hat{R}_y(\tau)e^{-j\omega\tau} \tag{2.91}$$

$$\hat{\phi}_{yu}(\omega) = \sum_{\tau=-\tau_m}^{\tau_m} \hat{R}_{yu}(\tau)e^{-j\omega\tau} \tag{2.92}$$

with a suitable τ_m like, *e.g.*, $\tau_m = N/10$. It can be shown that, for $N \to \infty$, these estimates will converge with probability one to the true $R_y(\tau)$, $R_{yu}(\tau)$, $\phi_y(\omega)$ and $\phi_{yu}(\omega)$, provided the signals are ergodic[7].

2.6.2 The Identification Method

The objective of system identification is to compute a parametrised model $\mathcal{M}(\theta)$ for the system:

$$\mathcal{M}(\theta) : \hat{y}(t) = G(z,\,\theta)u(t) + H(z,\,\theta)e(t) \tag{2.93}$$

[7] An ergodic stochastic process $x(t)$ is a process whose time average $\lim_{N\to\infty}\frac{1}{2N}\sum_{t=-N}^{N} x(t)$ tends to its ensemble average, *i.e.*, to its expectation $Ex(t)$.

This model lies in some model set $\mathcal{M}$ selected by the designer:

$$\mathcal{M} \triangleq \left\{ \mathcal{M}(\theta) \mid \theta \in D_\theta \subseteq \mathbb{R}^n \right\} \tag{2.94}$$

i.e., $\mathcal{M}$ is the set of all models with the same structure as $\mathcal{M}(\theta)$. The parameter vector θ ranges over a set $D_\theta \subseteq \mathbb{R}^n$ that is assumed to be compact and connected. We say that the true system is in the model set, which is denoted by $\mathcal{S} \in \mathcal{M}$, if

$$\exists \theta_0 \in D_\theta : G(z, \theta_0) = G_0,\ H(z, \theta_0) = H_0 \tag{2.95}$$

Otherwise, we say that there is *undermodelling* of the system dynamics. The case where the noise properties cannot be correctly described within the model set but where

$$\exists \theta_0 \in D_\theta : G(z, \theta_0) = G_0 \tag{2.96}$$

will be denoted by $G_0 \in \mathcal{G}$.

The prediction-error identification procedure uses a finite set of N input-output data

$$Z^N = \left\{ u(1),\ y(1),\ \ldots,\ u(N),\ y(N) \right\} \tag{2.97}$$

to compute the one-step-ahead prediction of the output signal at each time sample $t \in [1,\ N]$, using the available past data samples[8] and the model with its parameter vector θ:

$$\hat{y}(t \mid t-1,\ \theta) = H^{-1}(z,\ \theta)G(z,\ \theta)u(t) + \big(1 - H^{-1}(z,\ \theta)\big)y(t) \tag{2.98}$$

(Observe that, because $G(z,\ \theta)$ is strictly proper and $H(z,\ \theta)$ is monic, the two terms of the right-hand side depend only on past u's and y's.) The prediction error at time t is

$$\varepsilon(t,\ \theta) = y(t) - \hat{y}(t \mid t-1,\ \theta) = H^{-1}(z,\ \theta)\big(y(t) - G(z,\ \theta)u(t)\big) \tag{2.99}$$

Observe that it would be equal to the white noise $e(t)$, *i.e.*, to the only absolutely unpredictable part of $y(t)$, if $G(z,\ \theta)$ and $H(z,\ \theta)$ were equal to $G_0(z)$ and $H_0(z)$, respectively. The objective of prediction-error identification is to find a parameter vector $\hat{\theta}$ such that $\varepsilon(t,\ \hat{\theta})$ be whitened, meaning that all the useful information contained in the data Z^N (*i.e.*, everything but the stochastic white noise $e(t)$) has been exploited.

Given the chosen model structure (2.93) and measured data (2.97), the prediction-error estimate of θ is determined through

$$\hat{\theta} = \arg \min_{\theta \in D_\theta} V_N\big(\theta,\ Z^N\big) \tag{2.100}$$

[8]The difference between *simulation* and *prediction* is that the former only uses the measured input signal and filters it through the transfer function $G(z,\ \theta)$ of the model to compute an estimate of the output signal, while the latter uses all available information, including past output samples, to build an estimation, or prediction, of the future outputs.

where $V_N(\theta, Z^N)$ is a quadratic criterion:

$$V_N(\theta, Z^N) = \begin{cases} \frac{1}{N}\sum_{t=1}^{N} \varepsilon_F^T(t, \theta)\Lambda^{-1}\varepsilon_F(t, \theta) & \text{(MIMO case)} \\ \frac{1}{N}\sum_{t=1}^{N} \varepsilon_F^2(t, \theta) & \text{(SISO case)} \end{cases} \quad (2.101)$$

In this expression, Λ is a symmetric positive-definite weighting matrix and $\varepsilon_F(t, \theta)$ are the possibly filtered prediction errors:

$$\varepsilon_F(t, \theta) = L(z, \theta)\varepsilon(t, \theta) \quad (2.102)$$

where $L(z, \theta)$ is any linear, stable, monic and possibly parametrised prefilter. Since

$$\varepsilon_F(t, \theta) = L(z, \theta)H^{-1}(z, \theta)\big(y(t) - G(z, \theta)u(t)\big) \quad (2.103)$$

this filter can be included in the noise model structure and, without loss of generality, we shall make the assumption that $L(z, \theta) = I$ in the sequel. (Observe, apropos, that if $L(z, \theta) = H(z, \theta)$, then $\varepsilon_F(t, \theta) = y(t) - G(z, \theta)u(t)$, which is the simulation error at time t. The noise model disappears then from the identification criterion, meaning that no noise model is identified. This is the output-error case: see below.)

The following notation will often be used for the estimates $G(z, \hat{\theta})$, $H(z, \hat{\theta})$, *etc.*:

$$\hat{G}(z) = G(z, \hat{\theta}) \quad \text{and} \quad \hat{H}(z) = H(z, \hat{\theta}) \quad (2.104)$$

2.6.3 Usual Model Structures

Some commonly used polynomial model structures are the following.

- FIR (Finite Impulse Response model structure):

$$y(t) = B(z)z^{-k}u(t) + e(t) \quad (2.105)$$

- ARX (Auto-Regressive model structure with eXogenous inputs):

$$A(z)y(t) = B(z)z^{-k}u(t) + e(t) \quad (2.106)$$

- ARMAX (Auto-Regressive Moving-Average model structure with eXogenous inputs):

$$A(z)y(t) = B(z)z^{-k}u(t) + C(z)e(t) \quad (2.107)$$

- OE (Output-Error model structure):

$$y(t) = F^{-1}(z)B(z)z^{-k}u(t) + e(t) \quad (2.108)$$

- BJ (Box-Jenkins model structure):

$$y(t) = F^{-1}(z)B(z)z^{-k}u(t) + D^{-1}(z)C(z)e(t) \quad (2.109)$$

In these expressions, k is the length of the dead time of the transfer function (expressed in number of samples), $B(z)$ is a polynomial (SISO case) or a polynomial matrix (MIMO case) of order n_b in z^{-1}, and $A(z)$, $C(z)$, $D(z)$ and $F(z)$ are monic polynomials[9] or polynomial matrices in z^{-1}, respectively of orders n_a, n_c, n_d and n_f.

► **Remark.** Prediction-error identification only works with stable predictors, *i.e.*, the product $H^{-1}(z, \theta)G(z, \theta)$ in (2.98) is constrained to be stable (otherwise the prediction errors might not be bounded). This means that the only way to identify an unstable system is to use a model structure where the unstable poles of $G(z, \theta)$ are also in $H(z, \theta)$, *e.g.*, an ARX or ARMAX structure, allowing the predictor to be stable although the model is unstable. On the contrary, the use of an OE model structure will enforce stability of the estimate. ◄

2.6.4 Computation of the Estimate

Depending on the chosen model structure, $\hat{\theta}$ can be obtained algebraically or *via* an optimisation procedure.

A. The FIR and ARX cases. With a FIR or ARX model structure, the model is linear in the parameters:

$$\mathcal{M}(\theta) : \hat{y}(t) = \varphi^T(t)\theta + e(t) \tag{2.110}$$

hence

$$\hat{y}(t \mid t-1, \theta) = \varphi^T(t)\theta \tag{2.111}$$

where

$$\theta = \begin{bmatrix} a_1 & \dots & a_{n_a} & b_0 & \dots & b_{n_b} \end{bmatrix}^T \tag{2.112}$$

and

$$\varphi(t) = \begin{bmatrix} -y(t-1) & \dots & -y(t-n_a) & u(t-n_k) & \dots & u(t-n_b-n_k) \end{bmatrix}^T \tag{2.113}$$

are respectively a parameter vector and a regression vector. The minimising argument of $V_N(\theta, Z^N)$ is then obtained by the standard least-squares method:

$$\hat{\theta} = \left[\frac{1}{N} \sum_{t=1}^{N} \varphi(t)\varphi^T(t) \right]^{-1} \frac{1}{N} \sum_{t=1}^{N} \varphi(t)y(t) \tag{2.114}$$

---

[9] A polynomial is monic if its independent term is 1.

B. Other cases. Other model structures require numerical optimisation to find the estimate. A standard search routine is the following:

$$\hat{\theta}^{i+1} = \hat{\theta}^i - \mu^i \left[R_N^i\right]^{-1} V_N'(\hat{\theta}^i,\, Z^N) \tag{2.115}$$

where $\hat{\theta}^i$ is the estimate at iteration i,

$$V_N'(\hat{\theta}^i,\, Z^N) = -\frac{1}{N}\sum_{t=1}^{N} \psi(t,\, \hat{\theta}^i)\varepsilon(t,\, \hat{\theta}^i) \tag{2.116}$$

is the gradient of $V_N(\theta,\, Z^N)$ with respect to θ evaluated at $\hat{\theta}^i$, R_N^i is a matrix that determines the search direction, μ^i is a factor that determines the step size and

$$\psi(t,\, \theta) = -\frac{d}{d\theta}\varepsilon(t,\, \theta) = \frac{d}{d\theta}\hat{y}(t \mid t-1,\, \theta) \tag{2.117}$$

is the negative gradient of the prediction error. A common choice for the matrix R_N^i is

$$R_N^i = \frac{1}{N}\sum_{t=1}^{N} \psi(t,\, \hat{\theta}^i)\psi^T(t,\, \hat{\theta}^i) + \delta^i I \tag{2.118}$$

which gives the Gauss-Newton direction if $\delta^i = 0$; see, *e.g.*, (Dennis and Schnabel, 1983). δ^i is a regularisation parameter chosen so that $R_N^i > 0$.

2.6.5 Asymptotic Properties of the Estimate

The results of this subsection are given for both MIMO and SISO cases.

A. Asymptotic bias and consistency. Under mild conditions, in the MIMO case, there hold (Ljung, 1978)

$$V_N(\theta,\, Z^N) \to \bar{V}(\theta) = \bar{E}\varepsilon^T(t,\, \theta)\Lambda^{-1}\varepsilon(t,\, \theta) \quad \text{w.p. 1 as } N \to \infty \tag{2.119}$$

and

$$\hat{\theta} \to \theta^* = \arg\min_{\theta \in D_\theta} \bar{V}(\theta) \qquad \text{w.p. 1 as } N \to \infty \tag{2.120}$$

Note that we can write $\bar{V}(\theta)$ as

$$\begin{aligned} \bar{V}(\theta) &= \bar{E}\,\mathrm{trace}\big(\Lambda^{-1}\varepsilon(t,\, \theta)\varepsilon^T(t,\, \theta)\big) \\ &= \frac{1}{2\pi}\int_{-\pi}^{\pi} \mathrm{trace}\big(\Lambda^{-1}\phi_\varepsilon(\omega)\big)d\omega \end{aligned} \tag{2.121}$$

where $\phi_\varepsilon(\omega)$ is the power spectral density of the prediction error $\varepsilon(t, \theta)$. The second equality comes from Parseval's relationship. As a result, there holds[10]

$$\theta^* = \arg\min_{\theta\in D_\theta} \frac{1}{2\pi}\int_{-\pi}^{\pi} \text{trace}\Bigg(\begin{bmatrix} G_0 - G(\theta) & H_0 - H(\theta) \end{bmatrix} \begin{bmatrix} \phi_u & \phi_{ue} \\ \phi_{eu} & \Lambda_0 \end{bmatrix} \times \begin{bmatrix} (G_0 - G(\theta))^\star \\ (H_0 - H(\theta))^\star \end{bmatrix} \left(H(\theta)\Lambda H^\star(\theta)\right)^{-1} \Bigg) d\omega \quad (2.122)$$

which can be recast as

$$\theta^* = \arg\min_{\theta\in D_\theta} \frac{1}{2\pi}\int_{-\pi}^{\pi} \text{trace}\Bigg(\Big[\big(G_0 - G(\theta) + B(\theta)\big)\phi_u\big(G_0 - G(\theta) + B(\theta)\big)^\star + \big(H_0 - H(\theta)\big)\big(\Lambda_0 - \phi_{eu}\phi_u^{-1}\phi_{ue}\big)\big(H_0 - H(\theta)\big)^\star \Big] \times \left(H(\theta)\Lambda H^\star(\theta)\right)^{-1} \Bigg) d\omega \quad (2.123)$$

where

$$B(e^{j\omega}, \theta) = \big(H_0(e^{j\omega}) - H(e^{j\omega}, \theta)\big)\phi_{eu}(\omega)\phi_u^{-1}(\omega) \quad (2.124)$$

is a bias term that will vanish only if $\phi_{eu} = 0$, *i.e.*, if the data are collected in open loop so that u and e are uncorrelated, or if the noise model $H(z, \theta)$ is flexible enough so that $\mathcal{S} \in \mathcal{M}$; see (Forssell, 1999).

In the SISO case, these expressions become

$$\bar{V}(\theta) = \frac{1}{2\pi}\int_{-\pi}^{\pi} \phi_\varepsilon(\omega) d\omega \quad (2.125)$$

and

$$\theta^* = \arg\min_{\theta\in D_\theta} \left\{ \frac{1}{2\pi}\int_{-\pi}^{\pi} \frac{\left|G_0 - G(\theta) + B(\theta)\right|^2}{\left|H(\theta)\right|^2}\phi_u d\omega + \frac{1}{2\pi}\int_{-\pi}^{\pi} \frac{\left|H_0 - H(\theta)\right|^2}{\left|H(\theta)\right|^2}\frac{\lambda_0\phi_u^r}{\phi_u} d\omega \right\} \quad (2.126)$$

where the second term has been obtained by noting that $\lambda_0 - \frac{|\phi_{ue}|^2}{\phi_u} = \frac{\lambda_0\phi_u^r}{\phi_u}$.

Hence, under the condition that $\mathcal{S} \in \mathcal{M}$, we find that

$$G(\hat{\theta}) \to G_0 \quad \text{and} \quad H(\hat{\theta}) \to H_0 \quad \text{w.p. 1 as } N \to \infty \quad (2.127)$$

[10] The $e^{j\omega}$ or ω arguments of the transfer functions or spectra will often be omitted to simplify the notation.

B. Asymptotic variance in transfer function space. In the MIMO case, the asymptotic covariance of the estimate, as N and n both tend to infinity, is given by

$$\operatorname{cov}\operatorname{col}\left(\begin{bmatrix}\hat{G}(e^{j\omega}) & \hat{H}(e^{j\omega})\end{bmatrix}\right) \approx \frac{n}{N}\begin{bmatrix}\phi_u(\omega) & \phi_{ue}(\omega)\\ \phi_{eu}(\omega) & \Lambda_0\end{bmatrix}^{-T} \otimes \phi_v(\omega) \tag{2.128}$$

where $\otimes$ denotes the Kronecker product. In open loop, $\phi_{ue} = 0$ and

$$\operatorname{cov}\operatorname{col}\left(\hat{G}(e^{j\omega})\right) \approx \frac{n}{N}\phi_u^{-T}(\omega) \otimes \phi_v(\omega) \tag{2.129a}$$

$$\operatorname{cov}\operatorname{col}\left(\hat{H}(e^{j\omega})\right) \approx \frac{n}{N}\Lambda_0^{-1} \otimes \phi_v(\omega) \tag{2.129b}$$

In the SISO case, these expressions become

$$\operatorname{cov}\left(\hat{G}(e^{j\omega})\right) \approx \frac{n}{N}\frac{\phi_v(\omega)}{\phi_u(\omega)} \tag{2.130a}$$

$$\operatorname{cov}\left(\hat{H}(e^{j\omega})\right) \approx \frac{n}{N}\left|H_0(e^{j\omega})\right|^2 \tag{2.130b}$$

in open loop while, in closed loop, the covariance of $\hat{G}(e^{j\omega})$ is

$$\operatorname{cov}\left(\hat{G}(e^{j\omega})\right) \approx \frac{n}{N}\frac{\phi_v(\omega)}{\phi_u^r(\omega)} \tag{2.131}$$

where ϕ_u^r is the power spectral density of that part of the input $u(t)$ that originates from the reference $r(t)$. This shows the necessity to have a nonzero exogenous excitation at $r(t)$ to be able to identify the system. An input signal $u(t)$ that would only be generated by feedback of the disturbances through the controller would be useless, as it would yield an estimate with infinite variance.

These results were established in (Ljung, 1985) for the SISO case and in (Zhu, 1989) for the MIMO case. They show the importance of the experiment design (open-loop or closed-loop operation, spectrum of the excitation signal, *etc.*) in the tuning of the modelling error. This question will receive more attention in Chapter 3.

▶ **Remark.** These asymptotic variance expressions are widely used in practice, although they are not always reliable. It has been shown in (Ninness *et al.*, 1999) that their accuracy could depend on choices of fixed poles or zeroes in the model structure; alternative variance expressions with greatly improved accuracy and which make explicit the influence of any fixed poles or zeroes are given. Observe for instance that the use of a fixed prefilter $L(z)$ during identification amounts to impose fixed poles and/or zeroes in the noise model. In this case, the extended theory of (Ninness *et al.*, 1999) should be used, the more so if the number of fixed poles and zeroes is large with respect to the model order. ◀

C. Asymptotic variance in parameter space. The parameter vector estimate tends to a random vector with normal distribution as the number of data samples N tends to infinity (Ljung, 1999):

$$\hat{\theta} \to \theta^* \qquad \text{w.p. 1 as } N \to \infty \tag{2.132a}$$

$$\sqrt{N}(\hat{\theta} - \theta^*) \in AsN(0, P_\theta) \tag{2.132b}$$

where

$$P_\theta = R^{-1}QR^{-1} \tag{2.132c}$$

$$R = \bar{V}''(\theta^*) > 0 \tag{2.132d}$$

$$Q = \lim_{N\to\infty} N \cdot E\Big(\big[V_N'(\theta^*, Z^N)\big]\big[V_N'(\theta^*, Z^N)\big]^T\Big) \tag{2.132e}$$

The prediction-error identification algorithms of the MATLAB® Identification Toolbox, used with standard model structures, deliver an estimate $\hat{P}_\theta$ of P_θ.

2.6.6 Classical Model Validation Tools

Once a model $\mathcal{M}(\hat{\theta})$ has been identified, it is necessary to make it pass a range of validation tests to assess its quality. The classical validation tests are the following.

A. Model fit indicator. The simulated output and the one-step-ahead predicted output are respectively given by

$$\hat{y}_s\big(t \mid \mathcal{M}(\hat{\theta})\big) = G(z, \hat{\theta})u(t) \tag{2.133a}$$

and

$$\hat{y}_p\big(t \mid \mathcal{M}(\hat{\theta})\big) = H^{-1}(z, \hat{\theta})G(z, \hat{\theta})u(t) + \big(1 - H^{-1}(z, \hat{\theta})\big)y(t) \tag{2.133b}$$

Let us define the following model fit indicator:

$$J_x\big(\mathcal{M}(\hat{\theta})\big) = \frac{1}{N}\sum_{t=1}^{N}\Big|y(t) - \hat{y}_x\big(t \mid \mathcal{M}(\hat{\theta})\big)\Big|^2 \tag{2.134}$$

where x is either p or s depending on whether we are interested in one-step-ahead prediction or in simulation (several-steps-ahead prediction can also be considered: see (Ljung, 1999)). A normalised measure of this fit is given by

$$R_x^2\big(\mathcal{M}(\hat{\theta})\big) = 1 - \frac{J_x\big(\mathcal{M}(\hat{\theta})\big)}{\frac{1}{N}\sum_{t=1}^{N}|y(t)|^2} \tag{2.135}$$

A value of R_x close to 1 means that the predicted or simulated output fits the process output well, *i.e.*, that the observed output variations are well explained by the model, while a value close to 0 means that the model is unable to correctly explain the data.

The quality of the indicator J_x depends very much on the data set that is used to compute it. As a result, it would be nice to be able to evaluate $\bar{J}_x = EJ_x$, where the expectation is taken with respect to the data, the model $\mathcal{M}(\hat{\theta})$ being fixed. It can be shown that J_x will be an unbiased estimate of $\bar{J}_x$ only if it is computed from a different data set that the one used during the identification of the model. Therefore, it is always recommended to use two different data sets: one for estimation, the other for validation. See (Ljung, 1999) for more details.

Finally, note that a model with a good fit in prediction can have a bad fit in simulation. It is easier to find a good model for prediction than for simulation, because prediction takes the noise into account while simulation does not, meaning that the best model obtained by prediction-error identification will always produce a simulation error at least equal to $v(t) = H_0(z)e(t)$, which is usually an auto-correlated signal, while it will produce a prediction error close to $e(t)$ and approximately white provided an appropriate model structure is used. When $H(z, \theta) \equiv 1$, *i.e.*, in the output-error case, the simulation error and prediction error are of course the same. This means that the choice of an OE model structure is ideal when the objective is to find a good simulation model, although the resulting model can be far from optimal in prediction if $H_0(z) \neq 1$.

B. Residuals analysis. Since the objective of prediction-error identification is to make the prediction errors white, it is natural to test this whiteness in order to assess the quality of the model.

Let us define the residuals as

$$\varepsilon(t) = \varepsilon\big(t \mid \mathcal{M}(\hat{\theta})\big) \triangleq y(t) - \hat{y}_p\big(t \mid \mathcal{M}(\hat{\theta})\big) \tag{2.136}$$

The residuals are the prediction errors, ideally built from a validation data set different from the one used during identification as explained above. The experimental auto-correlation function of the residuals is given by

$$\hat{R}_\varepsilon^N(\tau) = \frac{1}{N}\sum_{t=1}^{N} \varepsilon(t)\varepsilon(t-\tau) \tag{2.137}$$

If $\varepsilon(t)$ is a real white noise sequence of variance λ, $\hat{R}_\varepsilon^N(\tau)$ will tend to 0 for $\tau \neq 0$ and to λ^2 for $\tau = 0$, as N tends to infinity. As a result, a good way to test the whiteness of $\varepsilon(t)$ is to check whether $\hat{R}_\varepsilon^N(\tau)$ is close enough to 0 for $\tau \neq 0$. The MATLAB® Identification Toolbox offers the possibility to plot $\hat{R}_\varepsilon^N(\tau)$ in function of τ, as well as the threshold beyond which it cannot be considered as 'sufficiently close' to zero, and which depends on a probability (confidence) level chosen by the user. More technically, the residuals will be considered as white with probability p if

$$\frac{N}{\big(\hat{R}_\varepsilon^N(0)\big)^2}\sum_{\tau=1}^{M}\big(\hat{R}_\varepsilon^N(\tau)\big)^2 < \chi_p^2(M) \tag{2.138}$$

where M is the number of time lags for which $\hat{R}_{\varepsilon}^{N}(\tau)$ is computed, N is the number of data samples used for the computation and $\chi_p^2(M)$ is the p level of the chi-square distribution with M degrees of freedom. Indeed,

$$\begin{aligned}\varepsilon(t) \in N(0, \lambda) \quad &\Longrightarrow \quad \sqrt{N}\hat{R}_{\varepsilon}^{N}(\tau) \in AsN(0, \lambda^2)\\ &\Longrightarrow \quad \frac{N}{\lambda^2}\sum_{\tau=1}^{M}\left(\hat{R}_{\varepsilon}^{N}(\tau)\right)^2 \in As\chi^2(M)\end{aligned} \tag{2.139}$$

See (Ljung, 1999) for details.

If the model fails to pass this test, it means that there is more information in the data than explained by the model, *i.e.*, that a higher-order or more flexible model structure should be used. This test will systematically fail if an output-error model structure is used while the real system output is subject to significantly nonwhite disturbances.

The cross-correlation between residuals and past inputs tells us if the residuals $\varepsilon(t)$ still contain information coming from the input signal $u(t)$. It is defined as

$$\hat{R}_{\varepsilon u}^{N}(\tau) = \frac{1}{N}\sum_{t=1}^{N}\varepsilon(t)u(t-\tau) \tag{2.140}$$

If it is significantly different of 0 for some time lag τ, then there is information coming from $u(t-\tau)$ that is present in $y(t)$ but not in $\hat{y}_p(t \mid \mathcal{M}(\hat{\theta}))$. This means that $G(z, \hat{\theta})$ is not representing the transfer from $u(t-\tau)$ to $y(t)$ correctly. This will typically be the case if the system delay or order is incorrectly estimated and it means that something has to be done with the structure of $G(z, \theta)$. Incidentally, note that there will never be any correlation between $\varepsilon(t)$ and future inputs $u(t+\tau)$, because of the causality of the system $G_0(z)$, except if the data are collected in closed loop, because $u(t)$ then depends on past outputs, hence on past disturbances $v(t)$ (and hence, $e(t)$). More technically, it can be shown (Ljung, 1999) that

$$\sqrt{N}\hat{R}_{\varepsilon u}^{N}(\tau) \in AsN(0, P), \quad P = \sum_{k=-\infty}^{\infty} R_{\varepsilon}(k)R_u(k) \tag{2.141}$$

and the uncorrelation test with probability p amounts to check if

$$\left|\hat{R}_{\varepsilon u}^{N}(\tau)\right| \leq \sqrt{\frac{P}{N}}N_p \tag{2.142}$$

where N_p denotes the p level of the $N(0, 1)$ distribution. $R_{\varepsilon}(k)$ and $R_u(k)$ are the unknown true auto-correlation functions of $\varepsilon(t)$ and $u(t)$, and they can only be approximated from a finite number of data. Once again, the MATLAB® Identification Toolbox offers the possibility to plot $\hat{R}_{\varepsilon u}^{N}(\tau)$ in function of τ, as well as the threshold beyond which it cannot be considered as 'sufficiently close' to zero and which depends on the confidence level p chosen by the user.

C. Pole-zero cancellations. While the residuals analysis tells us whether the model structure and order are respectively flexible and high enough to capture all the system dynamics, it does not detect a too high model order. In particular, it is important to avoid (near) pole-zero cancellations, as they do generally not represent the reality, are a cause of increased model variance by virtue of the n factor in (2.128) and lead to a (nearly) nonminimal model that may pose control design issues. Pole-zero cancellations can be detected by plotting the poles and zeroes of the model in the complex plane. Any such cancellation means that both the numerator and denominator orders of the model can be reduced by 1.

2.6.7 Closed-loop Identification

When the system G_0 operates in closed loop with some stabilising controller K as in Figure 2.1, it is possible to use closed-loop data for the identification of a model $\hat{G}$. This can be done using different approaches, among which the following three will be used in this book.

A. The indirect approach. This approach, proposed by (Söderström and Stoica, 1989), uses measurements of either $r_1(t)$ or $r_2(t)$ and either $y(t)$ or $u(t)$ to identify one of the four entries of the matrix $T(G_0, K)$ of (2.8). From this estimate, a model $\hat{G}$ for G_0 is derived, using knowledge of the controller, which has to be LTI. Such knowledge is required for this method.

B. The coprime-factor approach. This approach, proposed by (Van den Hof and Schrama, 1995), uses measurements of $r_1(t)$ or $r_2(t)$ and of $y(t)$ and $u(t)$ to identify the two entries of a column of $T(G_0, K)$. A model $\hat{G}$ for G_0 is then given by the ratio of these two entries. This method requires that the controller be LTI, but it can be unknown.

C. The direct approach. This approach consists in identifying a model $\hat{G}$ for G_0 directly from measurements of $u(t)$ and $y(t)$ collected in closed loop.

This is, of course, only a rough description of these approaches. Variants exist (*e.g.*, indirect identification with a tailor-made parametrisation), as well as other methods (*e.g.*, the dual Youla parametrisation method (Hansen, 1989; De Bruyne *et al.*, 1998), which will be used and described in Chapter 4). It has recently been shown that the qualitative properties of the different closed-loop identification methods are essentially equivalent, by observing that these methods can be seen as different parametrisations of one and the same prediction-error method (Forssell and Ljung, 1999).

2.6.8 Data Preprocessing

We cannot conclude this section on system identification without a word on data preprocessing, as it is a crucial phase upon which the whole identification procedure will rest.

As we have said in the beginning of this section, prediction-error identification is based on the assumption that the true system is LTI, which is generally acceptable if variations around a given operating point are considered. It is therefore important to remove the value of this operating point from the input and output signals that are used for identification, in such a way that only small variations around the setpoint are considered. Depending on the process, this operating point will be the initial value of the signals (*e.g.*, if the data used for identification are those of a step response), or its mean, or a slow trend. The removal of the continuous component can be done either algebraically, or by high-pass data filtering. The latter can also be used for data detrending.

The sampling frequency of the data must be chosen carefully and an anti-aliasing filter must be used during the acquisition procedure and during any possible subsequent step of downsampling. If the sampling period is too small with respect to the system time constant, the poles and zeroes of the identified model will drift towards the point $1 + 0\jmath$ of the unit circle in the complex plane, *i.e.*, the obtained model will be close to instability and numerically ill conditioned.

▶ **Example 2.5.** To illustrate this, consider for instance the first-order continuous-time system $G(s) = \frac{1}{s+1}$. Its time constant is 1 second. If it is sampled with a very small sampling period t_s, the derivative can be approximated by a finite difference, *i.e.*, by setting $s \approx \frac{1-z^{-1}}{t_s}$. Hence, $G(s)$ can be approximated by the discrete-time transfer function $G(z) = \frac{t_s}{(1+t_s)-z^{-1}}$, whose pole tends to $z = 1$ as t_s tends to 0. ◀

On the contrary, if the sampling period is too large, useful signal information will be lost. There are several rules for choosing the appropriate sampling period. (Zhu, 2001) proposes the following:

- to choose $t_s \approx \frac{T_{min}}{3}$, where T_{min} is the smallest time constant of interest;
- to choose $f_s = 10 f_o$, where f_s is the sampling frequency and f_o is the cut-off frequency of the process;
- to choose t_s in the range $\frac{T_{set}}{100} \leq t_s \leq \frac{T_{set}}{20}$, where T_{set} is the settling time of the process.

Other important operations in data preprocessing are peak shaving, removal of outliers, selection of a data set not affected by unmeasured disturbances or changes in the operating conditions, and division of the data into two subsets, respectively for parameter estimation and for validation.

2.7 Balanced Truncation

Here, we make the assumption that a strictly stable, high-order, continuous-time model $G_n(s)$ of the system is available (see Subsection 2.7.6 for some remarks on the discrete-time case), and that it has the following state-space representation:

$$G_n = \left[\begin{array}{c|c} A & B \\ \hline C & D \end{array}\right] \tag{2.143a}$$

i.e.,

$$G_n(s) = C(sI - A)^{-1}B + D \tag{2.143b}$$

The balanced truncation procedure, which was first proposed in (Moore, 1981), consists in computing a state-space representation of $G_n(s)$ in which the most controllable modes coincide with the most observable ones. The least observable and controllable modes, which have little influence on the input-output behaviour of the model, are then discarded. This procedure can be extended to the case where frequency weightings are used (Enns, 1984*a*, 1984*b*). We shall first briefly review the notions of controllability and observability and then describe the balanced truncation procedure without and with frequency weightings.

2.7.1 The Concepts of Controllability and Observability

A. Controllability.

Definition 2.6. (CONTROLLABILITY) *A state $x_0 \in \mathbb{R}^n$ of the system G_n is called controllable if there exists a control input signal $u(t)$, $t \in [0, T]$, that brings the system from initial state $x(0) = x_0$ to $x(T) = 0$ in a finite time T. The controllable subspace of G_n, $X^{cont} \subseteq \mathbb{R}^n$, is the set of all controllable states of G_n. G_n is called controllable if $X^{cont} = \mathbb{R}^n$.*

The controllability matrix of G_n is

$$\mathbf{C} \triangleq \begin{bmatrix} B & AB & \ldots & A^{n-1}B \end{bmatrix} \tag{2.144}$$

The columns of $\mathbf{C}$ generate X^{cont}. As a result, G_n (or, equivalently, the pair (A, B)) is controllable if and only if $\operatorname{rank} \mathbf{C} = n$. The controllability of G_n implies that it is always possible to find a finite input signal $u(t)$ that brings the state from any initial condition $x(0) = x_1$ to any desired value $x(T) = x_2$ in a finite time $T > 0$.

The controllability Gramian of G_n at time $0 \le t < \infty$ is defined as

$$\mathbf{P}(t) \triangleq \int_0^t e^{A\tau} BB^T e^{A^T\tau} d\tau, \qquad t \in \mathbb{R}^+ \tag{2.145}$$

It has the following properties:

- $\forall t \in \mathbb{R}^+ \quad \mathbf{P}(t) = \mathbf{P}^T(t) \ge 0$, *i.e.*, it is symmetric and positive semi-definite;
- $\forall t \in \mathbb{R}^+ \quad \operatorname{im} \mathbf{P}(t) = \operatorname{im} \mathbf{C}$, *i.e.*, the columns of $\mathbf{P}(t)$ generate X^{cont} for all positive t.

For a strictly stable system, the infinite controllability Gramian is defined as

$$\mathbf{P} \triangleq \int_0^\infty e^{A\tau} BB^T e^{A^T\tau} d\tau \tag{2.146}$$

It has the same properties as $\mathbf{P}(t)$. As a result, a stable system G_n is fully controllable if and only if $\mathbf{P} > 0$, *i.e.*, if it is positive definite. This infinite controllability Gramian, which we shall simply call the controllability Gramian in the sequel, can be computed by solving the following Lyapunov equation:

$$A\mathbf{P} + \mathbf{P}A^T + BB^T = 0 \tag{2.147}$$

The controllability Gramian $\mathbf{P}$ gives the minimum input energy that would be necessary to bring the system from the free equilibrium $x(-\infty) = 0$ to a state $x(0) = x_0$:

$$\min_u \left\{ \int_{-\infty}^0 u^T(t)u(t)dt \mid x(0) = x_0 \right\} = x_0^T \mathbf{P}^{-1} x_0 \quad \text{when } x(-\infty) = 0 \tag{2.148}$$

B. Observability.

Definition 2.7. (OBSERVABILITY) *A state $x_0 \in \mathbb{R}^n$ of the system G_n is called unobservable if the free response of the output of G_n to this state is identically zero for all $t \ge 0$, i.e., if this state cannot be distinguished from the zero state. The unobservable subspace of G_n, $\complement_{\mathbb{R}^n} X^{obs} \subseteq \mathbb{R}^n$, is the set of all unobservable states of G_n. G_n is called observable if $\complement_{\mathbb{R}^n} X^{obs} = \{0\}$ or, equivalently, if $X^{obs} = \mathbb{R}^n_0$.*

The observability matrix of G_n is

$$\mathbf{O} \triangleq \begin{bmatrix} C \\ CA \\ \vdots \\ CA^{n-1} \end{bmatrix} \tag{2.149}$$

The null space of $\mathbf{O}$ is $\complement_{\mathbb{R}^n} X^{obs}$, hence G_n (or, equivalently, the pair (A, C)) is observable if and only if $\operatorname{rank} \mathbf{O} = n$. G_n is observable means that equal outputs for equal inputs imply equal initial state conditions.

The observability Gramian of G_n at time $0 \leq t < \infty$ is defined as

$$\mathbf{Q}(t) \triangleq \int_0^t e^{A^T(t-\tau)} C^T C e^{A(t-\tau)} d\tau, \qquad t \in \mathbb{R}^+ \tag{2.150}$$

It has the following properties:

- $\forall t \in \mathbb{R}^+ \quad \mathbf{Q}(t) = \mathbf{Q}^T(t) \geq 0$, *i.e.*, it is symmetric and positive semi-definite;
- $\forall t \in \mathbb{R}^+ \quad \ker \mathbf{Q}(t) = \ker \mathbf{O}$, *i.e.*, the null space of $\mathbf{Q}(t)$ is $\complement_{\mathbb{R}^n} X^{obs}$ for all positive t.

For a strictly stable system, the infinite observability Gramian is defined as

$$\mathbf{Q} \triangleq \int_0^\infty e^{A^T\tau} C^T C e^{A\tau} d\tau \tag{2.151}$$

It has the same properties as $\mathbf{Q}(t)$. As a result, G_n is fully observable if and only if $\mathbf{Q} > 0$, *i.e.*, if it is positive definite. This infinite observability Gramian, which we shall simply call the observability Gramian in the sequel, can be computed by solving the following Lyapunov equation:

$$A^T\mathbf{Q} + \mathbf{Q}A + C^T C = 0 \tag{2.152}$$

The observability Gramian $\mathbf{Q}$ gives the total energy of the free output response of the system to an initial state $x(0) = x_0$:

$$\int_0^\infty y^T(t) y(t) dt = x_0^T \mathbf{Q} x_0 \quad \text{when } u(t) = 0 \ \forall t \geq 0 \text{ and } x(0) = x_0 \tag{2.153}$$

2.7.2 Balanced Realisation of a System

The following observations can be made regarding the controllability and observability Gramians:

- the controllability and observability Gramians $\mathbf{P}$ and $\mathbf{Q}$ depend on the state-space realisation of the system G_n;
- their eigenvalues give information about the 'level' of observability or controllability of the state variables;
- depending on the chosen realisation (2.143), some state variables (*i.e.*, some dynamics) can be very observable but little controllable, or vice-versa.

If the realisation (2.143) is minimal, *i.e.*, if it is both controllable and observable ($\mathbf{P} > 0$ and $\mathbf{Q} > 0$), it is possible to find a state transformation that brings the system to a form where the most observable dynamics are also the most controllable ones. This is called a *balanced realisation*. When the system is in balanced form, its Gramians are diagonal and equal:

$$\mathbf{P} = \mathbf{Q} = \Sigma \triangleq \operatorname{diag}(\varsigma_1, \ldots, \varsigma_n) \tag{2.154}$$

where the ς_i's are the Hankel singular values of G_n in decreasing order:

$$\varsigma_i = \sqrt{\lambda_i(\mathbf{PQ})} > 0, \quad \varsigma_i \geq \varsigma_{i+1} \tag{2.155}$$

Observe that, although the Gramians depend on the state realisation of the system, the Hankel singular values are independent of this realisation. Once a balanced realisation has been obtained, it is possible to reduce the order of the system by simply discarding the state variables that correspond to the least observable and controllable dynamics. Indeed, these are the dynamics that contribute the least to the global input-output behaviour of the system.

Starting from any realisation (2.143) of G_n, one can compute a balanced realisation as follows.

1. Compute the controllability and observability Gramians $\mathbf{P}$ and $\mathbf{Q}$ by solving the Lyapunov equations (2.147) and (2.152).
2. Compute
$$\Sigma = \text{diag}(\varsigma_1, \ldots, \varsigma_n) \tag{2.156}$$
where the ς_i's are given by (2.155).
3. Compute R such that
$$\mathbf{P} = R^T R \tag{2.157}$$
It can be obtained as
$$R = \Upsilon\sqrt{\Lambda}\Upsilon^T \tag{2.158}$$
where Λ is a diagonal matrix containing the eigenvalues of $\mathbf{P}$ and Υ is a matrix whose columns are the eigenvectors of $\mathbf{P}$ associated with the entries of Λ.
4. Make a singular-value decomposition of $R\mathbf{Q}R^T$:
$$R\mathbf{Q}R^T = U\Sigma^2 U^T \tag{2.159}$$
5. Then
$$T = \Sigma^{1/2} U^T R^{-T} \tag{2.160}$$
is the balancing transformation matrix and there holds
$$T\mathbf{P}T^T = T^{-T}\mathbf{Q}T^{-1} = \Sigma \tag{2.161}$$
6. Compute
$$\breve{A} = TAT^{-1} \qquad \breve{B} = TB \qquad \breve{C} = CT^{-1} \qquad \breve{D} = D \tag{2.162}$$
Then,
$$\breve{G}_n = \left[\begin{array}{c|c} \breve{A} & \breve{B} \\ \hline \breve{C} & \breve{D} \end{array}\right] \tag{2.163}$$
is a balanced realisation of $G_n(s)$.

2.7.3 Balanced Truncation

The diagonal of Σ contains the Hankel singular values of G_n in decreasing order. The state variables of the balanced realisation $\breve{G}_n$ follow the same order: x_i is more observable and controllable than x_j for $1 \leq i < j \leq n$. More precisely, $\breve{G}_n$ is more observable and controllable in the direction of $\vec{e}_i$ than in the direction of $\vec{e}_j$ for $1 \leq i < j \leq n$. Let us then consider the following partition:

$$\Sigma = \begin{bmatrix} \Sigma_{11} & 0 \\ 0 & \Sigma_{22} \end{bmatrix} \tag{2.164a}$$

$$\Sigma_{11} = \text{diag}(\varsigma_1, \ldots, \varsigma_r) \tag{2.164b}$$

$$\Sigma_{22} = \text{diag}(\varsigma_{r+1}, \ldots, \varsigma_n) \tag{2.164c}$$

The corresponding partition of $\breve{G}_n$ is

$$\breve{G}_n = \left[\begin{array}{cc|c} \breve{A}_{11} & \breve{A}_{12} & \breve{B}_1 \\ \breve{A}_{21} & \breve{A}_{22} & \breve{B}_2 \\ \hline \breve{C}_1 & \breve{C}_2 & \breve{D} \end{array}\right] \tag{2.165}$$

Let us define

$$\hat{G}_r = \left[\begin{array}{c|c} \breve{A}_{11} & \breve{B}_1 \\ \hline \breve{C}_1 & \breve{D} \end{array}\right]$$

$$\triangleq \text{bt}\,(G_n, r) \tag{2.166}$$

which is obtained by truncating the least observable and controllable dynamics of $\breve{G}_n$. The obtained reduced-order system $\hat{G}_r$ is a stable suboptimal solution to the following minimisation problem:

$$\min_{G_r(s) \text{ of order } r} \|G_n(s) - G_r(s)\|_\infty \tag{2.167}$$

which is hard to solve exactly. Upper and lower bounds of the reduction error in the $\mathcal{H}_\infty$ norm can easily be computed from the Hankel singular values of G_n (Glover, 1984), (Enns, 1984*a*, 1984*b*):

$$\varsigma_r \leq \left\| G_n(s) - \hat{G}_r(s) \right\|_\infty \leq 2 \sum_{i=r+1}^{n} \varsigma_i \tag{2.168}$$

The level r of truncation can be chosen by plotting the ς_i's: it is best to have $\varsigma_r >> \varsigma_{r+1}$, which means that the dynamics corresponding to $\varsigma_{r+1}, \ldots, \varsigma_n$ can really be neglected because of their relatively poor degree of observability and controllability. A tighter upper bound in (2.168) can be obtained by counting the ς_i's of multiplicity larger than 1 only once in the summation.

2.7.4 Numerical Issues

The balancing transformation of Subsection 2.7.2 is a frequent source of numerical difficulty, as is often the case with nonorthogonal transformations. There exist, however, algorithms that compute the reduced-order system $\hat{G}_r(s)$ starting from any realisation of the full-order system $G_n(s)$ without performing explicitly this balancing transformation. The following algorithm, for instance, has been proposed by (Safonov and Chiang, 1989).

1. Starting from the Gramians $\mathbf{P}$ and $\mathbf{Q}$ of any state-space realisation of $G_n(s)$, compute ordered Schur decompositions of the product $\mathbf{PQ}$:
$$V_A^T \mathbf{PQ} V_A = S_A \qquad V_D^T \mathbf{PQ} V_D = S_D \tag{2.169}$$
where S_A and S_D are upper triangular matrices with the eigenvalues of $\mathbf{PQ}$ on their diagonals, respectively in ascending and descending order and V_A and V_D are orthogonal.
2. Compute the following submatrices, where r is the order of the desired reduced-order system $\hat{G}_r(s)$:
$$V_a = V_A \begin{bmatrix} 0_{(n-r)\times r} \\ I_{r\times r} \end{bmatrix} \qquad V_d = V_D \begin{bmatrix} I_{r\times r} \\ 0_{(n-r)\times r} \end{bmatrix} \tag{2.170}$$
3. Compute a singular-value decomposition of $V_a^T V_d$:
$$U_L S U_R^T = V_a^T V_d \tag{2.171}$$
where S is diagonal with positive entries and U_L and U_R are orthogonal.
4. Transformation matrices are obtained as
$$L = S^{-\frac{1}{2}} U_L^T V_a^T \qquad R = V_d U_R S^{-\frac{1}{2}} \tag{2.172}$$
($LR = I$, meaning that the transformation is orthogonal) and a state-space realisation of $G_r(s)$ is given by
$$A_r = LAR \qquad B_r = LB \qquad C_r = CR \qquad D_r = D \tag{2.173}$$

2.7.5 Frequency-weighted Balanced Truncation

A very common case is when the reduction criterion contains stable input and/or output frequency-weighting filters W_r and/or W_l. The minimisation problem becomes

$$\min_{G_r(s) \text{ of order } r} \left\| W_l(s)\big(G_n(s) - G_r(s)\big)W_r(s) \right\|_\infty \tag{2.174}$$

to which a suboptimal solution can be computed by frequency-weighted balanced truncation.

Let us consider any minimal realisation of the stable system $G_n(s)$ defined in (2.143) and realisations of the two filters as follows:

$$W_l = \left[\begin{array}{c|c} A_l & B_l \\ \hline C_l & D_l \end{array}\right] \qquad W_r = \left[\begin{array}{c|c} A_r & B_r \\ \hline C_r & D_r \end{array}\right] \tag{2.175}$$

Then, a realisation of $W_l(s)G_n(s)W_r(s) \triangleq \tilde{G}_n(s) = \tilde{C}(sI - \tilde{A})^{-1}\tilde{B} + \tilde{D}$ is given by

$$\breve{G}_n = \left[\begin{array}{c|c} \tilde{A} & \tilde{B} \\ \hline \tilde{C} & \tilde{D} \end{array}\right] = \left[\begin{array}{ccc|c} A_l & B_l C & B_l D C_r & B_l D D_r \\ 0 & A & B C_r & B D_r \\ 0 & 0 & A_r & B_r \\ \hline C_l & D_l C & D_l D C_r & D_l D D_r \end{array}\right] \tag{2.176}$$

The controllability and observability Gramians of this input-output frequency-weighted system are respectively the solutions of the following Lyapunov equations:

$$\tilde{A}\tilde{\mathbf{P}} + \tilde{\mathbf{P}}\tilde{A}^T + \tilde{B}\tilde{B}^T = 0 \tag{2.177}$$

$$\tilde{A}^T\tilde{\mathbf{Q}} + \tilde{\mathbf{Q}}\tilde{A} + \tilde{C}^T\tilde{C} = 0 \tag{2.178}$$

These Gramians can be partitioned similarly to $\tilde{A}$ in (2.176):

$$\tilde{\mathbf{P}} = \begin{bmatrix} \mathbf{P}_{11} & \mathbf{P}_{12} & \mathbf{P}_{13} \\ \mathbf{P}_{21} & \mathbf{P}_{22} & \mathbf{P}_{23} \\ \mathbf{P}_{31} & \mathbf{P}_{32} & \mathbf{P}_{33} \end{bmatrix} \qquad \tilde{\mathbf{Q}} = \begin{bmatrix} \mathbf{Q}_{11} & \mathbf{Q}_{12} & \mathbf{P}_{13} \\ \mathbf{Q}_{21} & \mathbf{Q}_{22} & \mathbf{P}_{23} \\ \mathbf{Q}_{31} & \mathbf{Q}_{32} & \mathbf{P}_{33} \end{bmatrix} \tag{2.179}$$

$\mathbf{P}_{22}$ and $\mathbf{Q}_{22}$, which correspond to the A block in (2.176), are then the frequency-weighted controllability and observability Gramians of $G_n(s)$. If $W_r(s) \neq I$, then $\mathbf{P}_{22} \neq \mathbf{P}$ given by (2.147), which means that the input weighting filter modifies the controllability Gramian of the system. Similarly, if $W_l(s) \neq I$, then $\mathbf{Q}_{22} \neq \mathbf{Q}$ given by (2.152), which means that the output weighting filter modifies the observability Gramian of the system.

The rest of the procedure consists of finding a transformation matrix T such that $T\mathbf{P}_{22}T^T = T^{-T}\mathbf{Q}_{22}T^{-1} = \Sigma = \text{diag}(\varsigma_1, \ldots, \varsigma_n)$ where the ς_i's are the frequency-weighted Hankel singular values of $G_n(s)$ and to apply this transformation to the realisation (2.143). This produces a frequency-weighted balanced realisation $\breve{G}_n$ of the system $G_n(s)$. Its order can then be reduced, as in unweighted balanced truncation, by discarding the modes corresponding to the smallest Hankel singular values. In the sequel,

$$\hat{G}_r = \text{fwbt}\,(G_n, W_l, W_r, r) \tag{2.180}$$

will denote the r-th order system produced by frequency-weighted balanced truncation of $G_n(s)$. An upper bound of the approximation error in the $\mathcal{H}_\infty$ norm is given by (Kim *et al.*, 1995)

$$\left\| W_l(s)\big(G_n(s) - G_r(s)\big)W_r(s) \right\|_\infty \leq 2 \sum_{i=r+1}^{n} \sqrt{\varsigma_i^2 + (\alpha_i + \beta_i)\varsigma_i^{3/2} + \alpha_i\beta_i\varsigma_i} \tag{2.181}$$

where

$$\alpha_i = \|\Xi_{i-1}(s)\|_\infty \left\| C_r \Phi_r(s) \mathbf{P}_{33}^{1/2} \right\|_\infty \tag{2.182a}$$

$$\beta_i = \left\| \mathbf{Q}_{11}^{1/2} \Phi_l(s) B_l \right\|_\infty \|\Gamma_{i-1}(s)\|_\infty \tag{2.182b}$$

$$\Xi_{i-1}(s) = A_{21}^{i-1}(sI - A_{i-1})^{-1} B_{i-1} + b_i \tag{2.182c}$$

$$\Gamma_{i-1}(s) = C_{i-1}(sI - A_{i-1})^{-1} A_{12}^{i-1} + c_i \tag{2.182d}$$

$$\Phi_r(s) = (sI - A_r)^{-1} \tag{2.182e}$$

$$\Phi_l(s) = (sI - A_l)^{-1} \tag{2.182f}$$

$$A_i = \begin{bmatrix} A_{i-1} & A_{12}^{i-1} \\ A_{21}^{i-1} & a_{ii} \end{bmatrix} \tag{2.182g}$$

$$B_i = \begin{bmatrix} B_{i-1} \\ b_i \end{bmatrix} \tag{2.182h}$$

$$C_i = \begin{bmatrix} C_{i-1} & c_i \end{bmatrix} \tag{2.182i}$$

and b_i and c_i are the i-th row of B_i and the i-th column of C_i, respectively, and $A_n = A$, $B_n = B$, $C_n = C$.

Let us finally remark that the reduced-order model $\hat{G}_r$ is guaranteed to be stable if frequency weighting is used at only one side of the reduction criterion.

2.7.6 Balanced Truncation of Discrete-time Systems

The balanced truncation and frequency-weighted balanced truncation procedures are exactly the same in the discrete-time case, except that the controllability and observability Gramians are respectively the solutions of the discrete Lyapunov equations

$$A\mathbf{P}A^T - \mathbf{P} + BB^T = 0 \tag{2.183}$$

and

$$A^T\mathbf{Q}A - \mathbf{Q} + C^TC = 0 \tag{2.184}$$

Observe that, since the $\mathcal{H}_2$ norm of a stable system is upper bounded by its $\mathcal{H}_\infty$ norm in the discrete-time case (Boyd and Doyle, 1987), the $\mathcal{H}_\infty$ upper bound of the approximation error, computed from the Hankel singular values (both in the unweighted and single-side weighted cases), is also an $\mathcal{H}_2$ upper bound of it.

CHAPTER 3

Identification in Closed Loop for Better Control Design

3.1 Introduction

This chapter discusses a number of issues related to the problem of modelling and identification for control design. The goal is to provide insights and partial answers to the following central question:

> *How should we identify a model $(\hat{G}, \hat{H})$ in such a way that it is good for control design?*

These insights will constitute the thread of this book and each of the following chapters will investigate more deeply some of the questions raised here.

Obviously, a reasonable qualification of a good model $(\hat{G}, \hat{H})$ for control design would be

1. **simultaneous stabilisation:** the controller $K(\hat{G}, \hat{H})$ designed from this model stabilises both the model $\hat{G}$ and the plant G_0, and
2. **similar performance:** The performance achieved by the controller $K(\hat{G}, \hat{H})$ when it is applied to the real system (G_0, H_0) is close to the designed performance, *i.e.*, to the performance it achieves with the nominal model $(\hat{G}, \hat{H})$.

The characterisation of all models that satisfy (1), for a model reference control design criterion, was studied in (Blondel *et al.*, 1997) and (Gevers *et al.*, 1997).

Observe that the problem of modelling for control involves three players: the plant (G_0, H_0), the model $(\hat{G}, \hat{H})$ and the to-be-designed controller $K(\hat{G}, \hat{H})$.

(For the sake of simplicity, the latter will simply be denoted by K in the sequel of this chapter.)

Identification for control often involves one or several steps of closed-loop identification, by which we mean identification of a model $(\hat{G}, \hat{H})$ of the plant (G_0, H_0) using data collected on the closed-loop system formed by the feedback connection of G_0 and some to-be-replaced controller K_{id}. Thus, closed-loop identification also involves three players: the plant (G_0, H_0), the model $(\hat{G}, \hat{H})$ and the current controller K_{id}. The third player (the controller) is different from the one above. As a result, closed-loop identification for control involves four players: the plant (G_0, H_0), the model $(\hat{G}, \hat{H})$, the controller K_{id} applied during identification and the to-be-designed controller K. The focus of our discussion is on the interplay between these four players. The typical context is one in which a plant (G_0, H_0) is presently under closed-loop control with some controller K_{id} (*e.g.*, a PID controller with a suboptimal level of performance) and where it is desired to estimate a model $(\hat{G}, \hat{H})$ with the view of designing a new model-based controller K that would achieve a better performance on the plant (G_0, H_0) while guaranteeing stability robustness.

In particular we address the following issues.

- We examine the role of the controller K_{id} in changing the experimental conditions.
- We compare the effects of open-loop and closed-loop identification in terms of bias and variance errors in the context of identification for control.
- We motivate the need for cautious controller adjustment in iterative design.

3.2 The Role of Feedback

It is well known that two plants that present similar open-loop behaviours (in terms of their impulse responses or Nyquist diagrams, for instance) may present considerably different closed-loop behaviours when they are connected to the same controller. Conversely, one can attach a stabilising controller to two plants and observe what appears to be identical closed-loop behaviours, when the open-loop behaviours of the plants are considerably different.

▶ **Example 3.1.** These facts are illustrated in (Skelton, 1989) where it is shown, for instance, that the plants $G_1(s) = \frac{1}{s+1}$ and $G_2(s) = \frac{1}{s}$ have remarkably different open-loop behaviours, while a static output feedback $u(t) = r(t) - ky(t)$ will make their closed-loop behaviours almost indistinguishable for large k (at least if one considers the closed-loop step responses): see Figure 3.1.

It is also interesting to observe that

$$\delta_\nu(100G_1, 100G_2) = 0.01 << \delta_\nu(G_1, G_2) = 0.71$$

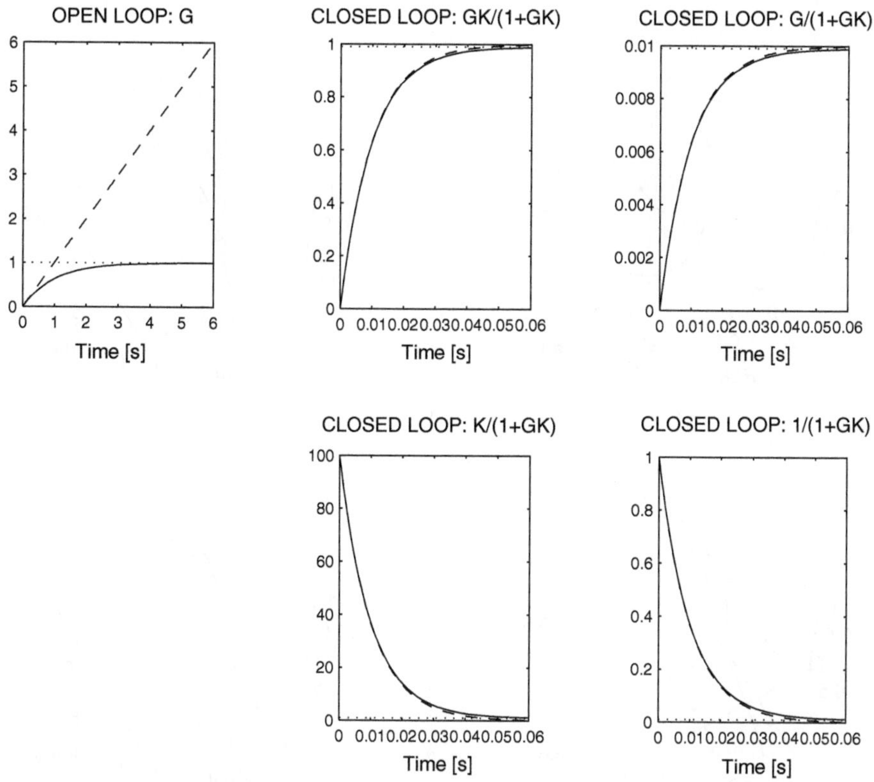

Figure 3.1. Step responses of $G_1(s) = \frac{1}{s+1}$ (—) and of $G_2(s) = \frac{1}{s}$ (−−) in open loop and in closed loop with $k = 100$

which shows quantitatively that G_1 and G_2 are much 'closer' to each other in closed loop with the static controller $k = 100$ than in open loop. The effect of k on $\delta_\nu(kG_1, kG_2)$ is illustrated in Figure 3.2. Although k only has an homothetic effect on the Nyquist plots of G_1 and G_2, it has a completely different effect on their projections onto the Riemann sphere, which become closer to each other as k increases. In the present case, $\delta_\nu(kG_1, kG_2) = \max_\omega \kappa(kG_1(j\omega), kG_2(j\omega))$ is determined at $\omega = 0$ rad/s. At this frequency, as k increases, the projection of kG_1 moves from the equator of the sphere (for $k = 1$) to its north pole (for $k = \infty$) while the projection of G_2 stays at the north pole. As a result, the ν-gap decreases from $\sqrt{0.5}$ to 0. ◀

We now return to the modelling game with the three players described in the introduction. Since the model $(\hat{G}, \hat{H})$ is used for the design, what is required of $(\hat{G}, \hat{H})$ is that the stability of the $(\hat{G}, K)$ loop implies that of the (G_0, K) loop (stability robustness) and that the closed-loop transfer functions of these two loops are close to each other (performance robustness).

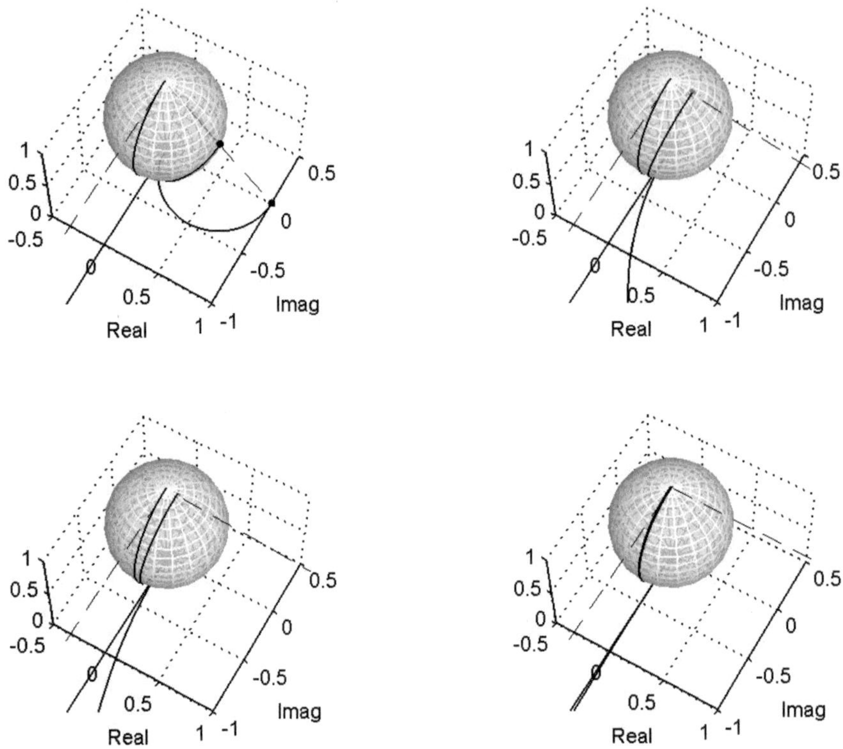

Figure 3.2. Effect of k on the ν-gap between $kG_1(s) = \frac{k}{s+1}$ and $kG_2(s) = \frac{k}{s}$. From left to right and from top to bottom: $k = 1$, $k = 5$, $k = 10$ and $k = 100$.

One of the main aims of feedback is to reduce the effects of model uncertainty on the open-loop plant. Feedback often has a sensitivity reduction objective. Therefore, any sensible feedback design will have the effect that the closed-loop systems $(\hat{G}, K)$ and (G_0, K) behave much closer to each other than the open-loop systems $\hat{G}$ and G_0.

What are the consequences of these observations on identification for control? The issue here is change of experimental conditions. Models can only have their quality evaluated for a particular set of experimental conditions. Changing from open-loop to closed-loop operation with a specified controller, or changing the controller, is of course one change of experimental conditions. Unless a plant model is exact, high accuracy under one set of experimental conditions (*e.g.*, open loop) does not guarantee its efficacy under changed experimental conditions. It is also intuitively clear that small changes of controller should probably avoid, or ameliorate, the problem of possible loss of efficacy of a model under changed experimental conditions.

As a consequence, the best way to evaluate the quality of the model $(\hat{G}, \hat{H})$ is to test it under the experimental conditions under which the plant (G_0, H_0) is due to operate, *i.e.*, in closed loop with the to-be-designed controller K. For the same reason, it should ideally be identified under those same feedback conditions; see, *e.g.*, (Hjalmarsson *et al.*, 1994*a*, 1996). This is of course impossible since knowledge of the model is required to design the controller K. The philosophy behind iterative design of models and controllers is to approximate these experimental conditions in successive steps. This allows one to successively reduce the uncertainty in the frequency bands of importance for the design of the next controller, even with reduced-order models (*i.e.*, models such that $\mathcal{S} \notin \mathcal{M}$ and/or $G_0 \notin \mathcal{G}$).

The alternative, advocated in (Ljung, 1997, 1998), is to identify in open loop a model of sufficiently high order that is validated by the data, and to subsequently perform a step of model or controller reduction. This will lead to much more conservative designs because such a validated model will have an uncertainty distribution that is not shaped for control design. In addition, this uncertainty region cannot be reduced by order reduction, whether frequency shaped or not. This will be investigated further in Section 3.4.

3.3 The Effect of Feedback on the Modelling Errors

We now analyse the effect of performing the identification of a system in closed loop on the bias and variance errors of the estimate.

3.3.1 The Effect of Feedback on the Bias Error

For the purpose of analysis, we consider that there exists a true LTI system that can be described by (2.79):

$$y(t) = G_0(z)u(t) + v(t) \tag{3.1}$$

We consider that this system is connected to a feedback controller K_{id}:

$$u(t) = r(t) - K_{id}(z)y(t) \tag{3.2}$$

where $r(t)$ denotes either $r_2(t)$ or $K_{id}(z)r_1(t)$ (see Figure 2.1 for the definitions of $r_1(t)$ and $r_2(t)$).

For simplicity, we restrict our analysis to the SISO case, but the results presented here can be readily extended to the MIMO case.

Consider that a parametrised model $G(z, \hat{\theta})$ is identified from N input-output data Z^N obtained on this system, using a prediction-error method. Define, as usual (see Section 2.6),

$$\varepsilon_F(t, \theta) = L(z, \theta)\big(y(t) - G(z, \theta)u(t)\big) \tag{3.3}$$

(which is equivalent to equation (2.103) since the noise model $H(z, \theta)$ can be included in the prefilter $L(z, \theta)$),

$$V_N(\theta, Z^N) = \frac{1}{N}\sum_{t=1}^{N} \varepsilon_F^2(t, \theta) \tag{3.4}$$

and

$$\hat{\theta} = \arg\min_{\theta} V_N(\theta, Z^N) \tag{3.5}$$

It follows easily that

$$\hat{\theta} \to \theta^* = \arg\min_{\theta} \bar{V}(\theta) \quad \text{as} \quad N \to \infty \tag{3.6}$$

where[1]

$$\bar{V}(\theta) = \frac{1}{2\pi}\int_{-\pi}^{\pi} |G_0 - G(\theta)|^2 \, |L(\theta)|^2 \, \phi_u^r d\omega + \frac{1}{2\pi}\int_{-\pi}^{\pi} |1 + G(\theta)K_{id}|^2 \, |S(G_0, K_{id})|^2 \, |L(\theta)|^2 \, \phi_v d\omega \tag{3.7}$$

Here $S(G_0, K_{id}) = \frac{1}{1+G_0K_{id}}$ is the closed-loop sensitivity function and ϕ_u^r is the spectral density of that part of the control signal that originates from the reference input. Indeed, since r and v are uncorrelated, one can split the input spectrum into

$$\phi_u(\omega) = \phi_u^r(\omega) + \phi_u^v(\omega) \tag{3.8}$$

where

$$\phi_u^r(\omega) = \left|S\big(G_0(e^{j\omega}), K_{id}(e^{j\omega})\big)\right|^2 \phi_r(\omega) \tag{3.9}$$

Observe now that if one takes a parametrised prefilter $L(\theta) = \frac{1}{1+G(\theta)K_{id}}$, the second term in (3.7) becomes independent of θ and the minimising argument of $\bar{V}(\theta)$ is the same as the minimising argument of

$$\tilde{V}(\theta) = \frac{1}{2\pi}\int_{-\pi}^{\pi} \left|\frac{G_0 - G(\theta)}{1 + G(\theta)K_{id}}\right|^2 |S(G_0, K_{id})|^2 \, \phi_r d\omega \tag{3.10}$$

As a result, the identification will tend to a model that realises the best compromise between fitting $G(\theta)$ to G_0 and making the model sensitivity function $\frac{1}{1+G(\theta)K_{id}}$ small. Furthermore, the weighting factor in (3.10) contains the actual sensitivity function $S(G_0, K_{id})$. Thus, the fit between $G(\hat{\theta})$ and G_0 will be tight where either the actual sensitivity function or the model sensitivity function is large, *i.e.* around the corresponding cross-over frequencies.

This last observation has been the main heuristic motivation for performing the identification in closed loop when low-order models are needed for the purpose of designing new controllers with higher performance. The heuristic reasoning is that, since the model is necessarily biased due to its low order, one should

[1]Under the standard assumption that all signals are ergodic, we find that $\bar{V}(\theta) = \bar{E}\,|\varepsilon_F(t, \theta)|^2$. Parseval's relationship (2.88) then yields $\bar{V}(\theta) = \frac{1}{2\pi}\int_{-\pi}^{\pi} \phi_{\varepsilon_F}(\omega)d\omega$ which can be expressed as (3.7).

tune the bias distribution so as to make it low around the cross-over frequency of the present control loop. A small error around this cross-over frequency then allows one to design a new controller that will expand the the existing bandwidth.

This heuristic motivation can be given more solid support by considering our observations of Section 3.2 on the assessment of the quality of models. Ideally the model $\hat{G} \triangleq G(\hat{\theta})$ should be evaluated by how well it mimics the behaviour of the actual system G_0 when both are connected in feedback with the to-be-designed controller K. In other words, if we denote respectively by $y(t)$ and $\hat{y}(t)$ the outputs of the feedback connections of G_0 and $\hat{G}$ with the same controller K, *i.e.*,

$$y(t) = \frac{G_0(z)}{1+G_0(z)K(z)}r(t) + \frac{1}{1+G_0(z)K(z)}v(t) \tag{3.11}$$

$$\hat{y}(t) = \frac{\hat{G}(z)}{1+\hat{G}(z)K(z)}r(t) \tag{3.12}$$

the model $\hat{G}$ will be judged to be good if the mean square error between the actual output and the simulated output, $E\left|y(t)-\hat{y}(t)\right|^2$, is small. Using Parseval's relationship (2.88), this quantity is given by

$$E\left|y(t)-\hat{y}(t)\right|^2 = \frac{1}{2\pi}\int_{-\pi}^{\pi}\left|\frac{G_0-\hat{G}}{1+\hat{G}K}\right|^2 \left|S(G_0,K)\right|^2 \phi_r d\omega + \frac{1}{2\pi}\int_{-\pi}^{\pi}\left|S(G_0,K)\right|^2 \phi_v d\omega \tag{3.13}$$

We now observe that, if $K_{id} = K$, then the minimisation of the asymptotic identification criterion (3.10) also minimises the performance criterion (3.13). Thus, when low-order models are used for control design, the bias should be minimised in the closed-loop sense of the criterion (3.10). This is achieved by closed-loop identification with the to-be-designed controller in the loop. Of course, this is impossible since the to-be-designed controller is unknown.

The concept of iterative identification and control design is an attempt to approach these ideal experimental conditions by slow adjustments from one controller to the next one. By slowly increasing the required performance, the to-be-designed controller is never very different from the controller K_{id} that is presently operating and therefore the experimental conditions are always close to the optimal ones. This philosophy has been called the 'windsurfer approach' in (Lee *et al.*, 1993). The reason for introducing small steps of controller adjustment is also justified by stability robustness arguments: see, *e.g.*, (Anderson *et al.*, 1998).

In this subsection we have justified the use of closed-loop identification for control on the basis of bias considerations. We observe that the matching of the criteria (3.13) and (3.10) is impossible if the data are collected in open loop, even if the to-be-designed controller K were known, because both criteria contain the actual sensitivity function S which depends on the unknown true system. This bias justification only holds when reduced-order models are used. Full-order models can be identified without bias and there is then no theoretical reason to prefer closed-loop or open-loop identification.

3.3.2 The Effect of Feedback on the Variance Error

It has been shown in (Gevers and Ljung, 1986) for minimum-variance control design and in (Hjalmarsson *et al.*, 1996) for other control design criteria, that closed-loop identification is optimal for the minimisation of the variance error on a to-be-designed controller when the system is in the model set ($\mathcal{S} \in \mathcal{M}$). In other words, the variance of the difference between the controller designed from an identified model and the optimal controller that would be designed if the true system were known is minimised when the model is obtained by closed-loop identification and is unbiased.

For some criteria (*e.g.*, minimum variance), the optimal design consists of closed-loop identification with the to-be-designed controller in the loop during data collection (*i.e.*, $K_{id} = K$), a similar conclusion of that reached in the previous subsection on the basis of bias considerations. Needless to say, this optimal design is again not applicable because computation of the optimal to-be-designed controller requires knowledge of the system being identified. However, this result gives once again heuristic support to iterative schemes.

There are two important limitations to the relevance of the optimal design results of (Gevers and Ljung, 1986) and (Hjalmarsson *et al.*, 1996):

1. they assume the true system $\mathcal{S}$ to be in the model set $\mathcal{M}$, *i.e.*, only variance errors are considered;
2. these results are based on performance considerations only, without regard for robust stability.

Here we focus on the robust stability issue and we explain why, on the basis of variance considerations (but bias errors are allowed too), closed-loop identification with a controller that is as close as possible to the to-be-designed controller is to be preferred to open-loop identification.

Recall that, in prediction-error identification, the asymptotic variance in transfer function space (*i.e.*, as both the number of data an the model order tend to infinity) is given by (2.128):

$$\operatorname{cov}\begin{bmatrix}\hat{G}(e^{j\omega})\\ \hat{H}(e^{j\omega})\end{bmatrix} \approx \frac{n}{N}\begin{bmatrix}\phi_u(\omega) & \phi_{ue}(\omega)\\ \phi_{eu}(\omega) & \lambda_0\end{bmatrix}^{-T}\phi_v(\omega) \tag{3.14}$$

In open loop, $\phi_{ue}(\omega) = \phi_{eu}(\omega) = 0$ and the variance of the input-output model is

$$\mathrm{var}\big(\hat{G}(e^{j\omega})\big) \approx \frac{n}{N}\frac{\phi_v(\omega)}{\phi_u(\omega)} \tag{3.15}$$

In closed loop, if we also focus on the asymptotic variance of the input-output model, we obtain

$$\begin{aligned}\mathrm{var}\big(\hat{G}(e^{j\omega})\big) &\approx \frac{n}{N}\frac{\phi_v(\omega)}{\phi_u^r(\omega)} \\ &= \frac{n}{N}\left|1 + G_0(e^{j\omega})K_{id}(e^{j\omega})\right|^2 \frac{\phi_v(\omega)}{\phi_r(\omega)}\end{aligned} \tag{3.16}$$

by noting that

$$\begin{bmatrix}\phi_u(\omega) & \phi_{ue}(\omega) \\ \phi_{eu}(\omega) & \lambda_0\end{bmatrix}^{-T} = \begin{bmatrix}\dfrac{1}{\phi_u^r(\omega)} & -\dfrac{\phi_{eu}(\omega)}{\lambda_0\phi_u^r(\omega)} \\ -\dfrac{\phi_{ue}(\omega)}{\lambda_0\phi_u^r(\omega)} & \dfrac{1}{\lambda_0 - \left|\phi_{ue}(\omega)\right|^2/\phi_u(\omega)}\end{bmatrix} \tag{3.17}$$

Thus, if white noise is used to excite the system in closed loop, *i.e.*, if $\phi_r(\omega)$ is uniform, the variance of the resulting estimate will be smaller around the cross-over frequency, *i.e.*, where the actual sensitivity function $\frac{1}{1+G_0K_{id}}$ is large (at least if $v(t)$ is white). This is not the case in open loop unless more excitation power is put in the frequency range of the desired closed-loop bandwidth, *i.e.*, if a tailor-made input spectrum $\phi_u(\omega)$ is chosen. This increased confidence in the model around the cross-over frequency results for instance in a tighter upper bound on the gain margin, which allows one to compute less conservative robust controllers. This is illustrated in the following example. Note however that this potential advantage of closed-loop identification only holds if the current controller K_{id} is not too different from the to-be-designed controller K, *i.e.*, if the desired closed-loop bandwidth is in the frequency range where the current sensitivity function is large.

These insights on variance errors are still preliminary and will receive more theoretical attention in the sequel of this book.

► **Example 3.2.** Assume that the true system (G_0, H_0) is described by the following ARX model:

$$\mathcal{S}: \big(1 - 1.4z^{-1} + 0.45z^{-2}\big)y(t) = \big(1 + 0.25z^{-1}\big)z^{-1}u(t) + e(t)$$

where $e(t)$ is unit-variance white noise. The experiment consists in identifying an unbiased model

$$\mathcal{M}(\theta): \big(1 + a_1z^{-1} + a_2z^{-2}\big)y(t) = \big(b_0 + b_1z^{-1}\big)z^{-1}u(t) + e(t)$$

where $\theta = \begin{bmatrix}a_1 & a_2 & b_0 & b_1\end{bmatrix}^T$ is the parameter vector and, using the variance of the parameter estimates, in computing the maximal gain of a proportional output feedback controller that leaves the closed-loop poles of all possible models within the stability region, with a probability level of 95%. The 95% uncertainty region in parameter space is defined as an ellipsoid

$$\mathcal{U} = \left\{ \theta \in \mathbb{R}^4 \mid (\theta - \hat{\theta})^T \hat{P}_\theta^{-1} (\theta - \hat{\theta}) < \alpha \right\}$$

where $\hat{\theta}$ is the estimated parameter vector, $\hat{P}_\theta$ is its estimated covariance matrix as delivered by prediction-error identification and $\alpha = 9.49$ is given by $\Pr(\chi^2(4) < \alpha) = 95\%$, *i.e.*, $\alpha = \chi^2_{0.95}(4)$ where $\chi^2(4)$ is a chi-square-distributed random variable with 4 degrees of freedom (this will be explained in more detail in Chapter 5).

We performed the two identification experiments described below and for each of these we computed the maximum gain k_{max} that would guarantee closed-loop stability for all models with parameters in $\mathcal{U}$, *i.e.* such that the characteristic polynomial $1 + (a_1 + b_0 k_{max})z^{-1} + (a_2 + b_1 k_{max})z^{-2}$ is stable for any $\begin{bmatrix} a_1 & a_2 & b_0 & b_1 \end{bmatrix}^T \in \mathcal{U}$.

The two experiments were as follows.

Experiment 1: open-loop identification. The system was simulated in open loop with unit-variance white noise as input signal $u(t)$. 1000 input-output data were collected and used to compute the estimates $\hat{\theta}$ and $\hat{P}_\theta$.

Experiment 2: closed-loop identification. A proportional controller $u(t) = r(t) - y(t)$ was used. The reference signal $r(t)$ was chosen as white noise with variance 18.38, leading to the same output variance as in the open-loop experiment[2]. 1000 input-output data were collected and used to compute the estimates $\hat{\theta}$ and $\hat{P}_\theta$.

Each of these two experiments was run 100 times, in order to get conclusions that would not depend on a particular noise realisation. For the true system, the smallest destabilising gain for the proportional controller is $k_{max} = 2.2$. The estimated limit gains $\hat{k}_{max}$ obtained by averaging the 100 Monte-Carlo runs with the two successive experiments were 1.36 and 2.04, respectively. This shows that the variance error is better tuned towards robust control design when the model is identified in closed loop. In addition, the experimental variances around the estimated limit gains were respectively 0.12 and 0.02, thus much smaller for the closed-loop experiments than for the open-loop experiments, which confirms the results of (Hjalmarsson *et al.*, 1996). ◄

3.4 The Effect of Model Reduction on the Modelling Errors

In the context of modelling for control, as we already mentioned in Section 3.2, it has been argued that a (wiser?) alternative to closed-loop identification of low-order models is to identify the plant in open loop using higher order, unbiased models that can be *validated* and to subsequently apply a model or

[2]Imposing the same output variance is a fair requirement when the objective is to compare different open-loop and/or closed-loop configurations of a given system since, in many practical applications, the output variance is constrained and the control objective is to reduce it.

controller reduction step if a low-order controller is desired. This approach has been strongly advocated in (Ljung, 1997). In a statistical framework a model is called validated if the uncertainty region around the estimated model contains the true system with some prescribed probability level (say 95%). For validated models, the variance error dominates the bias error (Ljung and Guo, 1997). We now briefly analyse this approach from the point of view of bias and variance errors and we compare it to closed-loop identification of a low-order model.

3.4.1 Using Model Reduction to Tune the Bias Error

When a reduced-order model is needed, an alternative, regarding the tuning of the bias error, is thus to first identify an unbiased (hence full-order) model (this can be done as well in open loop as in closed loop) and to subsequently use model reduction to reduce the order of the model to the desired value. Clearly, the discussion of Subsection 3.3.1 remains valid in this case. Therefore, the reduction step should also be performed in closed loop, *i.e.*, with frequency weightings that take the current controller (if such a controller is available) into account, so as to make the reduction criterion equivalent to the performance degradation criterion (3.13) if the to-be-designed controller were known. Such a procedure is proposed in Chapter 6, for an alternative performance degradation cost function of the form

$$J_{perf}(\hat{G}_r) = \left\| T_o(G_n, K) - T_o(\hat{G}_r, K) \right\|_\infty \tag{3.18}$$

where G_n is a high-order model obtained by identification or any other appropriate technique, $\hat{G}_r$ is the (biased) reduced-order model and K is the current controller.

3.4.2 Dealing with the Variance Error

A model reduction criterion like (3.18) reconciles the two approaches (closed-loop identification of a low-order model and open-loop identification of an unbiased model followed by order reduction) from a bias distribution point of view. However, as stressed in Subsection 3.3.2, the uncertainty distribution attached to a model obtained by open-loop identification is not ideally tuned for control design if no precautions are taken regarding the excitation spectrum. Therefore, we suggest to always perform the identification in closed loop, even if a full-order, unbiased model structure is used.

A question that arises, now, is how the variance of a low-order model obtained by (closed-loop) identification with a low-order (biased) model structure compares with the variance of a low-order model obtained by (closed-loop) identification of a full-order (unbiased) model followed by a step of model reduction. Few results have been published regarding this issue. It has been shown in (Tjärnström and Ljung, 1999) that, for a FIR system (but this result is also valid for other structures), the second approach could lead to a low-order model with smaller variance than the first approach, if $\mathcal{L}_2$ reduction is used with the exact input spectrum as frequency weighting. Let $(G_n(z, \theta), H(z, \theta))$ represent the full-order model with parameter vector $\theta \in \mathbb{R}^n$ and let $\hat{\theta}$ be the parameter estimate obtained by prediction-error identification. In this case, $\hat{\theta}$ tends asymptotically to

$$\theta^* = \arg \min_{\theta \in \mathbb{R}^n} \int_{-\pi}^{\pi} \left|G_0(e^{j\omega}) - G_n(e^{j\omega}, \theta)\right|^2 \frac{\phi_u(\omega)}{\left|H(e^{j\omega}, \theta)\right|^2} d\omega \tag{3.19}$$

as the number of data N tends to infinity ($H(e^{j\omega}, \theta) = 1$ in the FIR case). In the second step, the parameters $\eta^* \in \mathbb{R}^r$ ($r < n$) of the $\mathcal{L}_2$ reduced-order model $\hat{G}_r$ are determined through

$$\eta^* = \arg \min_{\eta \in \mathbb{R}^r} \int_{-\pi}^{\pi} \left|G_n(e^{j\omega}, \hat{\theta}) - G_r(e^{j\omega}, \eta)\right|^2 \frac{\phi_u(\omega)}{\left|H(e^{j\omega}, \hat{\theta})\right|^2} d\omega \tag{3.20}$$

using knowledge of $\phi_u(\omega)$.

The major limitation of this result is the fact that, in many practical situations, the exact input spectrum is unknown. This will for instance always be the case if the data samples are collected in closed loop, since in this case the input depends on the noise. Even in open loop, an imperfect knowledge of the actuator dynamics coupled with the presence of measurement noise will always result in an imperfect knowledge of the spectrum of the input signal that is actually applied to the plant.

When the exact input spectrum is unknown, the maximum-likelihood $\mathcal{L}_2$ reduced-order estimate of the high-order model is obtained by least-squares approximation (in the FIR case). In this case, the variance of the low-order estimate is the same for the two approaches considered here, as we now show.

Assume that the true FIR system is described by the following equation:

$$\begin{aligned} y(t) &= \sum_{k=1}^{n} b_k u(t-k) + e(t) \triangleq \theta_0^T \varphi(t) + e(t) \\ &= \sum_{k=1}^{r} b_k u(t-k) + \sum_{k=r+1}^{n} b_k u(t-k) + e(t) \\ &\triangleq \eta_0^T \varphi_1(t) + \xi_0^T \varphi_2(t) + e(t) \end{aligned} \tag{3.21}$$

where

$$\eta_0 = \begin{bmatrix} b_1 & \dots & b_r \end{bmatrix}^T \tag{3.22a}$$

$$\xi_0 = \begin{bmatrix} b_{r+1} & \dots & b_n \end{bmatrix}^T \tag{3.22b}$$

$$\theta_0 = \begin{bmatrix} \eta_0 \\ \xi_0 \end{bmatrix} \tag{3.22c}$$

$$\varphi_1(t) = \begin{bmatrix} u(t-1) & \dots & u(t-r) \end{bmatrix}^T \tag{3.22d}$$

$$\varphi_2(t) = \begin{bmatrix} u(t-r-1) & \dots & u(t-n) \end{bmatrix}^T \tag{3.22e}$$

$$\varphi(t) = \begin{bmatrix} u(t-1) & \dots & u(t-n) \end{bmatrix}^T \tag{3.22f}$$

and $e(t)$ is white noise with zero mean and variance λ_0.

If a low-order model of order r is directly identified using N input-output data, the estimate is given by

$$\begin{aligned} \hat{\eta}_{dir} &= \left[\frac{1}{N} \sum_{t=1}^{N} \varphi_1(t) \varphi_1^T(t) \right]^{-1} \frac{1}{N} \sum_{t=1}^{N} \varphi_1(t) y(t) \\ &= \eta_0 + \left[\frac{1}{N} \sum_{t=1}^{N} \varphi_1(t) \varphi_1^T(t) \right]^{-1} \frac{1}{N} \sum_{t=1}^{N} \varphi_1(t) \varphi_2^T(t) \xi_0 \\ &\quad + \left[\frac{1}{N} \sum_{t=1}^{N} \varphi_1(t) \varphi_1^T(t) \right]^{-1} \frac{1}{N} \sum_{t=1}^{N} \varphi_1(t) e(t) \end{aligned} \tag{3.23}$$

Its expectation (with respect to the noise) is

$$E\,\hat{\eta}_{dir} = \eta_0 + \left[\frac{1}{N} \sum_{t=1}^{N} \varphi_1(t) \varphi_1^T(t) \right]^{-1} \frac{1}{N} \sum_{t=1}^{N} \varphi_1(t) \varphi_2^T(t) \xi_0 \tag{3.24}$$

and its covariance matrix is

$$\hat{P}_{\eta_{dir}} = \left[\sum_{t=1}^{N} \varphi_1(t) \varphi_1^T(t) \right]^{-1} \lambda_0 \tag{3.25}$$

If a full-order model is first identified and then reduced by least-squares approximation, we have successively

$$\hat{\theta} = \left[\frac{1}{N} \sum_{t=1}^{N} \varphi(t) \varphi^T(t) \right]^{-1} \frac{1}{N} \sum_{t=1}^{N} \varphi(t) y(t) \tag{3.26}$$

and

$$\hat{\eta}_{red} = \left[\frac{1}{N} \sum_{t=1}^{N} \varphi_1(t) \varphi_1^T(t) \right]^{-1} \frac{1}{N} \sum_{t=1}^{N} \varphi_1(t) \hat{y}(t, \hat{\theta}) \tag{3.27}$$

where

$$\hat{y}(t, \hat{\theta}) = \varphi^T(t)\hat{\theta} \tag{3.28}$$

is the simulated output of the full-order estimate. It follows easily that

$$\begin{aligned}\hat{\eta}_{red} = &\left[\frac{1}{N}\sum_{t=1}^{N}\varphi_1(t)\varphi_1^T(t)\right]^{-1}\left[\frac{1}{N}\sum_{t=1}^{N}\varphi_1(t)\varphi^T(t)\right]\\ &\times\left[\frac{1}{N}\sum_{t=1}^{N}\varphi(t)\varphi^T(t)\right]^{-1}\frac{1}{N}\sum_{t=1}^{N}\varphi(t)y(t)\end{aligned} \tag{3.29}$$

where

$$\sum_{t=1}^{N}\varphi(t)y(t) = \sum_{t=1}^{N}\varphi(t)\varphi^T(t)\theta_0 + \sum_{t=1}^{N}\varphi(t)e(t) \tag{3.30}$$

and finally

$$\begin{aligned}\hat{\eta}_{red} = \eta_0 + &\left[\frac{1}{N}\sum_{t=1}^{N}\varphi_1(t)\varphi_1^T(t)\right]^{-1}\frac{1}{N}\sum_{t=1}^{N}\varphi_1(t)\varphi_2^T(t)\xi_0\\ +&\left[\frac{1}{N}\sum_{t=1}^{N}\varphi_1(t)\varphi_1^T(t)\right]^{-1}\left[\frac{1}{N}\sum_{t=1}^{N}\varphi_1(t)\varphi^T(t)\right]\\ \times&\left[\frac{1}{N}\sum_{t=1}^{N}\varphi(t)\varphi^T(t)\right]^{-1}\frac{1}{N}\sum_{t=1}^{N}\varphi(t)e(t)\end{aligned} \tag{3.31}$$

Its expectation and covariance matrix are respectively

$$E\,\hat{\eta}_{red} = \eta_0 + \left[\frac{1}{N}\sum_{t=1}^{N}\varphi_1(t)\varphi_1^T(t)\right]^{-1}\frac{1}{N}\sum_{t=1}^{N}\varphi_1(t)\varphi_2^T(t)\xi_0 \tag{3.32}$$

and

$$\hat{P}_{\eta_{red}} = \left[\sum_{t=1}^{N}\varphi_1(t)\varphi_1^T(t)\right]^{-1}\lambda_0 \tag{3.33}$$

which are equal to those of $\hat{\eta}_{dir}$. The same bias and variance errors are thus committed with both approaches if the exact input spectrum is unknown and the reduction is carried out on the basis of the same data samples that are used for identification.

It should also be noted that $\mathcal{L}_2$ model reduction is not an easy problem. Most of the time, it requires the use of iterative algorithms, with all the problems that such algorithms involve: local minima, huge computational cost, *etc.* Therefore, balanced truncation, or other analytic methods like Hankel-norm approximation, are generally preferred. The way the variance of a low-order model computed by balanced truncation is related to the variance of the initial high-order model is still an open problem, but there is no reason to think, *a priori*, that balanced truncation might have a systematic diminution or, on the contrary, increase effect on the model variance, at least in the unweighted case.

▶ **Example 3.3.** Consider the following 5th-order ARX system:

$$G_0(z) = \frac{0.017181z^{-1}(1-0.8669z^{-1})(1-0.8187z^{-1})(1-0.7164z^{-1})(1-0.3489z^{-1})}{(1-0.9512z^{-1})(1-0.8825z^{-1})(1-0.8465z^{-1})(1-0.7788z^{-1})(1-0.6065z^{-1})}$$

$$H_0(z) = \frac{1}{(1-0.9512z^{-1})(1-0.8825z^{-1})(1-0.8465z^{-1})(1-0.7788z^{-1})(1-0.6065z^{-1})}$$

in closed loop with the following PI controller:

$$K_{id}(z) = 4\left(1 + \frac{z^{-1}}{20(1-z^{-1})}\right)$$

This system is excited with a PRBS of amplitude 1 as reference signal $r_1(t)$ (see Figure 2.1) and its output is subject to disturbances $v(t) = H_0(z)e(t)$, where $e(t)$ is a sequence of Gaussian white noise with variance $\lambda_0 = 10^{-8}$. The following experiments are carried out:

Experiment 1: closed-loop identification of an unbiased model. Using a set of 1023 input-output data samples $Z^{1023} = \{u(t), y(t) \mid t = 1\ldots 1023\}$, an unbiased model $\hat{G}_5^{clid}$ with an ARX[5,5,1] structure is identified.

Experiment 2: closed-loop identification of a low-order model. Using a set of 1023 input-output data samples $Z^{1023} = \{u(t), y(t) \mid t = 1\ldots 1023\}$, a low-order model $\hat{G}_r^{clid}$ with an ARX[r,r,1] structure, $r < 5$, is identified.

Experiment 3: unweighted reduction of the unbiased model. The unbiased model $\hat{G}_5^{clid}$ obtained in Experiment 1 is reduced using unweighted normalised coprime-factor balanced truncation (see Subsection 6.4.1), producing a r-th-order model $\hat{G}_r^{unw}$.

Experiment 4: reduction of the unbiased model using closed-loop induced weightings. The unbiased model $\hat{G}_5^{clid}$ obtained in Experiment 1 is reduced by means of frequency-weighted normalised coprime-factor balanced truncation, using closed-loop induced frequency weightings (see Subsection 6.4.2), producing a r-th-order model $\hat{G}_r^{clw}$.

Following a Monte-Carlo scheme, these four experiments are repeated 100 times, *i.e.*, 100 realisations of each model are produced. Figures 3.3 and 3.4 show the experimental expectations and standard deviations of the magnitude of their frequency responses, respectively for $r = 2$ and $r = 1$. $\hat{G}_2^{unw}$ and $\hat{G}_1^{unw}$ have both a variance approximately equal to that of $\hat{G}_5^{clid}$, which seems to confirm the intuition that (unweighted) model reduction has little influence on the variance. For $r = 2$, the variance of the low-order model $\hat{G}_2^{clid}$ is approximately the same as that of the unbiased model $\hat{G}_5^{clid}$, while that of $\hat{G}_2^{clw}$ is lower. For $r = 1$, the variance of $\hat{G}_1^{clid}$ is significantly lower than that of $\hat{G}_5^{clid}$, while the variance of $\hat{G}_1^{clw}$ is much larger at low frequencies. This shows that conclusions are hard to draw regarding the choice between direct identification of a low-order model, or identification of a full-order model followed by model reduction with closed-loop induced frequency weightings. ◀

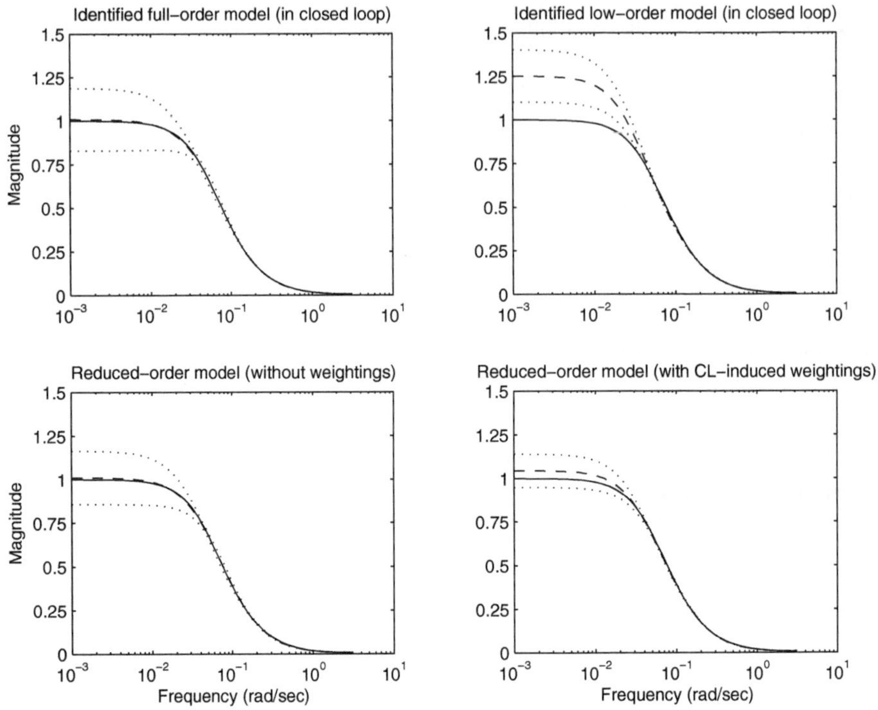

Figure 3.3. Magnitude of the frequency response of G_0 (—). Experimental expectation of the magnitude (– –) and experimental expectation of the magnitude $\pm$ 1 standard deviation ($\cdots$) of the frequency responses of $\hat{G}_5^{clid}$ (top left), $\hat{G}_2^{clid}$ (top right), $\hat{G}_2^{unw}$ (bottom left) and $\hat{G}_2^{clw}$ (bottom right).

Finally, note that the variance of the low-order model is not a very important issue, at least with respect to a robust control design objective. Indeed, this variance cannot be used for the computation of an uncertainty region containing the true system, since the model is biased. The only role it plays is that it determines the variance of the resulting controller: the higher the variance of the model, the higher the variance of the optimal controller designed from this model. Yet, from a robustness point of view, one must take into account the variance of an unbiased model in order to build an uncertainty region guaranteed to contain the true system (with a given probability): see Chapter 5.

3.5 Summary of the Chapter

We have shown that it is generally preferable to identify a (possibly low-order) model in closed loop when the purpose is to use this model for control design. Indeed, the bias and variance errors attached to the estimate are better tuned

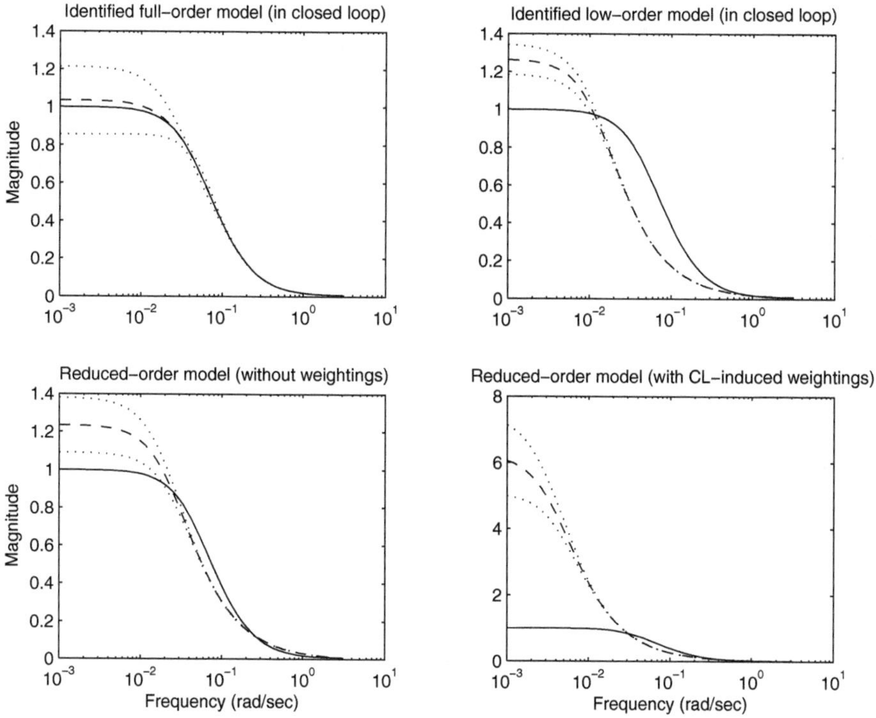

Figure 3.4. Magnitude of the frequency response of G_0 (—). Experimental expectation of the magnitude (– –) and experimental expectation of the magnitude ± 1 standard deviation (· · ·) of the frequency responses of $\hat{G}_5^{clid}$ (top left), $\hat{G}_1^{clid}$ (top right), $\hat{G}_1^{unw}$ (bottom left) and $\hat{G}_1^{clw}$ (bottom right).

for control design when the model is identified in closed loop rather than in open loop, at least if similar spectra are used to excite the system in both cases, and if the to-be-designed controller is close to the one used during identification. One could of course tune the open-loop excitation spectrum in order to put more energy around the desired closed-loop cross-over frequency. However, the claimed advantage of closed-loop identification is the fact that this tuning happens 'naturally' due to the filtering of the reference spectrum through the closed-loop sensitivity function.

CHAPTER 4

Dealing with Controller Singularities in Closed-loop Identification

4.1 Introduction

In the previous chapter, we have given several reasons for performing the identification of a system in closed loop when the objective is the computation of a model aimed for control design. Before we pursue the reasoning made there and extend it in the next chapters to the validation of a given model or controller, or to the reduction of a high-order model or controller, we address some problems that arise in closed-loop identification when the controller contains unstable poles or nonminimum-phase zeroes. Such singularities are likely to be encountered with controllers designed by optimal techniques like LQG or $\mathcal{H}_\infty$ synthesis, since the designer has little control on the poles and zeroes of these controllers, but it should be noted that even PI or PID controllers are unstable, since they contain an integrator.

We show that, with some of the commonly used closed-loop identification methods, the resulting model is not stabilised by the controller used during identification, even though the true system is stabilised by the same controller[1]. Furthermore, a model that is not stabilised by the present controller is intrinsically flawed for the design of a better controller, in that it deprives the control designer of some of their most important robust-stability tools for the design of this new controller.

The objective of this chapter is thus to answer the following questions:

[1]Of course, it is generally impossible to know with absolute certainty that a controller stabilises an imperfectly known plant. However, we can reasonably consider that closed-loop stability is achieved if no unstable behaviour has been detected during a sufficiently long period of closed-loop operation.

How do the different classical closed-loop identification methods compare with respect to this nominal instability issue? Which method should I use in function of the singularities of the controller and with which excitation signal?

We consider four different approaches to closed-loop identification: the indirect approach, the coprime-factor approach, the direct approach and the dual Youla parametrisation approach. They are all based on prediction-error identification. The first three have already been briefly presented in Subsection 2.6.7.

For the sake of simplifying the notations, we shall restrict our analysis to the SISO case. However, our results can easily be extended to the multivariable case.

4.2 The Importance of Nominal Closed-loop Stability for Control Design

When one identifies a model $\hat{G}$ of an unknown true system G_0 from data collected in closed loop with a stabilising controller K_{id}, it makes good sense to request that the nominal closed-loop system $(\hat{G},\, K_{id})$ be stable since we know that the actual closed-loop system $(G_0,\, K_{id})$ is stable. In iterative identification and control design, the model $\hat{G}$ is used for the design of a new controller K that must achieve a better performance on the actual system G_0 than the present controller K_{id}. We have just seen in the previous chapter that, for the model $\hat{G}$ to be suitable for the design of such a controller K, it is necessary that G_0 and $\hat{G}$ be close *in a closed-loop sense, i.e.*, $||T(G_0,\, K_{id}) - T(\hat{G},\, K_{id})||$ must be small. This is unlikely to be the case if one of these closed-loop systems is stable and the other is not.

There are also a number of technical grounds for requesting stability of the nominal closed-loop system in the framework of robust control analysis and design. The stability margin, $b_{\hat{G},K_{id}}$ and the maximal stability margin, $\sup_K b_{\hat{G},K}$, of the nominal model are very useful tools to ascertain that a new controller K, designed from $\hat{G}$, will stabilise the true system G_0. If $b_{\hat{G},K_{id}} = 0$ and/or if $\sup_K b_{\hat{G},K}$ is very small, it will be much harder – if not impossible – to design a new controller K with prior stability guarantees. Indeed, Proposition 2.9 tells us that such a stability guarantee is obtained for $(G_0,\, K)$ if

$$\delta_\nu(G_0,\, \hat{G}) < b_{\hat{G},K} \tag{4.1}$$

However, if K_{id} does not stabilise $\hat{G}$ while it stabilises G_0, we know that

$$b_{G_0,K_{id}} \le \delta_\nu(G_0,\, \hat{G}) \tag{4.2}$$

This implies that, in order to guarantee the stabilisation of G_0 by K using condition (4.1), we need to design a controller K with a nominal stability margin $b_{\hat{G},K}$ that is necessarily larger than the existing margin $b_{G_0,K_{id}}$. If K_{id} is a controller achieving small closed-loop bandwidth (a controller that one would like to replace by a better one), then $b_{G_0,K_{id}}$ will often already be large, meaning that $\delta_\nu(G_0, \hat{G})$ is also large. There is a risk, therefore, that

$$\sup_K b_{\hat{G},K} \leq \delta_\nu(G_0, \hat{G}) \tag{4.3}$$

In such a case, no controller K stabilising $\hat{G}$ is guaranteed to stabilise G_0 and it may even be impossible to find a controller that stabilises both $\hat{G}$ and G_0.

This reasoning may seem somewhat conservative, as condition (4.1) is only sufficient. We shall however see in Chapter 5 that it has high importance when the objective is to use the model $\hat{G}$ for robust control design. Indeed, in such a case, one does not only take the nominal model $\hat{G}$ into account, but also an uncertainty region $\mathcal{D}$ around it that is guaranteed to contain the true system G_0 (possibly with some probability). We shall see that the quantity

$$\delta_{WC}(\hat{G}, \mathcal{D}) \triangleq \sup_{G \in \mathcal{D}} \delta_\nu(\hat{G}, G) \tag{4.4}$$

which is a measure of the size of $\mathcal{D}$, should then be as small as possible. However, since

$$G_0 \in \mathcal{D} \Rightarrow \delta_{WC}(\hat{G}, \mathcal{D}) \geq \delta_\nu(\hat{G}, G_0) \tag{4.5}$$

it is of extreme importance to have $\delta_\nu(\hat{G}, G_0)$ as small as possible as well, therefore not larger than $b_{G_0,K_{id}}$. Hence $\hat{G}$ should be stabilised by K_{id}.

Nominal closed-loop instability can also be a handicap when one tries to assess *a priori* whether a new controller K (possibly designed using $\hat{G}$, but this is not a requirement) will stabilise the true system G_0. Such prior stability check is often based on the use of a 'small-gain theorem'-based result like Proposition 2.1. A sufficient condition for K to stabilise G_0 is (Vinnicombe, 1993*a*)

$$\delta_\nu(K_{id}, K) < b_{G_0,K_{id}} \tag{4.6}$$

Since $b_{G_0,K_{id}}$ is unknown, a possible solution is to replace it by an estimate $b_{\hat{G},K_{id}}$. This procedure is justified by Proposition 2.1 and has for instance been suggested in (De Bruyne and Kammer, 1999) as a tool for checking stability in Iterative Feedback Tuning. However, this method fails if $b_{\hat{G},K_{id}}$ is a poor estimate of $b_{G_0,K_{id}}$ and *a fortiori*, if $b_{\hat{G},K_{id}} = 0$ while the actual closed-loop system (G_0, K_{id}) is stable.

4.3 Poles and Zeroes of a System

Definition 4.1. (MINIMALITY) *A state-space realisation (A, B, C, D) of a system is called minimal if A has the smallest possible dimension, which is true if and only if the pair (A, B) is controllable and the pair (A, C) is observable.*

The notions of observability and controllability are defined in Subsection 2.7.1.

Definition 4.2. (POLE) *Let* $P(z) = C(zI-A)^{-1}B+D$ *be the transfer function of a discrete-time system* P *with minimal state-space realisation* (A, B, C, D). $z_0 \in \mathbb{C}$ *is a pole of* P *if* $P(z_0) = \infty$.

Alternatively, the poles of a system can also be defined as the eigenvalues (counted with multiplicity) of the matrix A in the minimal state-space realisation (A, B, C, D) of P. This definition is also valid in the MIMO case. It should be noted that any pole of a MIMO system is necessarily also a pole of at least one of its SISO subsystems. This can be useful when applying the results of this chapter to the multivariable case.

Definition 4.3. (TRANSMISSION ZERO) *Let* $P(z) = C(zI-A)^{-1}B+D$ *be the transfer function of a discrete-time system* P *with minimal state-space realisation* (A, B, C, D). $z_0 \in \mathbb{C}$ *is a transmission zero of* P *if* $P(z_0) = 0$.

Alternatively, the (transmission) zeroes can be defined as those values of z for which the matrix $\begin{bmatrix} zI-A & B \\ -C & D \end{bmatrix}$ does not have full rank. This definition is also valid in the MIMO case. It should be noted that, contrary to what happened with the poles, the zeroes of a MIMO system do not necessarily coincide with the zeroes of its SISO subsystems. This should be taken into account when applying the results of this chapter to the multivariable case.

A zero at $z = z_0$ is referred to as a 'transmission zero' because $P(z_0)e^{z_0 t} = 0 \ \forall t \geq 0$ provided the initial state $x(0)$ is unobservable, *i.e.*, the forced response of the system to the input signal $u(t) = e^{z_0 t}$ is zero. We shall also often talk of 'blocking zeroes' and 'blocking poles', which we define as a particular class of zeroes and poles.

Definition 4.4. (BLOCKING ZERO OR POLE) *A zero (resp. a pole)* $z_0 \in \mathbb{C}$ *of a discrete-time system* $P(z)$ *will be called a blocking zero (resp. a blocking pole) if it is located on the unit circle* ($|z_0| = 1$), *i.e., if* $z_0 = e^{j\omega_0}$ *for some* $\omega_0 \in \begin{bmatrix} 0 & 2\pi \end{bmatrix}$.

A blocking zero is thus a transmission zero with respect to a real sinusoidal excitation at frequency ω_0: $P(e^{j\omega_0})\sin(\omega_0 t) = 0 \ \forall t \geq 0$ if the initial state $x(0)$ is unobservable. A unit-circle pole of a controller is called a blocking pole because it becomes a blocking zero of the closed-loop transfer functions T_{12} and T_{22} defined in (2.8).

Finally, we define the notions of (strict) instability and (strictly) nonminimum phase as follows.

Definition 4.5. (INSTABILITY) *A pole* $z_0 \in \mathbb{C}$ *of a discrete-time system* $P(z)$ *will be called 'unstable' (resp. 'strictly unstable') if* $|z_0| \geq 1$ *(resp. if* $|z_0| > 1$*).* $P(z)$ *will be called (strictly) unstable if it has at least one (strictly) unstable pole.*

Definition 4.6. (NONMINIMUM PHASE) *A zero $z_0 \in \mathbb{C}$ of a discrete-time system $P(z)$ will be said to have a 'nonminimum phase' (resp. a 'strictly nonminimum phase') if $|z_0| \geq 1$ (resp. if $|z_0| > 1$). $P(z)$ will be said to have a (strictly) nonminimum phase if it has at least one (strictly) nonminimum-phase zero.*

The terminology 'nonminimum phase' is explained by the fact that such zeroes introduce additional negative phase displacements in the Bode diagram of the transfer function.

4.4 Loss of Nominal Closed-loop Stability with Unstable or Nonminimum-phase Controllers

Let us consider the identification of an unknown plant G_0 in closed loop with some stabilising controller K_{id} as in Figure 4.1.

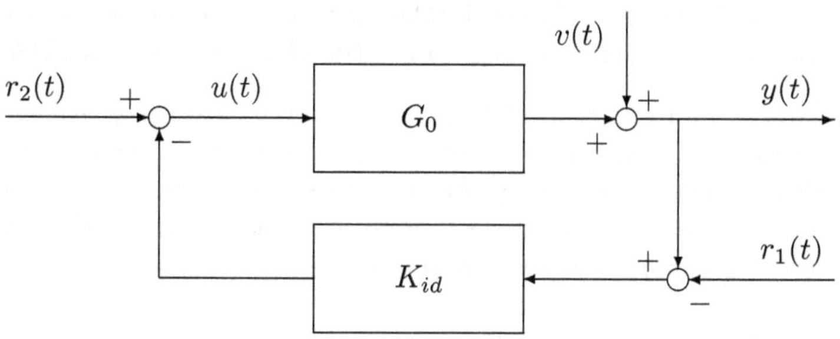

Figure 4.1. Closed-loop configuration during identification

The generalised closed-loop transfer matrix is

$$T(G_0, K_{id}) = \begin{bmatrix} \dfrac{G_0 K_{id}}{1 + G_0 K_{id}} & \dfrac{G_0}{1 + G_0 K_{id}} \\ \dfrac{K_{id}}{1 + G_0 K_{id}} & \dfrac{1}{1 + G_0 K_{id}} \end{bmatrix} \triangleq \begin{bmatrix} T_{11} & T_{12} \\ T_{21} & T_{22} \end{bmatrix} \tag{4.7}$$

and the noise transfer matrix is

$$N(G_0, K_{id}) = \begin{bmatrix} \dfrac{1}{1 + G_0 K_{id}} \\ \dfrac{-K_{id}}{1 + G_0 K_{id}} \end{bmatrix} = \begin{bmatrix} T_{22} \\ -T_{21} \end{bmatrix} \triangleq \begin{bmatrix} N_1 \\ N_2 \end{bmatrix} \tag{4.8}$$

We now analyse the effect of unstable poles or nonminimum-phase zeroes in K_{id} on the stability of the $(\hat{G}, K_{id})$ loop and on the possibility to use $\hat{G}$ for the design of a new controller K with a better performance. Four different approaches to closed-loop identification are studied.

4.4.1 The Indirect Approach

The indirect approach consists in first estimating one of the four entries of the closed-loop transfer matrix $T(G_0, K_{id})$ using an appropriate combination of measurements of either $r_1(t)$ or $r_2(t)$ and either $y(t)$ or $u(t)$. From this estimate, a model $\hat{G}$ is derived, using knowledge of the controller K_{id} and one of the following algebraic relations:

$$\begin{aligned} \hat{G}(\hat{T}_{11}) &= \frac{\hat{T}_{11}}{K_{id} - K_{id}\hat{T}_{11}} & \hat{G}(\hat{T}_{12}) &= \frac{\hat{T}_{12}}{1 - K_{id}\hat{T}_{12}} \\ \hat{G}(\hat{T}_{21}) &= \frac{1}{\hat{T}_{21}} - \frac{1}{K_{id}} & \hat{G}(\hat{T}_{22}) &= \frac{1}{K_{id}}\left(\frac{1}{\hat{T}_{22}} - 1\right) \end{aligned} \tag{4.9}$$

We can thus consider four different cases, depending on which of the four entries of $T(G_0, K_{id})$ is identified in the first step of the procedure. The following theorem gives conditions under which the obtained model $\hat{G}$ will be guaranteed to be either stabilised or destabilised by the controller K_{id}. These conditions are related to the poles and zeroes of K_{id} and to the identified entry of $T(G_0, K_{id})$.

Theorem 4.1. (Codrons *et al.*, 2002) *Consider the closed-loop set-up of Figure 4.1. Assume that an indirect closed-loop identification method is used to obtain a stable estimate of one of the four closed-loop transfer functions T_{ij} ($i, j \in \{1, 2\}$), from a finite number of noisy data samples, followed by an estimate $\hat{G}$ of G_0 using the corresponding formula in (4.9). Then*

1. *the nominal closed-loop system $(\hat{G}, K_{id})$ will be unstable if K_{id} has one or more nonminimum-phase zeroes and $\hat{G}$ is computed from $\hat{T}_{11}$, $\hat{T}_{21}$ or $\hat{T}_{22}$;*
2. *the nominal closed-loop system $(\hat{G}, K_{id})$ will be unstable if K_{id} has one or more unstable poles and $\hat{G}$ is computed from $\hat{T}_{11}$, $\hat{T}_{12}$ or $\hat{T}_{22}$.*
3. *the nominal closed-loop system $(\hat{G}, K_{id})$ will be stable if K_{id} has a minimum phase and $\hat{G}$ is computed from $\hat{T}_{21}$;*
4. *the nominal closed-loop system $(\hat{G}, K_{id})$ will be stable if K_{id} is stable and $\hat{G}$ is computed from $\hat{T}_{12}$.*

Proof. We provide the proof for the cases where either T_{11} or T_{12} is estimated in the first step of the procedure. The analysis of the other two cases is similar.

The generalised closed-loop transfer matrix of the loop $(\hat{G}(\hat{T}_{11}), K_{id})$ is given by

$$T\big(\hat{G}(\hat{T}_{11}), K_{id}\big) = \begin{bmatrix} \hat{T}_{11} & \dfrac{\hat{T}_{11}}{K_{id}} \\ K_{id}(1 - \hat{T}_{11}) & 1 - \hat{T}_{11} \end{bmatrix} \tag{4.10}$$

The following observations can be made about its stability:

(i) if K_{id} has a nonminimum-phase zero at some z_0, then $T_{11}(z_0) = 0$: see (4.7). However, the estimate $\hat{T}_{11}$ has an unavoidable variance error and possibly a bias error as well. As a result, $\hat{T}_{11}(z_0) \neq 0$ and $\frac{\hat{T}_{11}}{K_{id}}$ will be unstable because it contains an imperfect cancellation of an unstable pole-zero pair. The stability of the other entries of $T(\hat{G}(\hat{T}_{11}), K_{id})$ is unaffected by nonminimum-phase zeroes in K_{id};

(ii) if K_{id} has an unstable pole at some z_0, then $1 - T_{11}(z_0) = 0$, but $1 - \hat{T}_{11}(z_0) \neq 0$ due to estimation error. Hence, $K_{id}(1-\hat{T}_{11})$ will be unstable for the same reason of imperfect cancellation of an unstable pole-zero pair. The stability of the other entries of $T(\hat{G}(\hat{T}_{11}), K_{id})$ is not affected by unstable poles in K_{id}.

The generalised closed-loop transfer matrix of the loop $(\hat{G}(\hat{T}_{12}), K_{id})$ is given by

$$T(\hat{G}(\hat{T}_{12}), K_{id}) = \begin{bmatrix} \hat{T}_{12}K_{id} & \hat{T}_{12} \\ (1 - \hat{T}_{12}K_{id})K_{id} & 1 - \hat{T}_{12}K_{id} \end{bmatrix} \tag{4.11}$$

The following observations can be made about its stability:

(i) the presence of nonminimum-phase zeroes in K_{id} does not affect the stability of $T(\hat{G}(\hat{T}_{12}), K_{id})$;

(ii) if K_{id} has unstable poles, however, the three reconstructed entries of $T(\hat{G}(\hat{T}_{12}), K_{id})$ will be unstable. □

Conclusion: With the indirect approach one must

- identify $\hat{G}$ from $\hat{T}_{12}$ using $\{y(t), r_2(t)\}$ data if K_{id} has *nonminimum-phase zeroes but no unstable poles*;
- identify $\hat{G}$ from $\hat{T}_{21}$ using $\{u(t), r_1(t)\}$ data if K_{id} has *unstable poles but no nonminimum-phase zeroes*.

All other strategies lead to an unstable nominal closed-loop system $(\hat{G}, K_{id})$ if K_{id} has nonminimum-phase zeroes or unstable poles. If K_{id} is both unstable and nonminimum phase, the nominal closed-loop system $(\hat{G}, K_{id})$ will be unstable whichever entry of $T(G_0, K_{id})$ is identified.

▶ **Remark.** The following additional comments can be made for the case where the controller has poles or zeroes on the unit circle and when T_{11} or T_{12} is estimated in the first step of the procedure; similar remarks hold for the other two cases. These remarks concern the quality of the estimate in terms of modelling error and not in terms of nominal closed-loop stability.

- Assume that K_{id} has a blocking zero at some frequency ω_0 and that T_{11} is identified. Because of this unit-circle zero $T_{11}(e^{j\omega_0}) = 0$, while $N_1(e^{j\omega_0}) \neq 0$. Hence, the output Signal-to-Noise Ratio (SNR) will be zero at this frequency and very low around it, yielding a bad estimate of T_{11} at this frequency[2]. An exception is when $\phi_v(\omega_0) = 0$, *i.e.*, when $H_0(e^{j\omega_0}) = 0$. In this case, because of the absence of noise at frequency ω_0, the SNR of $y(t)$ is undetermined at ω_0 but, as it is not impaired in the neighbourhood of ω_0, the blocking zero has no impact on the quality of the estimate at this frequency (however, because of the noise acting at all other frequencies, a perfect matching of $T_{11}(e^{j\omega_0}) = 0$ will never happen in practice).

- Assume that K_{id} has a blocking zero at some frequency ω_0 and that T_{12} is identified. Since $T_{12}(e^{j\omega_0}) \neq 0$, this blocking zero will not impair the output SNR and the quality of the estimate will not be affected.
- Assume that K_{id} has a blocking pole at some frequency ω_0 (it is often the case, since most controllers contain an integrator, *i.e.*, a pole at $\omega_0 = 0$ rad/s) and that T_{11} is identified. Since $T_{11}(e^{j\omega_0}) \neq 0$ while $N_1(e^{j\omega_0}) = 0$, the output SNR will be infinite, hence, the quality of the estimate will be improved. However, because the model parameters also depend on measurements at other frequencies, a perfect matching of $T_{11}(e^{j\omega_0})$ will never happen in practice (this confirms, if necessary, that nominal closed-loop instability happens in this case).
- Assume that K_{id} has a blocking pole at some frequency ω_0 and that T_{12} is identified. Because of this unit-circle pole, $T_{12}(e^{j\omega_0}) = 0$ and $N_1(e^{j\omega_0}) = 0$, meaning that the noise is also generally cut off at this frequency. As a result, this blocking pole usually has no impact on the quality of the estimate at this frequency. An exception occurs when $H_0(e^{j\omega_0}) = \infty$, *i.e.*, when the noise process has a pole at the same frequency. A typical case is when the controller has an integrator and the system is subject to step disturbances, since these can be modelled as $v(t) = H_0(z)e(t)$ where $e(t)$ is a white noise sequence and $H_0(z)$ is, or contains, an integrator. In this case, the output SNR will be zero at frequency ω_0 and very low around it, yielding a bad estimate of T_{12} at this frequency. ◀

4.4.2 The Coprime-factor Approach

The coprime-factor approach consists in first estimating the two entries of one column of $T(G_0, K_{id})$ using measurements of $r_1(t)$, $y(t)$ and $u(t)$, or of $r_2(t)$, $y(t)$ and $u(t)$. A model $\hat{G}$ of G_0 is then given by the ratio of these two entries:

$$\hat{G}(\hat{T}_{11}, \hat{T}_{21}) = \frac{\hat{T}_{11}}{\hat{T}_{21}} \quad \text{or} \quad \hat{G}(\hat{T}_{12}, \hat{T}_{22}) = \frac{\hat{T}_{12}}{\hat{T}_{22}} \tag{4.12}$$

Since the closed-loop system (G_0, K_{id}) is stable, it is natural to request that the estimates of its entries be stable as well. As a result, $\hat{T}_{11}, \hat{T}_{21} \in \mathcal{RH}_\infty$ or $\hat{T}_{12}, \hat{T}_{22} \in \mathcal{RH}_\infty$, hence they define coprime factors of $\hat{G}$. This method requires that the controller be LTI, but it does not have to be known. Two cases can be considered, depending on whether $r_1(t)$ or $r_2(t)$ is used as excitation signal.

[2] If a nonparametric model was identified using spectral analysis methods, this model would be undetermined at the frequencies of the blocking zeroes. Here, in the case of a finite dimensional parametric model, the parameters are estimated by the measurements at other frequencies, but we can expect that the estimate of the transfer function at frequencies close to a blocking zero will be poor.

The following theorem gives conditions under which the obtained model $\hat{G}$ will be guaranteed to be either stabilised or destabilised by the controller K_{id}. These conditions are related to the poles and zeroes of K_{id} and to the identified entries of $T(G_0, K_{id})$ in the first step of the procedure.

Theorem 4.2. (Codrons *et al.*, 2002) *Consider the closed-loop set-up of Figure 4.1. Assume that a coprime-factor closed-loop identification method is used to obtain a model $\hat{G}$ for G_0 from a finite number of noisy data samples and that the estimates of the closed-loop transfer functions $\hat{T}_{ij}$ $(i, j \in \{1, 2\})$, identified during the first step of the procedure, are stable. Then*

1. *in the case where $\hat{G}$ is computed from $\hat{G}(\hat{T}_{11}, \hat{T}_{21}) = \frac{\hat{T}_{11}}{\hat{T}_{21}}$ using measurements of $r_1(t)$, $y(t)$ and $u(t)$, the nominal closed-loop system $(\hat{G}, K_{id})$ will be stable if and only if K_{id} has no strictly nonminimum-phase zeroes and $\frac{\hat{T}_{21}}{K_{id}} + \hat{T}_{11}$ has no nonminimum-phase zeroes;*
2. *in the case where $\hat{G}$ is computed from $\hat{G}(\hat{T}_{12}, \hat{T}_{22}) = \frac{\hat{T}_{12}}{\hat{T}_{22}}$ using measurements of $r_2(t)$, $y(t)$ and $u(t)$, the nominal closed-loop system $(\hat{G}, K_{id})$ will be stable if and only if K_{id} has no strictly unstable poles and $\hat{T}_{22} + \hat{T}_{12}K_{id}$ has no nonminimum-phase zeroes.*

Proof. We provide a full proof for case (1) only; case (2) is very similar. The generalised closed-loop transfer matrix of $\hat{G}(\hat{T}_{11}, \hat{T}_{21})$ is given by

$$T(\hat{G}, K_{id}) = \begin{bmatrix} \dfrac{\hat{T}_{11}}{\dfrac{\hat{T}_{21}}{K_{id}} + \hat{T}_{11}} & \dfrac{\dfrac{\hat{T}_{11}}{K_{id}}}{\dfrac{\hat{T}_{21}}{K_{id}} + \hat{T}_{11}} \\ \\ \dfrac{\hat{T}_{21}}{\dfrac{\hat{T}_{21}}{K_{id}} + \hat{T}_{11}} & \dfrac{\dfrac{\hat{T}_{21}}{K_{id}}}{\dfrac{\hat{T}_{21}}{K_{id}} + \hat{T}_{11}} \end{bmatrix} \neq \begin{bmatrix} \hat{T}_{11} & \dfrac{\hat{T}_{11}}{K_{id}} \\ \\ \hat{T}_{21} & \dfrac{\hat{T}_{21}}{K_{id}} \end{bmatrix} \tag{4.13}$$

We first note that, even though the coprime-factor procedure first computes estimates of $\hat{T}_{11}$ and $\hat{T}_{21}$, the resulting estimate of the first column of $T(\hat{G}, K_{id})$ is not $\begin{bmatrix} \hat{T}_{11} \\ \hat{T}_{21} \end{bmatrix}$. This is because $\hat{T}_{11} \neq \frac{\hat{G}K_{id}}{1+\hat{G}K_{id}}$ and $\hat{T}_{21} \neq \frac{K_{id}}{1+\hat{G}K_{id}}$ when $\hat{G}$ is computed as $\hat{G} = \frac{\hat{T}_{11}}{\hat{T}_{21}}$; as a result $\frac{\hat{T}_{21}(z)}{K_{id}(z)} + \hat{T}_{11}(z) \neq 1$, contrary to what should be. However, if $\hat{T}_{11}$ and $\hat{T}_{21}$ are reasonable estimates of T_{11} and T_{21}, then they must have the property that

$$\frac{\hat{T}_{21}(z)}{K_{id}(z)} + \hat{T}_{11}(z) \approx 1 \quad \forall z \tag{4.14}$$

Now observe that

(i) if K_{id} has a strictly nonminimum-phase zero at z_0, then (4.14) cannot hold at z_0. In addition, $\frac{\hat{T}_{11}}{K_{id}}$ and $\frac{\hat{T}_{21}}{K_{id}}$ will be unstable since, in practice, the nonminimum-phase zero of K_{id} will never be exactly cancelled by a corresponding zero in the estimated $\hat{T}_{11}$ and $\hat{T}_{21}$, because of the unavoidable errors attached to these estimates;

(ii) if K_{id} has a blocking zero, *i.e.*, a zero on the unit circle at some $z_0 = e^{j\omega_0}$, then $T_{11}(z_0) = T_{21}(z_0) = N_2(z_0) = 0$, while $N_1(z_0) \neq 0$. This will result in a very bad SNR of $y(t)$ at (and around) frequency ω_0, hence $\hat{T}_{11}$ will be a very poor estimate of T_{11} around this frequency. In particular, $\hat{T}_{11}(z_0) \neq 0$. Furthermore, because of the unavoidable variance error (and the possible bias error) attached to any model identified from noisy data, we shall normally have $\hat{T}_{21}(z_0) \neq 0$ as well. Hence

$$\left| \frac{\hat{T}_{21}(z_0)}{K_{id}(z_0)} + \hat{T}_{11}(z_0) \right| = \infty \tag{4.15}$$

will hold, while (4.14) must hold at frequencies where the quality of the estimates is not affected by blocking zeroes. This yields, for the four entries of $T(\hat{G}, K_{id})$ in (4.13),

$$\frac{\hat{T}_{11}(z_0)}{\frac{\hat{T}_{21}(z_0)}{K_{id}(z_0)} + \hat{T}_{11}(z_0)} = 0 = T_{11}(z_0) \tag{4.16a}$$

$$\frac{\frac{\hat{T}_{11}(z_0)}{K_{id}(z_0)}}{\frac{\hat{T}_{21}(z_0)}{K_{id}(z_0)} + \hat{T}_{11}(z_0)} = \frac{\hat{T}_{11}(z_0)}{\hat{T}_{21}(z_0)} = \hat{G}(z_0) \tag{4.16b}$$

$$\frac{\hat{T}_{21}(z_0)}{\frac{\hat{T}_{21}(z_0)}{K_{id}(z_0)} + \hat{T}_{11}(z_0)} = 0 = T_{21}(z_0) \tag{4.16c}$$

$$\frac{\frac{\hat{T}_{21}(z_0)}{K_{id}(z_0)}}{\frac{\hat{T}_{21}(z_0)}{K_{id}(z_0)} + \hat{T}_{11}(z_0)} = 1 = T_{22}(z_0) \tag{4.16d}$$

which are all finite and, for three of them, exact. Therefore a blocking zero does not cause nominal closed-loop instability;

(iii) unstable poles of K_{id} do not cause instability of the nominal closed-loop system;

(iv) if $\frac{\hat{T}_{21}(z)}{K_{id}(z)} + \hat{T}_{11}(z)$ has a nonminimum-phase zero, then $T(\hat{G}, K_{id})$ is unstable.

To summarise, when $\hat{G}$ is computed from $\hat{G} = \frac{\hat{T}_{11}}{\hat{T}_{21}}$ using measurements of $r_1(t)$, $y(t)$ and $u(t)$, the nominal closed-loop transfer matrix $T(\hat{G}, K_{id})$ will be stable if and only if K_{id} has no strictly nonminimum-phase zeroes and condition (4.14) holds, except at possible blocking zeroes of K_{id} where (4.15) will hold. Condition (4.14) serves as a way of validating the quality of the estimates $\hat{T}_{11}$ and $\hat{T}_{21}$; it also implies that $\frac{\hat{T}_{21}(z)}{K_{id}(z)} + \hat{T}_{11}(z)$ has a minimum phase.

In the case where $\hat{G}$ is computed from $\hat{G} = \frac{\hat{T}_{12}}{\hat{T}_{22}}$, the nominal closed-loop transfer matrix becomes

$$T(\hat{G}, K_{id}) = \begin{bmatrix} \dfrac{\hat{T}_{12}K_{id}}{\hat{T}_{22} + \hat{T}_{12}K_{id}} & \dfrac{\hat{T}_{12}}{\hat{T}_{22} + \hat{T}_{12}K_{id}} \\ \dfrac{\hat{T}_{22}K_{id}}{\hat{T}_{22} + \hat{T}_{12}K_{id}} & \dfrac{\hat{T}_{22}}{\hat{T}_{22} + \hat{T}_{12}K_{id}} \end{bmatrix} \tag{4.17}$$

In this case we need

$$\hat{T}_{22}(z) + \hat{T}_{12}(z)K_{id}(z) \approx 1 \quad \forall z \tag{4.18}$$

which cannot hold at possible poles of K_{id} on the unit circle. The same reasoning as before leads to the conclusion that $T(\hat{G}, K_{id})$ will be stable in this case if and only if K_{id} has no strictly unstable poles and condition (4.18) holds, except at possible unit-circle poles of K_{id}, where $|\hat{T}_{22} + \hat{T}_{12}K_{id}| = \infty$. Condition (4.18) serves as a way of validating the quality of the estimates $\hat{T}_{12}$ and $\hat{T}_{22}$; it also implies that $\hat{T}_{22}(z) + \hat{T}_{12}(z)K_{id}(z)$ has a minimum phase. □

Conclusion: With the coprime-factor approach, one must

- identify $\hat{G}$ from $\hat{G} = \hat{T}_{11}\hat{T}_{21}^{-1}$, if K_{id} has *strictly unstable poles but no strictly nonminimum-phase zeroes*, using $\{y(t), u(t), r_1(t)\}$ data. Condition (4.14) must be checked *a posteriori*;
- identify $\hat{G}$ from $\hat{G} = \hat{T}_{12}\hat{T}_{22}^{-1}$, if K_{id} has *strictly nonminimum-phase zeroes but no strictly unstable poles*, using $\{y(t), u(t), r_2(t)\}$ data. Condition (4.18) must be checked *a posteriori*.

If K_{id} has both poles and zeroes strictly outside the unit circle, the coprime-factor approach cannot be used to obtain a model $\hat{G}$ stabilised by K_{id}. Poles and/or zeroes on the unit circle are not an issue regarding nominal closed-loop stability.

► **Remark.** Although possible unit-circle poles and zeroes in K_{id} have no influence on the stability of $\hat{G}$ in closed loop, they can have a significant influence on the quality of the estimate, as we now show.

- If K_{id} has a blocking zero at some frequency ω_0, then $T_{11}(e^{j\omega_0}) = 0$, $T_{21}(e^{j\omega_0}) = 0$, $N_1(e^{j\omega_0}) \neq 0$ and $N_2(e^{j\omega_0}) = 0$. As a result, if $r_1(t)$ is used as excitation source, the SNR of $y(t)$ will be zero while that of $u(t)$ will be undetermined at ω_0, but not impaired by this blocking zero in the neighbourhood of this frequency. Therefore, we can expect a good estimation of T_{21} at ω_0, but a poor estimate of T_{11} at the same frequency, yielding a poor estimate of $\hat{G}$. It is therefore recommended to excite the system with $r_2(t)$ and to identify T_{12} and T_{22} in this case.
- If K_{id} has a blocking pole at some frequency ω_0, then $T_{12}(e^{j\omega_0}) = 0$, $T_{22}(e^{j\omega_0}) = 0$, $N_1(e^{j\omega_0}) = 0$ and $N_2(e^{j\omega_0}) \neq 0$. As a result, if $r_2(t)$ is used as excitation source, the SNR of $u(t)$ will be zero while that of $y(t)$ will be undetermined at ω_0, but not

impaired by this blocking pole in the neighbourhood of this frequency. Therefore, we can expect a good estimation of T_{22} at ω_0, but a poor estimate of T_{12} at the same frequency, yielding again a bad estimate of $\hat{G}$. It is therefore recommended to excite the system with $r_1(t)$ and to identify T_{11} and T_{21} in this case. ◄

4.4.3 The Direct Approach

In this approach, the model $\hat{G}$ of the system is directly identified using measurements of $u(t)$ and $y(t)$. It has the advantage that the structure and the order of the model are easy to tune and that the controller can be unknown, nonlinear and time-varying. However, it also suffers the following drawbacks:

- the input signal used for identification is correlated with the output noise, with the result that a correct noise model is required to avoid bias (Forssell and Ljung, 1999);
- the bias error distribution is difficult to tune since it is influenced by the unknown closed-loop sensitivity function (Ansay *et al.*, 1999).

The nominal closed-loop transfer matrix obtained by the direct approach is simply

$$T(\hat{G}, K_{id}) = \begin{bmatrix} \dfrac{\hat{G}K_{id}}{1+\hat{G}K_{id}} & \dfrac{\hat{G}}{1+\hat{G}K_{id}} \\ \dfrac{K_{id}}{1+\hat{G}K_{id}} & \dfrac{1}{1+\hat{G}K_{id}} \end{bmatrix} \triangleq \begin{bmatrix} \hat{T}_{11} & \hat{T}_{12} \\ \hat{T}_{21} & \hat{T}_{22} \end{bmatrix} \tag{4.19}$$

Its stability does not hinge on the cancellation of unstable poles or nonminimum-phase zeroes in K_{id}. Hence, the direct approach can be used with any external excitation, $r_1(t)$ or $r_2(t)$, even if K_{id} has unstable poles and/or nonminimum-phase zeroes. However, there is no prior guarantee of nominal closed-loop stability, which will have to be checked after identification.

► **Remark.** Once again, blocking zeroes and poles deserve particular attention.

- Assume that K_{id} has a blocking zero at some frequency ω_0 and that $r_1(t)$ is used to excite the system. The signals used for identification are then $y(t) = T_{11}(z)r_1(t) + N_1(z)v(t)$ and $u(t) = T_{21}(z)r_1(t) + N_2(z)v(t)$, with $T_{11}(e^{j\omega_0}) = 0$, $T_{21}(e^{j\omega_0}) = 0$, $N_2(e^{j\omega_0}) = 0$ and $N_1(e^{j\omega_0}) \neq 0$. Hence, the input power spectral density will be zero at this frequency ($\phi_u(\omega_0) = 0$), while $y(t)$ will only contain noise at this frequency ($\phi_y^{r_1}(\omega_0) = 0 \Rightarrow \phi_y(\omega_0) = \phi_y^v(\omega_0) = |N_1(e^{j\omega_0})|^2\phi_v(\omega_0)$). In the absence of input excitation and with a zero output SNR, the resulting model $\hat{G}(e^{j\omega})$ is allowed to take any value at $\omega = \omega_0$. This value will have no effect on the nominal closed-loop transfer functions $\hat{T}_{11}$ and $\hat{T}_{21}$ which, in any case, are identically zero at this frequency. However, the other two entries of $T(\hat{G}, K_{id})$ depend on $\hat{G}$ at ω_0, with the result that, since $\hat{G}(e^{j\omega_0})$ can be anything, these entries might be unstable, *i.e.*, the nominal closed-loop system might be unstable. Therefore, it is recommended to use $r_2(t)$ to excite the system if K_{id} has a zero on the unit circle. An exception occurs, however, when $\phi_v(\omega_0) = 0$, *i.e.*, when the noise process also has a unit-circle zero at ω_0: $H_0(e^{j\omega_0}) = 0$. In this case, because of the absence of noise at frequency ω_0, the SNR of $y(t)$ is undetermined at ω_0 but, as it is determined and not impaired in the neighbourhood of ω_0, the blocking zero will have no impact on the quality of the estimate at this frequency, the parameters of $\hat{G}$ being determined by the spectral content of the data at other frequencies. As H_0 is generally unknown in advance, a rule of good practice is, however, to always use $r_2(t)$ when the controller has a blocking zero.

- Assume that K_{id} has a blocking pole at some frequency ω_0 and that $r_2(t)$ is used to excite the system. The signals used for identification are then $y(t) = T_{12}(z)r_2(t) + N_1(z)v(t)$ and $u(t) = T_{22}(z)r_2(t) + N_2(z)v(t)$, with $T_{12}(e^{j\omega_0}) = 0$, $T_{22}(e^{j\omega_0}) = 0$, $N_1(e^{j\omega_0}) = 0$ and $N_2(e^{j\omega_0}) \neq 0$. Hence, in the general case where $\phi_v(\omega_0)$ is finite, the output power spectral density will be zero at this frequency ($\phi_y(\omega_0) = 0$), while $u(t)$ will only contain noise at this frequency ($\phi_u^{r_2}(\omega_0) = 0 \Rightarrow \phi_u(\omega_0) = \phi_u^v(\omega_0) = |N_2(e^{j\omega_0})|^2\phi_v(\omega_0)$). Note that, even though $\phi_u(\omega_0) \neq 0$, $u(t)$ bears no useful information at frequency ω_0[3]. So, we are in a situation where the useful part of the input spectrum and the output spectrum are both zero at ω_0 and the output SNR is undetermined at this frequency. Yet, this does not impair the quality of the estimate $\hat{G}(e^{j\omega})$ at $\omega = \omega_0$. Indeed, as there is useful input excitation and as the output SNR is determined and not affected by the blocking pole of K_{id} in the neighbourhood of this pole, the parameters of $\hat{G}$ will be determined from the spectral content of the data at frequencies close to ω_0. An exception occurs, however, when $\phi_v(\omega_0) = \infty$, *i.e.*, when H_0 also has a pole at ω_0. In this case, the output SNR is 0 at ω_0 and very low in its neighbourhood due to the cancellation of the blocking zero in N_1 by this pole. As a result, correct values of the parameters of $\hat{G}$ cannot longer be obtained from the spectral content of the signals around ω_0. $\hat{G}(e^{j\omega})$ can then take any value at $\omega = \omega_0$. This value will have no effect on the nominal closed-loop transfer functions $\hat{T}_{12}$ and $\hat{T}_{22}$ which, in any case, are identically zero at this frequency. However, the other two entries of $T(\hat{G}, K_{id})$ depend on $\hat{G}$ at ω_0, with the result that, since $\hat{G}(e^{j\omega_0})$ can be anything, these entries might again be unstable, *i.e.*, the nominal closed-loop system might be unstable. As a result and since H_0 is generally unknown in advance, it is always preferable to use $r_1(t)$ as excitation source if K_{id} has a blocking pole.

Note also that, with the direct approach, both excitation signals $r_1(t)$ and $r_2(t)$ can be used simultaneously to excite the system. This can be useful when the controller has both blocking zeroes and blocking poles, or when the control structure involves a feedback part (with $r_1(t)$ as reference signal for $y(t)$) and a feed-forward part (producing a feed-forward signal $r_2(t)$) that must both be kept in use during the identification tests because of operational constraints. ◀

---

[3] This is shown by (3.7) where it can be seen that only $\phi_u^r(\omega_0)$ (*i.e.*, the power spectral density of that part of $u(t)$ that originates from a reference signal and not from the noise) would play a role in the tuning of $\hat{G}(e^{j\omega_0})$. The second term of (3.7) is zero at ω_0 since $S(G_0(e^{j\omega_0}), K_{id}(e^{j\omega_0})) = N_1(e^{j\omega_0}) = 0$.

4.4.4 The Dual Youla Parametrisation Approach

(Hansen, 1989) has proposed a closed-loop identification method using the dual Youla parametrisation of all LTI plants that are stabilised by a given controller.

Consider any auxiliary system G_{aux} known to be stabilised by K_{id} (typically a known but imperfect model of the true plant G_0, although this is not required), and define the following coprime factorisations:

$$G_{aux} = \frac{N}{M} \qquad \text{and} \qquad K_{id} = \frac{U}{V} \tag{4.20}$$

where N, M, U and V belong to $\mathcal{RH}_\infty$ and satisfy the following Bezout identity:

$$VM + UN = 1 \tag{4.21}$$

Remember (Lemma 2.2) that this equation expresses the fact that the feedback loop formed by G_{aux} and K_{id} is internally stable.

Then, the set of all LTI plants stabilised by K_{id} is given by (Vidyasagar, 1985)

$$\mathcal{G}_{K_{id}} = \left\{ G(R) = \frac{N + VR}{M - UR} \mid R \in \mathcal{RH}_\infty \right\} \tag{4.22}$$

Since we have assumed that G_0 is stabilised by K_{id}, *i.e.*, $G_0 \in \mathcal{G}_{K_{id}}$, there exists $R_0 \in \mathcal{RH}_\infty$ such that

$$G_0 = \frac{N + VR_0}{M - UR_0} \tag{4.23}$$

and the feedback system (G_0, K_{id}) of Figure 4.1 can thus be redrawn as in Figure 4.2.

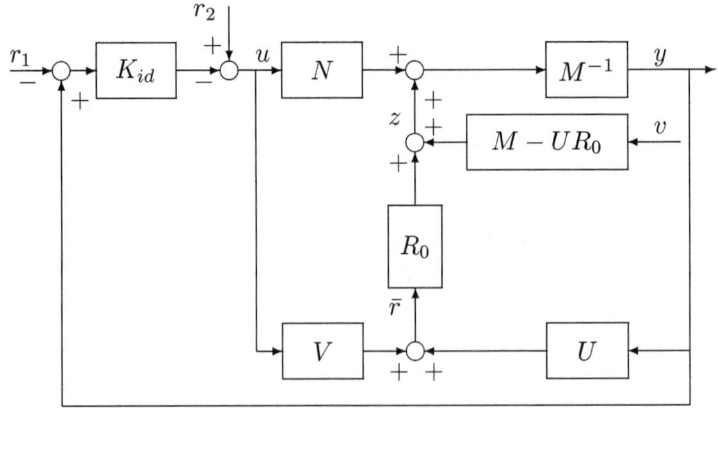

Figure 4.2. Alternative representation of the closed-loop system of Figure 4.1 using coprime factors and a dual Youla parametrisation of the plant

Let us now define the following auxiliary signals, as shown in Figure 4.2:

$$z(t) = My(t) - Nu(t) \tag{4.24}$$

and

$$\begin{aligned}\bar{r}(t) &= Uy(t) + Vu(t) \\ &= U\big(NUr_1(t) + NVr_2(t)\big) + V\big(MUr_1(t) + MVr_2(t)\big) \\ &= Ur_1(t) + Vr_2(t)\end{aligned} \tag{4.25}$$

where the second equality is obtained by observing that

$$\begin{bmatrix} y(t) \\ u(t) \end{bmatrix} = \begin{bmatrix} N \\ M \end{bmatrix} \begin{bmatrix} U & V \end{bmatrix} \begin{bmatrix} r_1(t) \\ r_2(t) \end{bmatrix} \tag{4.26}$$

which results from (2.62) applied to the SISO case ($\tilde{U} = U$, $\tilde{V} = V$) and from (4.21) and where the third equality results, again, from (4.21). These auxiliary signals can thus be reconstructed from the data. We then have

$$z(t) = R_0\bar{r}(t) + (M - UR_0)H_0e(t) \tag{4.27}$$

A model $\hat{G}$ of G_0 can be obtained through the identification of a model $\hat{R}$ of the 'true' dual Youla parameter R_0 using data

$$Z^N = \Big\{\bar{r}(1),\, z(1),\, \dots,\, \bar{r}(N),\, z(N)\Big\} \tag{4.28}$$

This is an open-loop identification problem since $\bar{r}(t)$ and $e(t)$ are uncorrelated, which is made obvious by the third equality in (4.25).

Due to the properties of the dual Youla parametrisation, a necessary and sufficient condition for the resulting model

$$\hat{G} = \frac{\hat{N}}{\hat{M}} \triangleq \frac{(N + V\hat{R})}{(M - U\hat{R})} \tag{4.29}$$

to be stabilised by K_{id} is that $\hat{R}$ be stable, which can be guaranteed by an appropriate choice of the model structure (*e.g.*, OE or BJ). As a result, the nominal generalised closed-loop transfer matrix is given by

$$T(\hat{G}, K) = \begin{bmatrix} \hat{N} \\ \hat{M} \end{bmatrix} \hat{\Phi}^{-1} \begin{bmatrix} U & V \end{bmatrix} = \begin{bmatrix} \hat{N}U & \hat{N}V \\ \hat{M}U & \hat{M}V \end{bmatrix} \tag{4.30}$$

where the second equality follows from the fact that $\hat{\Phi} = \hat{N}U + \hat{M}V = NU + V\hat{R}U + MV - U\hat{R}V = NU + MV = 1$. Hence, $T(\hat{G}, K)$ is stable if and only if $\hat{N}$ and $\hat{M}$ (or, equivalently, $\hat{R}$) are stable. Thus, with this method, the stability of the nominal closed-loop system is guaranteed by the structure of the solution, disregarding possible unstable poles and nonminimum-phase zeroes in K_{id}.

► **Remark.** Although $\hat{G}$ is guaranteed to be stabilised by K_{id} if $\hat{R}$ is stable, it should be noted that the factorisation of K_{id} into coprime factors may put constraints on the design of $r_1(t)$ and $r_2(t)$. Indeed, the coprime factors U and V may have a low gain at some frequencies, either because of the particular factorisation of K_{id} (*e.g.*, if the coprime factors are chosen normalised, meaning that $|U|^2 + |V|^2 = 1$, a low gain of one of these factors at some frequency always corresponds to a high gain of the other factor at the same frequency) or because of blocking poles or zeroes in K_{id} (indeed, any blocking zero of K_{id} is also a blocking zero of U, while any blocking pole of K_{id} is a blocking zero of V).

- If U has a low gain at some frequencies, $r_1(t)$ will be cut off and $\bar{r}(t)$ will be poorly (or not at all) excited at these frequencies. $r_2(t)$ should then be nonzero in order to have a good SNR and avoid a poor estimation of R_0 at these frequencies;
- If V has a low gain at some frequencies, $r_2(t)$ will be cut off and $\bar{r}(t)$ will be poorly (or not at all) excited at these frequencies. $r_1(t)$ should then be nonzero in order to have a good SNR and avoid a poor estimation of R_0 at these frequencies.

As in the direct approach, it is possible to use both sources of excitation $r_1(t)$ and $r_2(t)$ simultaneously, which is an advantage in comparison with the indirect and coprime-factor approaches. ◀

4.5 Guidelines for an Appropriate Closed-loop Identification Experiment Design

4.5.1 Guidelines for the Choice of an Identification Method

The effects of the presence of unstable poles or nonminimum-phase zeroes in the controller on the stability of the nominal closed-loop systems obtained by the four closed-loop identification techniques considered in this chapter can be summarised in Table 4.1. This table gives guidelines for the design of an identification experiment that are appropriate regarding the closed-loop stability of the obtained model. These guidelines are given in terms of the closed-loop identification methods and of the excitation signals that can be used in function of the controller singularities.

The following example illustrates how this table must be interpreted.

▶ **Example 4.1.** Assume that K_{id} has a unit-circle zero. Then,

- the indirect method will deliver a model $\hat{G}$ that is stabilised by K_{id} if and only if a stable T_{12} is identified using measurements of $r_2(t)$ and $y(t)$ (horizontal reading of the table: '+') and K_{id} has no unit-circle or strictly unstable poles (vertical reading of the table: '–');
- the coprime-factor method can deliver a model $\hat{G}$ that is stabilised by K_{id} (the stability must be checked *a posteriori*)
  - using measurement of $r_1(t)$ (horizontal reading of the table: '·') except if K_{id} also has strictly nonminimum-phase zeroes (vertical reading of the table: '–');

  - using measurement of $r_2(t)$, which is better than using $r_1(t)$ (horizontal reading of the table: '⊙'), except if K_{id} also has strictly unstable poles (vertical reading of the table: '−');
- the dual Youla parametrisation method will always deliver a model $\hat{G}$ that is stabilised by K_{id}, but it is better to use $r_2(t)$, possibly with $r_1(t)$ (horizontal reading of the table: '⊕'), than $r_1(t)$ alone (horizontal reading of the table: '+') for excitation. ◀

Table 4.1. Stability of the nominal closed-loop model w.r.t. the identification method, the excitation signal and the singularities of K_{id}: stability is guaranteed if a stable model structure is used (+); stability has to be checked *a posteriori* (·); instability is guaranteed (−). When the use of some signal or combination of signals should be favoured within a given method, the corresponding mark is encircled (⊕ or ⊙). When K_{id} has several listed singularities, the most unfavourable one prevails.

Singularities of K_{id}	**Indirect**				**Coprime-factor**	
	r_1, y	r_1, u	r_2, y	r_2, u	r_1	r_2
Strictly unstable poles	−	+	−	−	·	−
Unit-circle poles	−	+	−	−	⊙	·
Strictly nonminimum-phase zeroes	−	−	+	−	−	·
Unit-circle zeroes	−	−	+	−	·	⊙

Singularities of K_{id}	**Direct**			**Dual-Youla param.**		
	r_1	r_2	r_1 & r_2	r_1	r_2	r_1 & r_2
Strictly unstable poles	·	·	·	+	+	+
Unit-circle poles	⊙	·	⊙	⊕	+	⊕
Strictly nonminimum-phase zeroes	·	·	·	+	+	+
Unit-circle zeroes	·	⊙	⊙	+	⊕	⊕

4.5.2 Remark on High-order Models Obtained by Two-stage Methods

The models delivered by two-stage techniques like the indirect, coprime-factor or dual Youla parametrisation approaches, will generally have a much higher order than the plant G_0 if unbiased transfer functions are identified in the first step. For instance, if G_0 has order n and K_{id} has order m, then the closed-loop transfer functions T_{ij}, $i, j \in \{1, 2\}$ will generally have order $n + m$. If $\hat{G}$ is obtained by an indirect identification method, then, by virtue of the mapping (4.9), its order will typically be that of the identified transfer function plus that of the controller, *i.e.*, $n + 2m$. If the coprime-factor method is used, the order of $\hat{G}$ will typically be $2(n+m)$. The situation is even worse with the dual Youla parametrisation approach.

The additional modes introduced during the reconstruction of $\hat{G}$ from the identified transfer function(s) are, of course, not representative of the plant but are caused by the modelling errors, which prevent exact cancellation of poles and zeroes that would otherwise perfectly match. Hence, if the identification is done properly, most of these additional poles and zeroes will present near mutual cancellations, meaning that the realisation of $\hat{G}$ will be nearly nonminimal.

However, in order to be suitable for control design, a model $\hat{G}$ must generally have a minimal realisation, *i.e.*, it must be observable and controllable (this is the case, for instance, with control design techniques like LQG or $\mathcal{H}_\infty$). The near nonminimality of the high-order models obtained *via* two-stage closed-loop identification methods may then be the cause of troubles during control design. It is therefore recommended to remove the quasi-nonminimal modes before using the model for control design. This can be done by means of various model reduction techniques. If the user has no intention to reduce the order of the model beyond that of the true system, *i.e.*, if no bias is introduced, open-loop reduction techniques will generally suffice. Otherwise, closed-loop techniques like those proposed in Chapter 6 should be considered.

4.6 Numerical Illustration

4.6.1 Problem Description

We consider as 'true system' the following ARX model:

$$G_0(z) = \frac{0.1028z + 0.1812}{z^4 - 1.992z^3 + 2.203z^2 - 1.841z + 0.8941} \tag{4.31a}$$

$$H_0(z) = \frac{z^4}{z^4 - 1.992z^3 + 2.203z^2 - 1.841z + 0.8941} \tag{4.31b}$$

It describes a flexible transmission system that was used in (Landau *et al.*, 1995*b*) and references therein as a benchmark for testing various control design methods. The system is normally excited using $r_2(t)$ with the following two-degree-of-freedom control law:

$$u(t) = F_{id}(z)r_2(t) - K_{id}(z)y(t) \tag{4.32}$$

where $F_{id}(z)$ is the feed-forward part of the controller. Here, we consider the optimal controller that was obtained by means of Iterative Feedback Tuning in (Hjalmarsson *et al.*, 1995):

$$F_{id}(z) = \frac{0.104z^4 - 0.0857z^3 - 0.04274z^2 + 0.03793z + 0.03612}{z^3(z-1)} \tag{4.33a}$$

$$K_{id}(z) = \frac{0.5517z^4 - 1.765z^3 + 2.113z^2 - 1.296z + 0.4457}{z^3(z-1)} \tag{4.33b}$$

As $F_{id}(z)$ does not influence the stability of $T(G_0,\, K_{id})$, we shall replace it by 1 in the sequel. For the purpose of illustration, we shall also consider that it is possible to excite the system with a reference signal $r_1(t)$[4].

With this controller, the achieved generalised stability margin is

$$b_{G_0,K_{id}} = 0.2761 \tag{4.34}$$

The maximum value that could be reached for this system is

$$\sup_K b_{G_0,K} = 0.4621 \tag{4.35}$$

Notice that $K_{id}(z)$ has a unit-circle pole located in $z = 1$, which will have a blocking effect on $r_2(t)$. $K_{id}(z)$ also has a pair of strictly nonminimum-phase complex zeroes in $z = 1.2622 \pm 0.2011j$, which may pose problems with some methods if $r_1(t)$ is used alone.

We now test the various closed-loop identification techniques described in this chapter and we analyse their results in the light of Section 4.2. Therefore, each of the models $\hat{G}$ obtained by these methods will be used to compute a two-degree-of-freedom controller $C(z) = \begin{bmatrix} F(z) & K(z) \end{bmatrix}$ stabilising $\hat{G}$ and presenting a good nominal performance (we want a zero static error and a faster response than with the current controller K_{id}, namely a closed-loop bandwidth located between the two resonance peaks of the open-loop system, while keeping the step response overshoot under 10%. Due to the increased closed-loop bandwidth, the new controllers will normally have smaller stability margins than the current one). The suitability for control design of each model $\hat{G}$ will be assessed by checking the sufficient conditions of Section 4.2 and the performance with the true system G_0 of the corresponding designed controller.

The signals $r_1(t)$ and/or $r_2(t)$ and $e(t)$ used to excite the closed-loop system $(G_0,\, K_{id})$ for identification are chosen as mutually independent Gaussian sequences with zero mean and variances 1, 1 and 0.05, respectively. No prefilter is used for identification (*i.e.*, $L(z,\, \theta)$=1) and the model structures we use are all unbiased[5] in order to show that the variance error that is always committed during identification from noisy data is sufficient to produce the instability

[4]In (Landau *et al.*, 1995*b*), it was demanded that the controllers $\begin{bmatrix} F(z) & K(z) \end{bmatrix}$ produced in the benchmark study all had an R-S-T structure, meaning that $F(z)$ and $K(z)$ had to share the same denominator. In the present case, they both have an unstable pole at $z = 1$, hence the feed-forward $F_{id}(z)$ cannot actually be implemented. However, since this pole also appears in $K_{id}(z)$, it would be easy to replace the equation for $u(t)$ by $u(t) = K_{id}(z)(L(z)r_1(t) - y(t))$, where $L(z)$ would be a stable feed-forward filter. This justifies, if necessary, the replacement of $F_{id}(z)$ by 1, or the use of $r_1(t)$ to excite the system, in our experiments.

[5]Observe that the indirect, coprime-factor and dual Youla parametrisation methods are actually 'open-loop' methods since they use the reference signals $r_1(t)$, $r_2(t)$ or $\bar{r}(t)$, which are not correlated with the noise, as input signals. Therefore, output-error model structures can be used with these methods without introducing a bias error.

problems described in this chapter. It should be clear that biased models would also present at least the same problems.

The control design method used with each identified model $\hat{G}$ is the following. First, coprime-factor-based $\mathcal{H}_\infty$ control design (Skogestad and Postlethwaite, 1996) is used to compute a stabilising single-degree-of-freedom controller $\tilde{K}$ for an augmented model

$$\tilde{G}(z) \triangleq W_1(z)\hat{G}(z)W_2(z) \tag{4.36}$$

where W_1 and W_2 are loop-shaping filters aimed to obtain a good nominal performance, so that the following $\mathcal{H}_\infty$ constraint is satisfied:

$$\left\| T(\tilde{G}, \tilde{K}) \right\|_\infty \leq \frac{1}{\sup_K b_{\tilde{G},K}} + \varepsilon \tag{4.37}$$

where $\varepsilon = 10^{-6}$. Then, a two-degree-of-freedom controller $C = \begin{bmatrix} F & K \end{bmatrix}$ for the nominal model $\hat{G}$ is obtained by setting

$$K(z) = W_2(z)\tilde{K}(z)W_1(z) \qquad \text{and} \qquad F(z) = W_2(1)\tilde{K}(1)W_1(z) \tag{4.38}$$

W_1 will typically contain an integrator in order to ensure a zero static error, while W_2 will be adjusted in order to reduce the effect of the second resonant mode.

Sometimes, in the light of Subsection 4.5.2, the order of $\hat{G}$ will be reduced using coprime-factor balanced truncation[6] before control design, the goal being the cancellation of the nearly nonminimal modes of high-order models produced by the coprime-factor or dual Youla parametrisation identification methods. The reduced-order models will be denoted by $\check{G}$ in the sequel, and the two-degree-of-freedom controllers computed from such models will be denoted by $\check{C} = \begin{bmatrix} \check{F} & \check{K} \end{bmatrix}$.

4.6.2 The Indirect Approach

Here, we consider two alternatives among the four possible ones: to identify T_{12} using $r_2(t)$ or to identify T_{21} using $r_1(t)$.

A. Identification of T_{12}**.** The closed-loop system was simulated with $r_2(t) \in N(0, 1)$ and $e(t) \in N(0, 0.05)$. Using 1000 measurements of $r_2(t)$ and $y(t)$, an output-error model $\hat{T}_{12}$ with exact structure OE[3,8,3] was estimated, from which a model $\hat{G}$ of the plant was derived using (4.9). Figure 4.3 shows the Bode diagrams of G_0 and $\hat{G}$. As expected, $\hat{G}$ is a very bad estimate at low frequencies, since it has an incorrect zero at the blocking pole of K_{id}. Furthermore, the three reconstructed entries of the nominal closed-loop transfer matrix (4.11) all have a pole in $z = 1$, meaning that the nominal closed-loop system is unstable ($b_{\hat{G},K_{id}} = 0$). This is because the zero at $z = 1$ of T_{12} is imperfectly estimated

[6]This procedure is explained in Subsection 6.4.1.

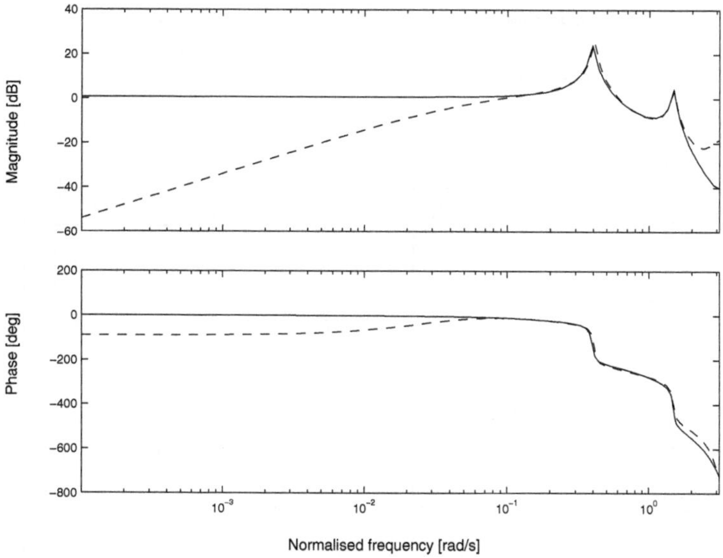

Figure 4.3. Indirect approach using T_{12}: Bode diagrams of G_0 (—) and $\hat{G}$ (– –)

as a zero at $z = 0.9590$ in $\hat{T}_{12}$, which does not cancel the integrator of K_{id}. Finally, note that

$$\delta_\nu(G_0, \hat{G}) = 1 > \sup_K b_{\hat{G},K} = 0.2389 \tag{4.39}$$

where the first equality results from a violation of the winding number condition (see Section 2.5). Thus, any controller K stabilising $\hat{G}$, even the most conservative one, cannot be guaranteed to also stabilise the true system G_0. Clearly, the model we have obtained here is not suitable for control design.

However, we did design a controller for $\hat{G}$ and we tried to apply it to G_0. The following loop-shaping filters were used for the design:

$$W_1(z) = \left(\frac{z}{z-1}\right)^2 \qquad \text{and} \qquad W_2(z) = \frac{2.44z^2 - 1.017z + 0.167}{z^2 - 0.1519z + 0.7423} \tag{4.40}$$

Here, a double integrator in $W_1(z)$ was necessary to ensure a zero static error, because of the differentiator contained in $\hat{G}$ (zero at $z = 1$). The resulting two-degree-of-freedom controller $C = \begin{bmatrix} F & K \end{bmatrix}$ has K of degree 14. Its performance with $\hat{G}$ and G_0 is depicted in Figure 4.4, where it can be seen that the nominal closed loop is stable and has a good performance while, as expected, the achieved closed loop is unstable.

B. Identification of T_{21}. The closed-loop system was simulated with $r_1(t) \in N(0, 1)$ and $e(t) \in N(0, 0.05)$. Using 1000 measurements of $r_1(t)$ and $y(t)$, an output-error model $\hat{T}_{21}$ with exact structure OE[9,8,0] was estimated, from

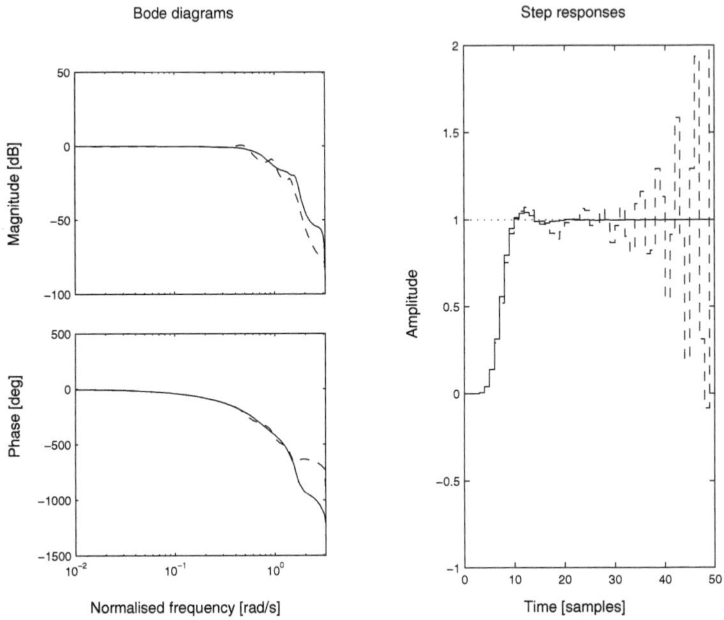

Figure 4.4. Control design *via* indirect identification using T_{12}: Bode diagrams (left) and step responses (right) of $\frac{F\hat{G}}{1+K\hat{G}}$ (—) and $\frac{FG_0}{1+KG_0}$ (−−)

which a model $\hat{G}$ of the plant was derived using (4.9). Figure 4.5 shows the Bode diagrams of G_0 and $\hat{G}$. It also shows the true T_{21}, and one can see that its modulus is very low around 0.4 rad/s, which results in a low SNR around that frequency and explains the bad quality of $\hat{G}$ around the first resonance peak.

The three reconstructed entries of the nominal closed-loop transfer matrix all have a pair of complex unstable poles in $z = 1.2622 \pm 0.2011\jmath$, meaning that the nominal closed-loop system is unstable ($b_{\hat{G},K_{id}} = 0$). This is because the zeroes at $z = 1.2622 \pm 0.2011\jmath$ of T_{21} are imperfectly estimated as zeroes at $z = 1.2650 \pm 0.1757\jmath$ in $\hat{T}_{21}$.

Finally, note that

$$\delta_\nu(G_0, \hat{G}) = 1 > \sup_K b_{\hat{G},K} = 0.0032 \tag{4.41}$$

The extremely small value of $\sup_K b_{\hat{G},K}$ indicates that, even if one can compute a controller that stabilises $\hat{G}$, this controller will have an extremely small generalised stability margin and, given inequality (4.41), it will have little chance of also stabilising G_0. Clearly, the model we have obtained here is not suitable for control design.

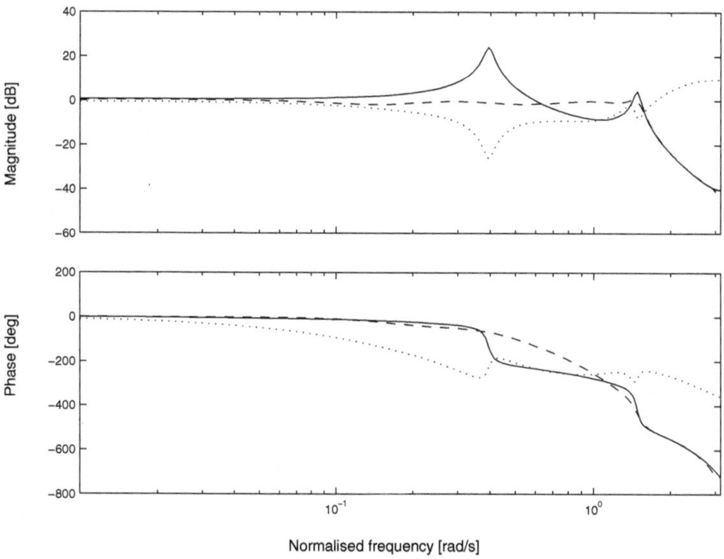

Figure 4.5. Indirect approach using T_{21}: Bode diagrams of G_0 (—), $\hat{G}$ (−−) and T_{21} ($\cdots$)

Although $\sup_K b_{\hat{G},K}$ is very small, we managed to design a stabilising controller for $\hat{G}$ with the following loop-shaping filters:

$$W_1(z) = \frac{z}{z-1} \qquad \text{and} \qquad W_2(z) = \frac{2.44z^2 - 1.017z + 0.167}{z^2 - 0.1519z + 0.7423} \tag{4.42}$$

The resulting two-degree-of-freedom controller $C = \begin{bmatrix} F & K \end{bmatrix}$ has K of degree 14. Its performance with $\hat{G}$ and G_0 is depicted in Figure 4.6, where it can be seen that the nominal closed loop is stable but the performance is poor, due to the small value of $\sup_K b_{\hat{G},K}$, while the achieved closed loop is unstable as expected.

4.6.3 The Coprime-factor Approach

Since the controller has a pole on the unit circle and zeroes outside the unit circle, we can expect that the model $\hat{G}$ obtained *via* the first alternative, *i.e.*, from estimates of T_{11} and T_{21}, will be destabilised by K_{id} and unsuitable for control design, while the one obtained *via* the second alternative, *i.e.*, from estimates of T_{12} and T_{22}, should be stabilised by K_{id} and suited to control design.

A. Identification of T_{11} and T_{21}. The closed-loop system was simulated with $r_1(t) \in N(0, 1)$ and $e(t) \in N(0, 0.05)$. $\hat{T}_{11}$ was identified using an exact OE[6,8,3] model structure and 1000 samples of $r_1(t)$ and $y(t)$. $\hat{T}_{21}$ was identified

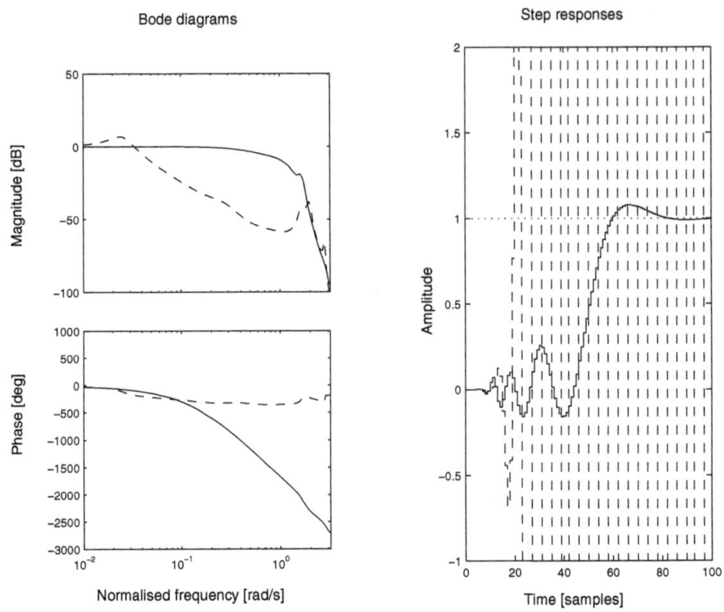

Figure 4.6. Control design *via* indirect identification using T_{21}: Bode diagrams (left) and step responses (right) of $\frac{F\hat{G}}{1+K\hat{G}}$ (—) and $\frac{FG_0}{1+KG_0}$ (−−)

using an exact OE[9,8,0] model structure and 1000 samples of $r_1(t)$ and $u(t)$. A model $\hat{G}$ of order 16 was then computed from $\hat{G} = \hat{T}_{11}/\hat{T}_{21}$. Figure 4.7 shows the Bode diagrams of G_0 and $\hat{G}$.

As expected, the nominal closed-loop transfer matrix (4.13) is unstable and

$$\delta_\nu(G_0, \hat{G}) = 1 > \sup_K b_{\hat{G},K} = 0.0299 \tag{4.43}$$

This model is again unsuitable for control design. This is confirmed by our attempt to compute a controller for G_0 using $\hat{G}$. A two-degree-of-freedom $\mathcal{H}_\infty$ controller $C = \begin{bmatrix} F & K \end{bmatrix}$, with K of degree 22, stabilising $\hat{G}$, was obtained with the following weightings:

$$W_1(z) = \frac{z}{z-1} \quad \text{and} \quad W_2(z) = \frac{1.09z^2 - 0.3616z + 0.2327}{z^2 - 0.3072z + 0.2683} \tag{4.44}$$

Its performance with $\hat{G}$ and G_0 is illustrated in Figure 4.8. The nominal closed loop is stable but the performance is poor, due to the small value of $\sup_K b_{\hat{G},K}$, while the achieved closed loop is unstable as expected.

An attempt was then made to use coprime-factor balanced truncation to obtain a 6th-order estimate $\check{G}$ of $\hat{G}$, from which a new $\check{C} = \begin{bmatrix} \check{F} & \check{K} \end{bmatrix}$ was produced using the same technique and loop-shaping filters as with $\hat{G}$. However, this new controller also destabilised the true system G_0.

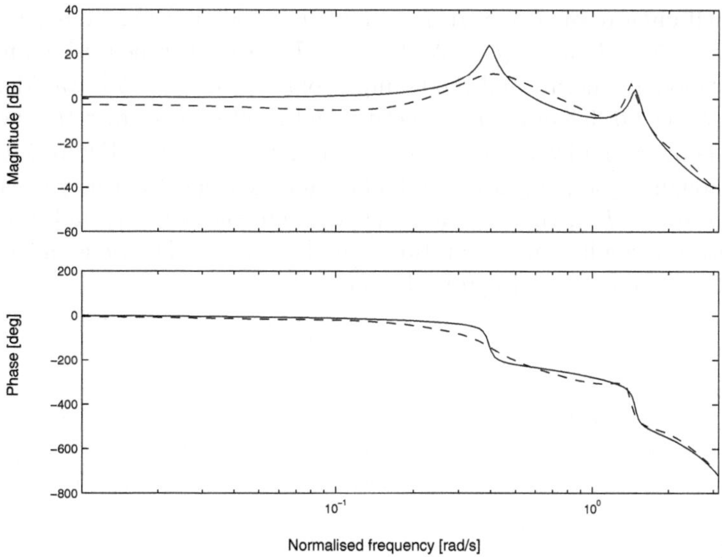

Figure 4.7. Coprime-factor approach using $r_1(t)$: Bode diagrams of G_0 (—) and $\hat{G}$ (−−)

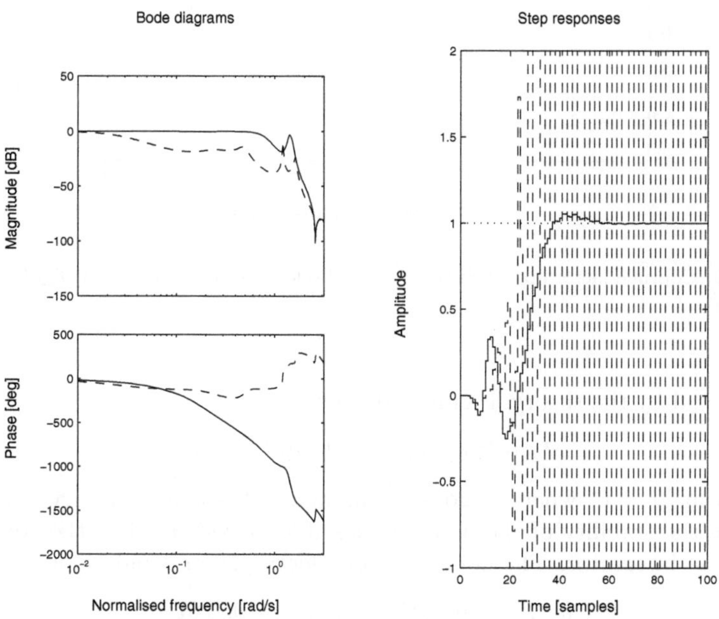

Figure 4.8. Control design *via* coprime-factor identification using $r_1(t)$: Bode diagrams (left) and step responses (right) of $\frac{F\hat{G}}{1+K\hat{G}}$ (—) and $\frac{FG_0}{1+KG_0}$ (−−)

B. Identification of T_{12} and T_{22}. The closed-loop system was simulated with $r_2(t) \in N(0, 1)$ and $e(t) \in N(0, 0.05)$. $\hat{T}_{12}$ was identified using an exact OE[3,8,3] model structure and 1000 samples of $r_2(t)$ and $y(t)$. $\hat{T}_{22}$ was identified using an exact OE[6,8,0] model structure and 1000 samples of $r_2(t)$ and $u(t)$. A model $\hat{G}$ of order 13 was then computed from $\hat{G} = \hat{T}_{12}/\hat{T}_{22}$. Figure 4.9 shows that, although $\hat{T}_{12}$ and $\hat{T}_{22}$ are very bad estimates at low frequency because of the integrator in K_{id}, the way they deviate from the true T_{12} and T_{22} is the same, with the result that their ratio, $\hat{G}$, is close to G_0. The Bode diagrams of G_0 and $\hat{G}$ are depicted in Figure 4.13 below.

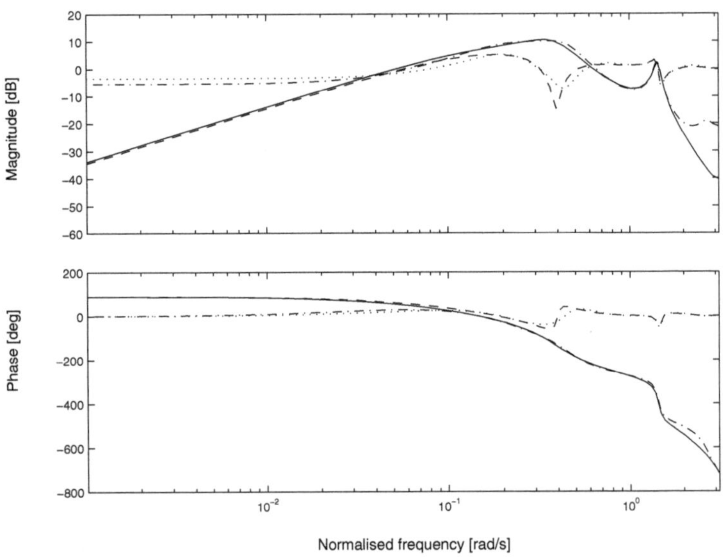

Figure 4.9. Coprime-factor approach using $r_2(t)$: Bode diagrams of T_{12} (—), T_{22} (−−), $\hat{T}_{12}$ (−·) and $\hat{T}_{22}$ (···)

As predicted by the theory, Figure 4.10 shows that

$$\hat{T}_{22}(e^{j\omega}) + \hat{T}_{12}(e^{j\omega})K_{id}(e^{j\omega}) \approx 1 \tag{4.45}$$

except near $\omega = 0$, where it goes to infinity. This is a necessary condition to ensure the stability of the nominal closed-loop system when the controller contains an integrator. The nominal closed-loop system is indeed stable with

$$b_{\hat{G},K_{id}} = 0.2517 \approx b_{G_0,K_{id}} = 0.2761 \tag{4.46}$$

Furthermore,

$$\delta_\nu(G_0, \hat{G}) = 0.1881 < \sup_K b_{\hat{G},K} = 0.4560 \tag{4.47}$$

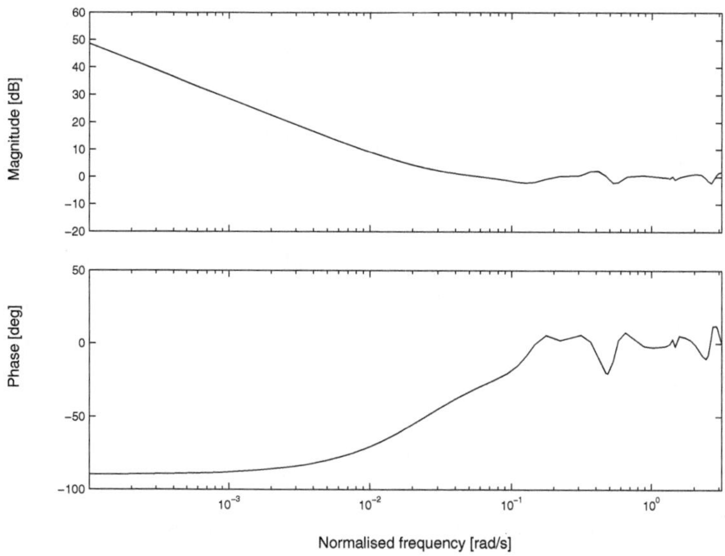

Figure 4.10. Coprime-factor approach using $r_2(t)$: Bode diagram of $\hat{T}_{22} + \hat{T}_{12}K_{id}$

which means that $\hat{G}$ is a good model for control design since every controller K such that $b_{\hat{G},K} > 0.1881$ will stabilise G_0.

A two-degree-of-freedom controller $C = \begin{bmatrix} F & K \end{bmatrix}$, with K of degree 20, stabilising $\hat{G}$, was obtained with the following weightings:

$$W_1(z) = \frac{z}{z-1} \qquad \text{and} \qquad W_2(z) = \frac{1.54z^2 - 0.5912z + 0.208}{z^2 - 0.2672z + 0.424} \tag{4.48}$$

Its performance with $\hat{G}$ and G_0 is illustrated in Figure 4.11. Although the achieved performance is not as good as expected, the nominal and actual closed loops are both stable, with respective generalised stability margins

$$b_{\hat{G},K} = 0.1175 \qquad \text{and} \qquad b_{G_0,K} = 0.0106 \tag{4.49}$$

As in Case A, we made an attempt to cancel the quasi-nonminimal modes of $\hat{G}$ by model reduction. Therefore, we computed a normalised coprime factorisation $\hat{G} = \hat{N}\hat{M}^{-1}$ and we reduced the order of $\begin{bmatrix} \hat{N} \\ \hat{M} \end{bmatrix}$ by balanced truncation. The huge gap between the 4th and 5th Hankel singular values (see Figure 4.12) is an indication that the state variables 5 to 13 are little observable and controllable, *i.e.*, the corresponding modes are quasi-nonminimal. Consequently, the order 4 was selected for the reduced-order model. This order is that of the plant G_0, which means that the identification was realised correctly in the first

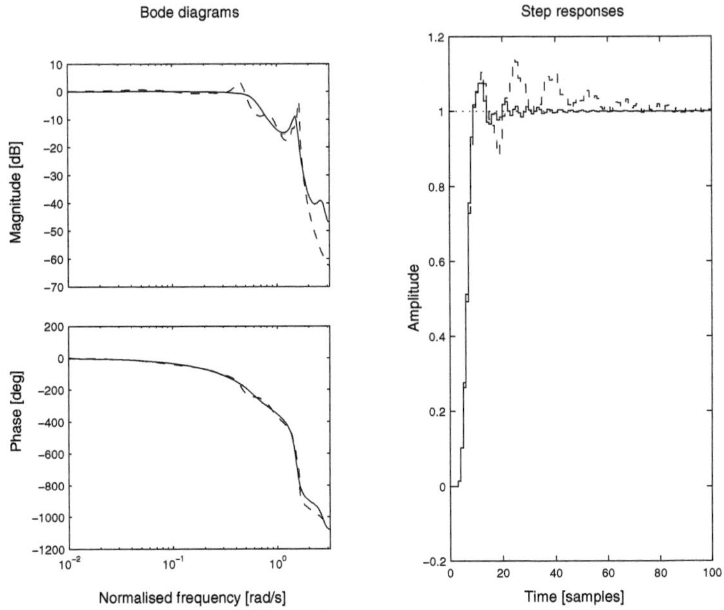

Figure 4.11. Control design *via* coprime-factor identification using $r_2(t)$: Bode diagrams (left) and step responses (right) of $\frac{F\hat{G}}{1+K\hat{G}}$ (—) and $\frac{FG_0}{1+KG_0}$ (– –)

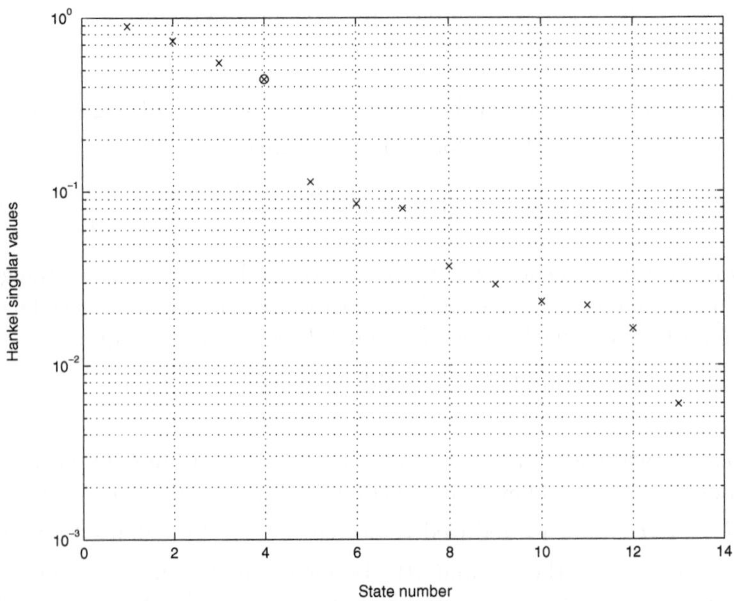

Figure 4.12. Reduction of the 13th-order model $\hat{G}$ obtained by coprime-factor identification using $r_2(t)$: Hankel singular values of $\begin{bmatrix} \hat{N} \\ \hat{M} \end{bmatrix}$

step of the procedure and did not introduce significant additional modes. This was not so in Case A, where the reduction had to be stopped at order 6.

The reduced-order model $\check{G}$ has properties close to those of $\hat{G}$:

$$b_{\check{G},\, K_{id}} = 0.2846 \tag{4.50}$$

and

$$\delta_\nu(G_0,\, \check{G}) = 0.2016 < \sup_K b_{\check{G},K} = 0.4585 \tag{4.51}$$

Its Bode diagram is plotted in Figure 4.13. Observe that it is a better approximation of G_0 than $\hat{G}$.

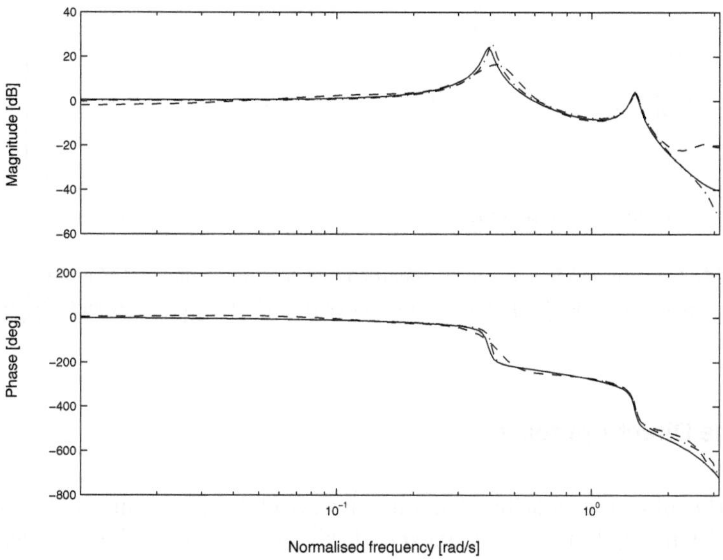

Figure 4.13. Coprime-factor approach using $r_2(t)$: Bode diagrams of G_0 (—), $\hat{G}$ (−−) and $\check{G}$ (−·)

A stabilising controller $\check{C} = \begin{bmatrix} \check{F} & \check{K} \end{bmatrix}$ was computed for $\check{G}$ using the following weightings:

$$W_1(z) = \frac{z}{z-1} \qquad \text{and} \qquad W_2(z) = \frac{2.382z^2 - 1.068z + 0.1816}{z^2 - 0.2556z + 0.7513} \tag{4.52}$$

In this case, the achieved performance is much closer to the nominal one (see Figure 4.14) and the nominal and actual closed loops are both stable with respective generalised stability margins

$$b_{\check{G},\check{K}} = 0.0700 \qquad \text{and} \qquad b_{G_0,\check{K}} = 0.0105 \tag{4.53}$$

This shows the importance of the reduction step to cancel interfering poles and zeroes.

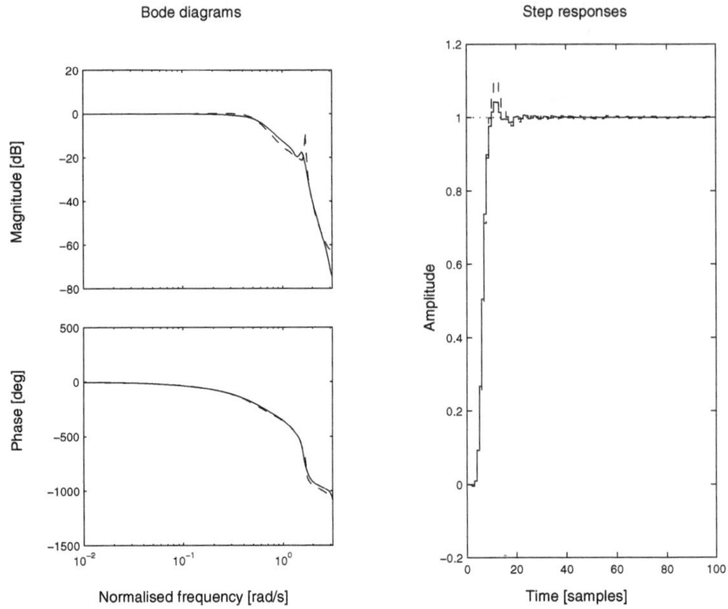

Figure 4.14. Control design *via* coprime-factor identification using $r_2(t)$ followed by model reduction: Bode diagrams (left) and step responses (right) of $\frac{\check{F}\check{G}}{1+\check{K}\check{G}}$ (—) and $\frac{\check{F}G_0}{1+\check{K}G_0}$ (−−)

4.6.4 The Direct Approach

Due to the blocking pole in K_{id}, we can expect better results if the system is excited with $r_1(t)$ than if it is excited with $r_2(t)$ alone. We consider three different cases, depending on the reference signals used. In all cases, 1000 samples of $u(t)$ and $y(t)$ were used to identify a model $\hat{G}$ with the correct ARX[4,2,3] structure.

A. Excitation *via* $r_1(t)$ alone. The closed-loop system was simulated with $r_1(t) \in N(0, 1)$ and $e(t) \in N(0, 0.05)$. The model $\hat{G}$ obtained by direct identification is close to the true system G_0 as shown in Figure 4.15. It is stabilised by K_{id} and the achieved nominal stability margin is close to that of G_0:

$$b_{\hat{G},K_{id}} = 0.2751 \approx b_{G_0,K_{id}} = 0.2761 \tag{4.54}$$

Furthermore,

$$\delta_\nu(G_0, \hat{G}) = 0.0620 < \sup_K b_{\hat{G},K} = 0.4760 \tag{4.55}$$

Clearly, this is a good model for control design.

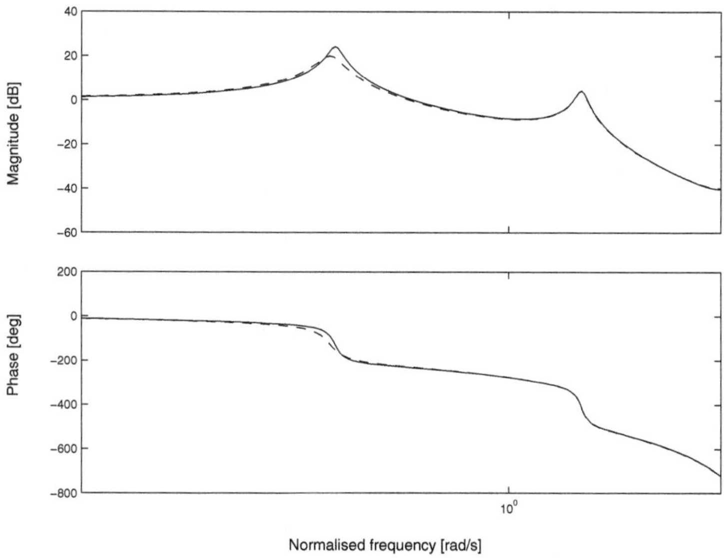

Figure 4.15. Direct approach with $r_1(t)$: Bode diagrams of G_0 (—) and $\hat{G}$ (– –)

We computed a stabilising two-degree-of-freedom controller $C = \begin{bmatrix} F & K \end{bmatrix}$ for $\hat{G}$ using the following loop-shaping filters:

$$W_1(z) = \frac{z}{z-1} \quad \text{and} \quad W_2(z) = \frac{2.44z^2 - 1.017z + 0.167}{z^2 - 0.1519z + 0.7423} \tag{4.56}$$

Its performance with $\hat{G}$ and G_0 is illustrated in Figure 4.16. The achieved closed-loop performance is as good as the designed one and the nominal and achieved stability margins are, respectively,

$$b_{\hat{G},K} = 0.0648 \quad \text{and} \quad b_{G_0,K} = 0.0665 \tag{4.57}$$

B. Excitation *via* $r_2(t)$ alone. The closed-loop system was simulated with $r_2(t) \in N(0, 1)$ and $e(t) \in N(0, 0.05)$. Because of the blocking pole in K_{id}, we could expect less satisfactory results than in Case A. The ν-gap distance between the obtained model $\hat{G}$ and G_0 has indeed doubled ($\delta_\nu(G_0, \hat{G}) = 0.1269$), but $\hat{G}$ is still very close to the true system (see Figure 4.17) and has a nominal stability margin close to the actual one:

$$b_{\hat{G},K_{id}} = 0.2725 \approx b_{G_0,K_{id}} = 0.2761 \tag{4.58}$$

We can explain the – perhaps surprisingly – good quality of $\hat{G}$ in the low frequency range by the fact that not only T_{12} but also N_1 has a blocking zero at $z = 1$ (see Figure 4.17). Therefore, the output SNR does not tend to 0 but remains constant as the frequency decreases. This would not be true if the

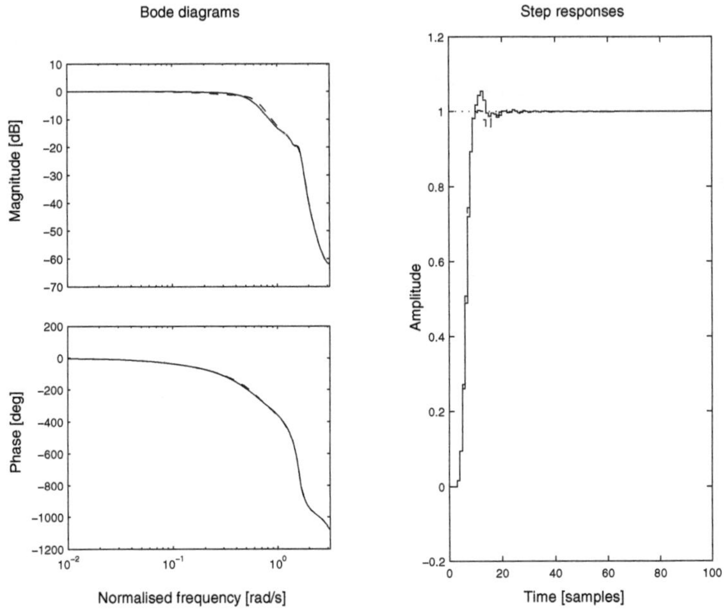

Figure 4.16. Control design *via* direct identification using $r_1(t)$: Bode diagrams (left) and step responses (right) of $\frac{F\hat{G}}{1+K\hat{G}}$ (—) and $\frac{FG_0}{1+KG_0}$ (−−)

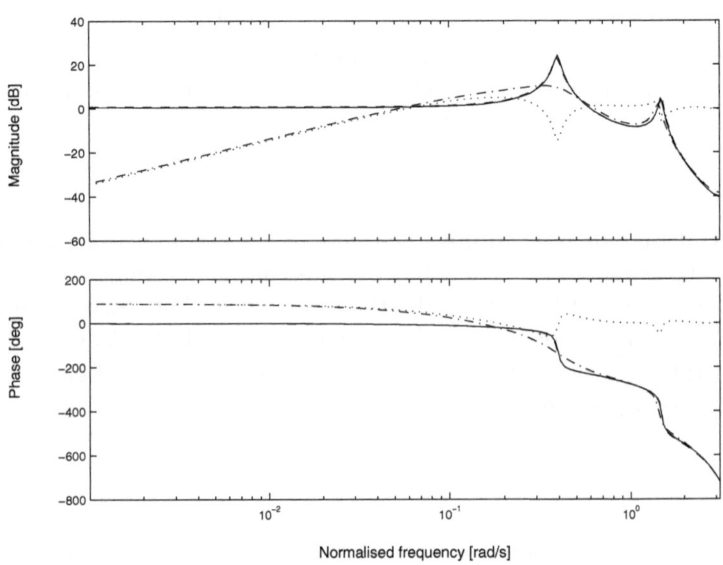

Figure 4.17. Direct approach with $r_2(t)$: Bode diagrams of G_0 (—), $\hat{G}$ (−−), T_{12} (−·) and N_1H_0 (···)

system was subject to step disturbances. In this case, indeed, H_0 would have a pole at $z = 1$ which would cancel the corresponding zero in N_1: see the remark in Subsection 4.4.3.

Finally, note that

$$\delta_\nu(G_0, \hat{G}) = 0.1269 < \sup_K b_{\hat{G},K} = 0.4603 \tag{4.59}$$

which shows that this model is good for control design.

We computed a stabilising two-degree-of-freedom controller $C = \begin{bmatrix} F & K \end{bmatrix}$ for $\hat{G}$ using the same loop-shaping filters as in Case A. Its performance with $\hat{G}$ and G_0 is illustrated in Figure 4.18. The actual performance is as good as the designed one and the nominal and achieved stability margins are, respectively,

$$b_{\hat{G},K} = 0.0650 \qquad \text{and} \qquad b_{G_0,K} = 0.0780 \tag{4.60}$$

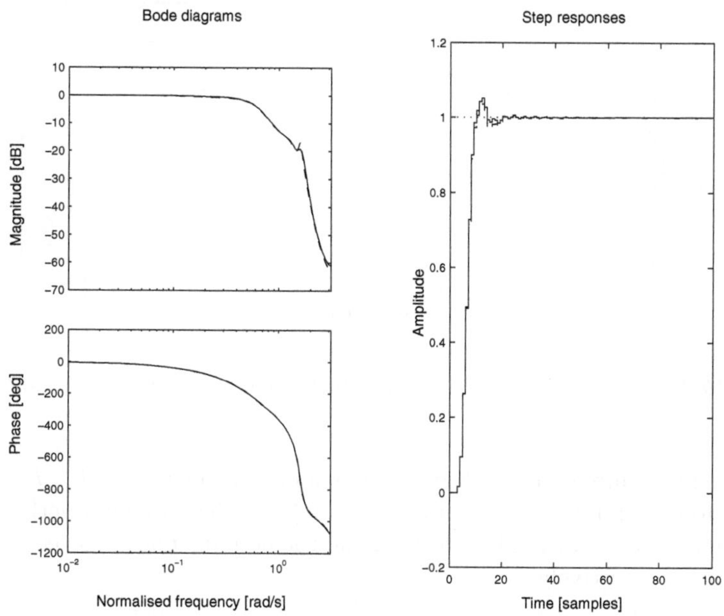

Figure 4.18. Control design *via* direct identification using $r_2(t)$: Bode diagrams (left) and step responses (right) of $\frac{F\hat{G}}{1+K\hat{G}}$ (—) and $\frac{FG_0}{1+KG_0}$ (– –)

C. Excitation *via* $r_1(t)$ and $r_2(t)$. Here, all three signals $r_1(t)$, $r_2(t)$ and $e(t)$ defined above were used to excite the system. As expected, this case yields the best results with a nominal stability margin very close to the actual one:

$$b_{\hat{G},K_{id}} = 0.2776 \approx b_{G_0,K_{id}} = 0.2761 \tag{4.61}$$

and an even smaller ν-gap between G_0 and $\hat{G}$ than in Case A:

$$\delta_\nu(G_0, \hat{G}) = 0.0507 < \sup_K b_{\hat{G},K} = 0.4651 \tag{4.62}$$

The Bode diagram of $\hat{G}$ cannot be distinguished from that of G_0 (see Figure 4.19).

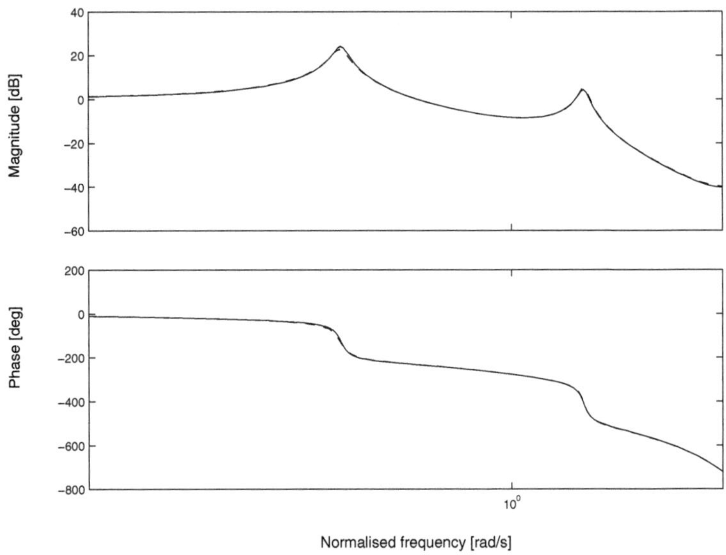

Figure 4.19. Direct approach with $r_1(t)$ and $r_2(t)$: Bode diagrams of G_0 (—) and $\hat{G}$ (– –)

$\hat{G}$ was used to compute a two-degree-of-freedom controller $C = \begin{bmatrix} F & K \end{bmatrix}$, using the same loop-shaping filters as in Cases A and B. It has similar a performance with $\hat{G}$ and G_0, as shown in Figure 4.20, and the nominal and achieved stability margins are, respectively,

$$b_{\hat{G},K} = 0.0646 \qquad \text{and} \qquad b_{G_0,K} = 0.0688 \tag{4.63}$$

4.6.5 The Dual Youla Parametrisation Approach

For the auxiliary model G_{aux}, we chose a model that represents the plant with zero load, while G_0 represents the plant for a 50% load (Landau *et al.*, 1995*b*):

$$G_{aux}(z) = \frac{0.2826z + 0.5067}{z^4 - 1.418z^3 + 1.589z^2 - 1.316z + 0.8864} \tag{4.64}$$

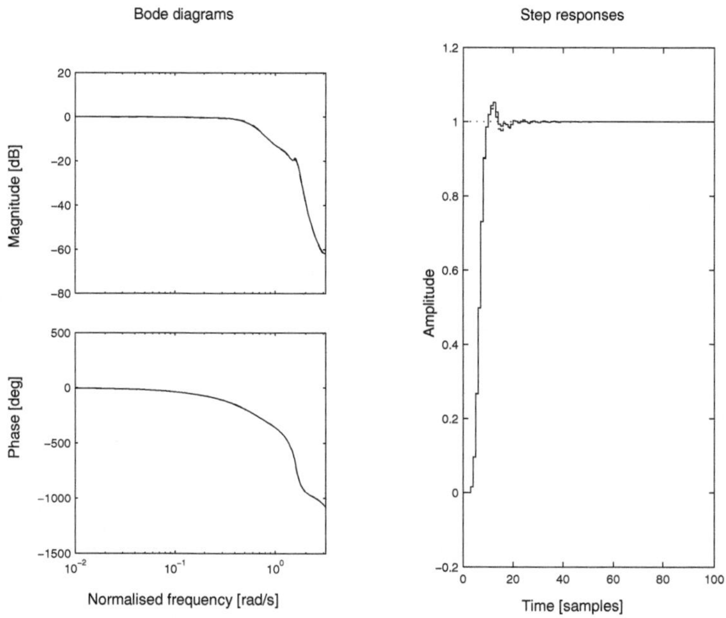

Figure 4.20. Control design *via* direct identification using $r_1(t)$ and $r_2(t)$: Bode diagrams (left) and step responses (right) of $\frac{F\hat{G}}{1+K\hat{G}}$ (—) and $\frac{FG_0}{1+KG_0}$ (−−)

It also has two resonance peaks, but at different frequencies than G_0 (see Figure 4.30 on page 108) and it is stabilised by K_{id}.

We built coprime factors of G_{aux} and K_{id} satisfying the Bezout identity (4.21) and the coprime factors U and V of K_{id} were chosen normalised, which makes the factorisation of G_{aux} and K_{id} unique. Figure 4.21 shows the Bode diagrams of U and V. Since V has a blocking zero at $\omega = 0$, using $r_2(t)$ alone may not produce a good estimate. But the same problem may occur when using $r_1(t)$ alone, because of the low gain of U between 0.1 rad/s and 2 rad/s. Hence, it might be necessary to use both sources of excitation simultaneously to obtain a good estimate of G_0. We shall now consider the three possible scenarios.

A. Excitation *via* $r_1(t)$ alone. The closed-loop system was simulated with $r_1(t) \in N(0, 1)$ and $e(t) \in N(0, 0.05)$. An output-error model structure OE[14,16,3] was used for the Youla parameter $\hat{R}$. It was identified using the auxiliary signals defined in (4.24) and (4.25) and a high (28th) order model $\hat{G}$ of G_0 was obtained using (4.29).

The nominal stability margin of this model with the current controller is

$$b_{\hat{G},K_{id}} = 0.0612 \tag{4.65}$$

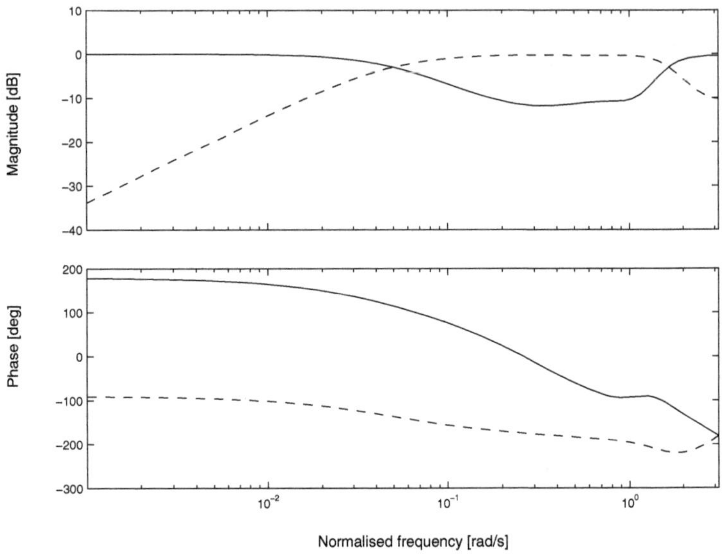

Figure 4.21. Bode diagrams of the normalised coprime factors of K_{id}: U (—) and V (−−)

and the ν-gap between the model and the true system is

$$\delta_\nu(G_0, \hat{G}) = 0.6472 > \sup_K b_{\hat{G},K} = 0.5414 \tag{4.66}$$

Hence, although the obtained model $\hat{G}$ is stabilised by K_{id}, it has a very low generalised stability margin when connected to it and a large ν-gap with respect to the true system G_0, which make it unsuitable for control design: no controller that stabilises $\hat{G}$ would be guaranteed to stabilise G_0. The Bode diagram of $\hat{G}$ is shown in Figure 4.23 below. Clearly, the quality of the estimate is poor in the frequency range where the gain of U is low.

$\hat{G}$ was used to compute a two-degree-of-freedom controller $C = \begin{bmatrix} F & K \end{bmatrix}$ using the following weightings:

$$W_1(z) = \frac{z}{z-1} \quad \text{and} \quad W_2(z) = \frac{2.411z^2 - 1.042z + 0.1741}{z^2 - 0.2037z + 0.7468} \tag{4.67}$$

This controller stabilises $\hat{G}$ and exhibits a good nominal performance but, as expected, it destabilises G_0: see Figure 4.22.

Since the model $\hat{G}$ has a very high order, we made an attempt to reduce it as we did after coprime-factor identification. The inspection of the Hankel singular values of the coprime factors of $\hat{G}$ led again to the selection of order 4 for the reduced-order model $\check{G}$. Its Bode diagram is shown in Figure 4.23.

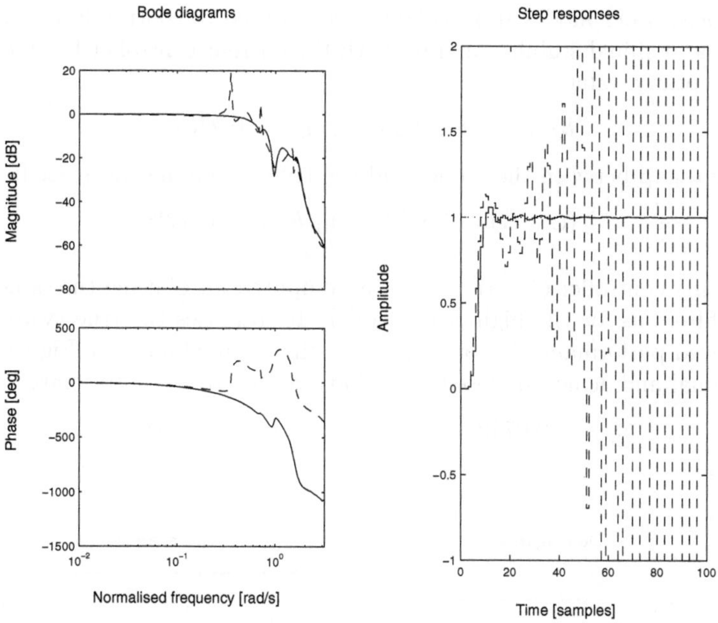

Figure 4.22. Control design *via* dual Youla parameter identification using $r_1(t)$: Bode diagrams (left) and step responses (right) of $\frac{F\hat{G}}{1+K\hat{G}}$ (—) and $\frac{FG_0}{1+KG_0}$ (– –)

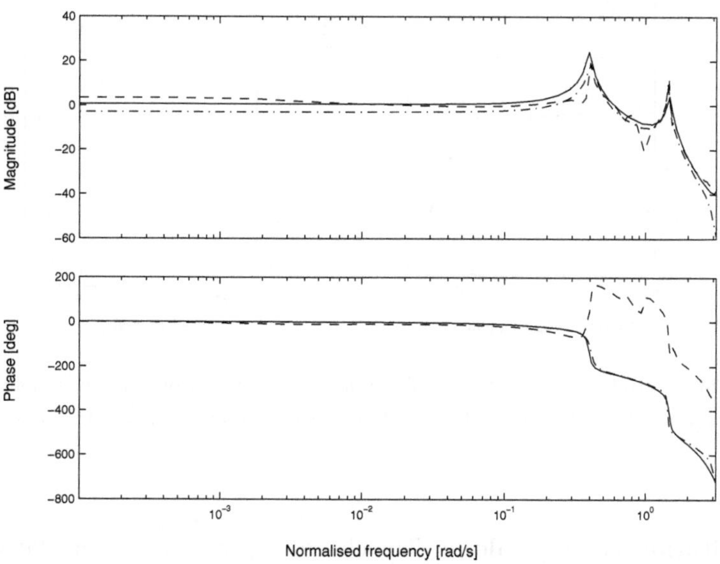

Figure 4.23. Dual Youla parameter identification with $r_1(t)$: Bode diagrams of G_0 (—), $\hat{G}$ (– –) and $\breve{G}$ (–·)

This reduced-order model appears to be more suited to control design than $\hat{G}$. Indeed, the nominal stability margin with the current controller has increased to an acceptable value:

$$b_{\check{G},\,K_{id}} = 0.2795 \approx b_{G_0,K_{id}} = 0.2761 \tag{4.68}$$

while the ν-gap between the model and the true system has decreased:

$$\delta_\nu(G_0,\,\check{G}) = 0.5121 < \sup_K b_{\check{G},K} = 0.5523 \tag{4.69}$$

A stabilising controller $\check{C} = \begin{bmatrix} \check{F} & \check{K} \end{bmatrix}$ was computed for $\check{G}$ using the same loop-shaping filters as for the high-order model. It stabilises the true system, but the achieved performance is not as good as the nominal one (see Figure 4.24). The nominal and achieved closed loops have respective stability margins

$$b_{\check{G},\check{K}} = 0.0742 \qquad \text{and} \qquad b_{G_0,\check{K}} = 0.0558 \tag{4.70}$$

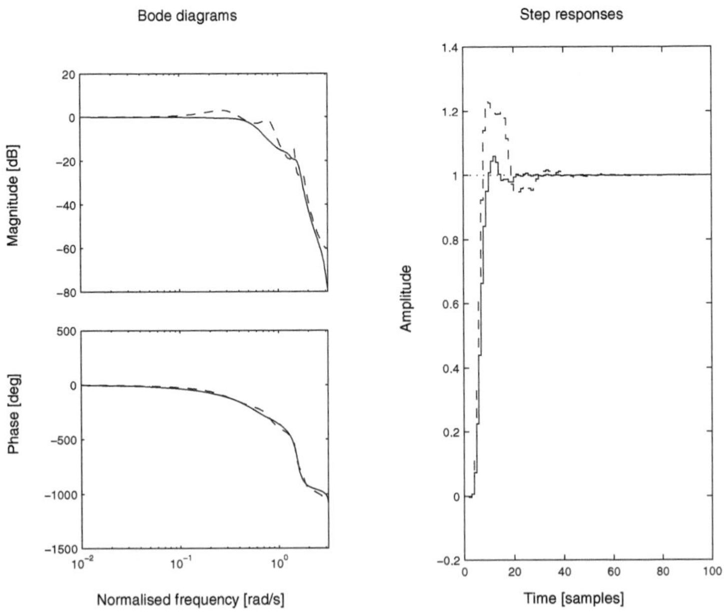

Figure 4.24. Control design *via* dual Youla parameter identification using $r_1(t)$ followed by model reduction: Bode diagrams (left) and step responses (right) of $\frac{\check{F}\check{G}}{1+\check{K}\check{G}}$ (—) and $\frac{\check{F}G_0}{1+\check{K}G_0}$ (-- -)

B. Excitation *via* $r_2(t)$ alone. The closed-loop system was simulated with $r_2(t) \in N(0,\,1)$ and $e(t) \in N(0,\,0.05)$. The same output-error model structure OE[14,16,3] as in the previous case was used for the Youla parameter $\hat{R}$, which led to a 28th-order model $\hat{G}$.

The nominal stability margin of this model with the current controller is

$$b_{\hat{G},K_{id}} = 0.1298 \tag{4.71}$$

and the ν-gap between the model and the true system is

$$\delta_\nu(G_0, \hat{G}) = 0.2366 < \sup_K b_{\hat{G},K} = 0.4493 \tag{4.72}$$

This is a better model for control design than the high-order one obtained in Case A; it also matches G_0 much better, as can be seen in Figure 4.26 below.

A stabilising controller $C = \begin{bmatrix} F & K \end{bmatrix}$ was computed for $\hat{G}$ using the following loop-shaping filters:

$$W_1(z) = \frac{z}{z-1} \qquad \text{and} \qquad W_2(z) = \frac{1.558z^2 - 0.5329z + 0.1935}{z^2 - 0.1904z + 0.409} \tag{4.73}$$

Although its actual performance is not as good as its nominal one (see Figure 4.25), the nominal and achieved closed loops are both stable, with respective stability margins

$$b_{\hat{G},K} = 0.1171 \qquad \text{and} \qquad b_{G_0,K} = 0.0041 \tag{4.74}$$

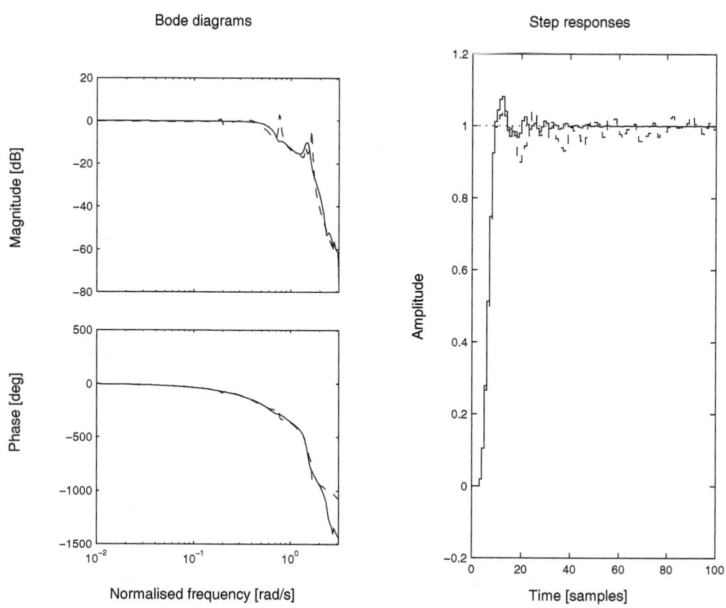

Figure 4.25. Control design *via* dual Youla parameter identification using $r_2(t)$: Bode diagrams (left) and step responses (right) of $\frac{F\hat{G}}{1+K\hat{G}}$ (—) and $\frac{FG_0}{1+KG_0}$ (– –)

Once again, the nominal model $\hat{G}$ has a very high order and we reduced it using coprime-factor balanced truncation to obtain a 4th-order model $\check{G}$. It achieves

$b_{\breve{G},\,K_{id}} = 0.2512$ and

$$\delta_\nu(G_0, \breve{G}) = 0.2224 < \sup_K b_{\breve{G},K} = 0.4504 \tag{4.75}$$

Its Bode diagram is shown in Figure 4.26.

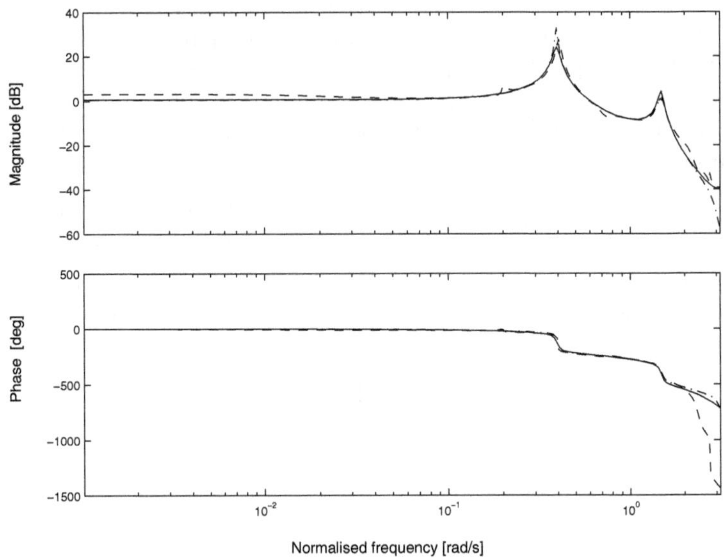

Figure 4.26. Dual Youla parameter identification with $r_2(t)$: Bode diagrams of G_0 (—), $\hat{G}$ (– –) and $\breve{G}$ (–·)

A stabilising controller $\breve{C} = \begin{bmatrix} \breve{F} & \breve{K} \end{bmatrix}$ was computed for $\breve{G}$ using the following loop-shaping filters:

$$W_1(z) = \frac{z}{z-1} \qquad \text{and} \qquad W_2(z) = \frac{2.44z^2 - 1.017z + 0.167}{z^2 - 0.1519z + 0.7423} \tag{4.76}$$

In this case, the actual closed loop is stable and the achieved performance is very close to the nominal one (see Figure 4.27). The nominal and achieved closed loops have respective stability margins

$$b_{\breve{G},\breve{K}} = 0.0627 \qquad \text{and} \qquad b_{G_0,\breve{K}} = 0.0133 \tag{4.77}$$

C. Excitation *via* $r_1(t)$ and $r_2(t)$. The closed-loop system was simulated with $r_1(t)$, $r_2(t)$ and $e(t)$ defined above. The same output-error model structure OE[14,16,3] was used for the Youla parameter $\hat{R}$. Once again, a 28th-order model $\hat{G}$ was produced.

The nominal stability margin of this model with the current controller is

$$b_{\hat{G},K_{id}} = 0.2847 \tag{4.78}$$

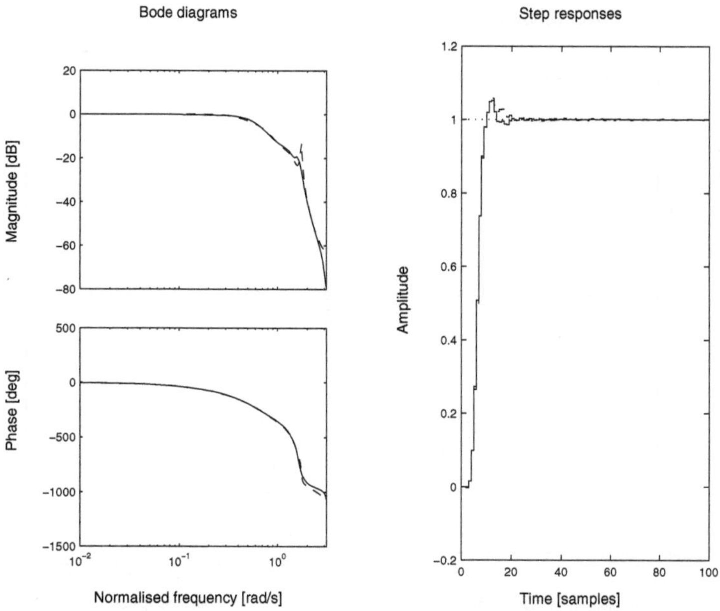

Figure 4.27. Control design *via* dual Youla parameter identification using $r_2(t)$ followed by model reduction: Bode diagrams (left) and step responses (right) of $\frac{\check{F}\hat{G}}{1+\check{K}\hat{G}}$ (—) and $\frac{\check{F}G_0}{1+\check{K}G_0}$ (– –)

and the ν-gap between the model and the true system is

$$\delta_\nu(G_0, \hat{G}) = 0.1581 < \sup_K b_{\hat{G},K} = 0.4574 \tag{4.79}$$

As expected, this is the best model for control design obtained by the dual Youla parametrisation approach. Its Bode diagram is depicted in Figure 4.30 below; one can see that it is very close to the true system G_0. For the purpose of illustration, the Bode diagram of the initial auxiliary model G_{aux} is also plotted and one can see that it is very different from G_0.

A stabilising controller $C = \begin{bmatrix} F & K \end{bmatrix}$ was computed for $\hat{G}$ using the following loop-shaping filters:

$$W_1(z) = \frac{z}{z-1} \qquad \text{and} \qquad W_2(z) = \frac{2.411z^2 - 1.042z + 0.1741}{z^2 - 0.2037z + 0.7468} \tag{4.80}$$

Although its actual performance is not as good as expected (see Figure 4.28), the nominal and achieved closed loops are both stable, with respective stability margins

$$b_{\hat{G},K} = 0.0663 \qquad \text{and} \qquad b_{G_0,K} = 0.0394 \tag{4.81}$$

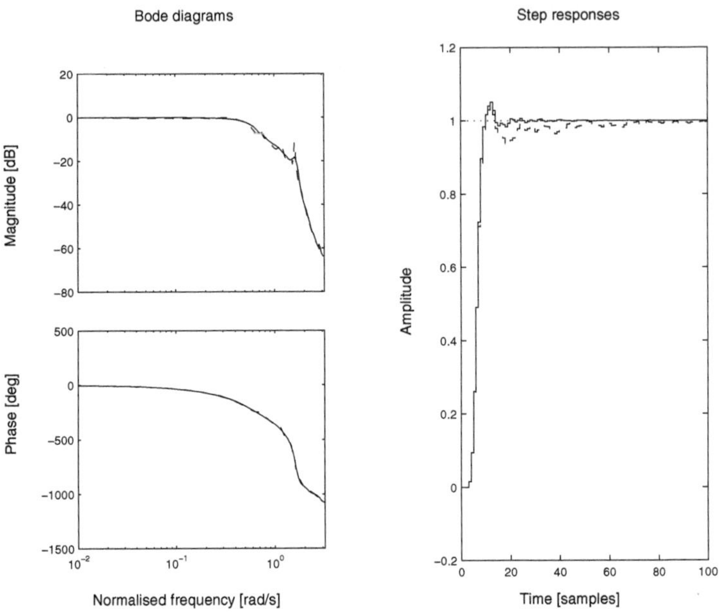

Figure 4.28. Control design *via* dual Youla parameter identification using $r_1(t)$ and $r_2(t)$: Bode diagrams (left) and step responses (right) of $\frac{F\hat{G}}{1+K\hat{G}}$ (—) and $\frac{FG_0}{1+KG_0}$ (−−)

As before, a step of model reduction was applied to obtain a 4th-order model $\check{G}$. Figure 4.29 shows that the reduction procedure has cancelled the quasi-nonminimal modes of $\hat{G}$.

As shown in Figure 4.30, the two models $\hat{G}$ and $\check{G}$ are nearly indistinguishable from the plant G_0.

The reduced-order model $\check{G}$ has the following properties:

$$b_{\check{G},\,K_{id}} = 0.2761 \tag{4.82}$$

and

$$\delta_\nu(G_0,\,\check{G}) = 0.0290 < \sup_K b_{\check{G},K} = 0.4580 \tag{4.83}$$

Observe that

$$\delta_\nu(G_0,\,\check{G}) = 0.0290 << \delta_\nu(G_0,\,\hat{G}) = 0.1581 \tag{4.84}$$

which, again, illustrates the importance of model reduction to eliminate the quasi-nonminimal modes of the model.

A stabilising controller $\check{C} = \begin{bmatrix}\check{F} & \check{K}\end{bmatrix}$ was computed for $\check{G}$ using the same loop-shaping filters as for the high-order model. In this case, the achieved performance is as good as the nominal one (see Figure 4.31), and the nominal

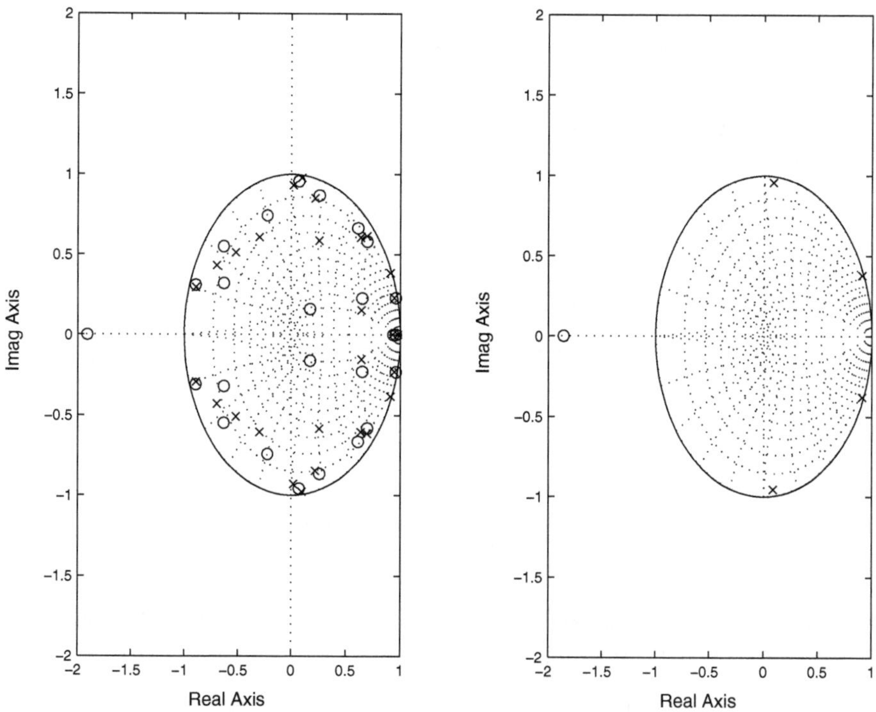

Figure 4.29. Dual Youla parameter identification with $r_1(t)$ and $r_2(t)$ followed by a step of model reduction: poles ($\times$) and zeroes ($\circ$) of $\hat{G}$ (left) and $\breve{G}$ (right)

and achieved closed loops are both stable with respective stability margins

$$b_{\breve{G},\breve{K}} = 0.0662 \qquad \text{and} \qquad b_{G_0,\breve{K}} = 0.0645 \tag{4.85}$$

4.6.6 Comments on the Numerical Example

This example confirmed all the theoretical results of this chapter and clearly showed that, unless an appropriate method is selected together with an appropriate experiment design, the model that results from closed-loop identification of a system with an unstable or nonminimum-phase controller can be a very poor estimate of the true system. Furthermore, this model will often not be stabilised by the controller used during identification, thereby causing problems if it is used for a subsequent control design.

In the present situation, it was possible to use the coprime-factor approach with excitation $r_2(t)$ because the unstable pole of the controller was located on the unit circle. However, it should be clear that this approach would not have

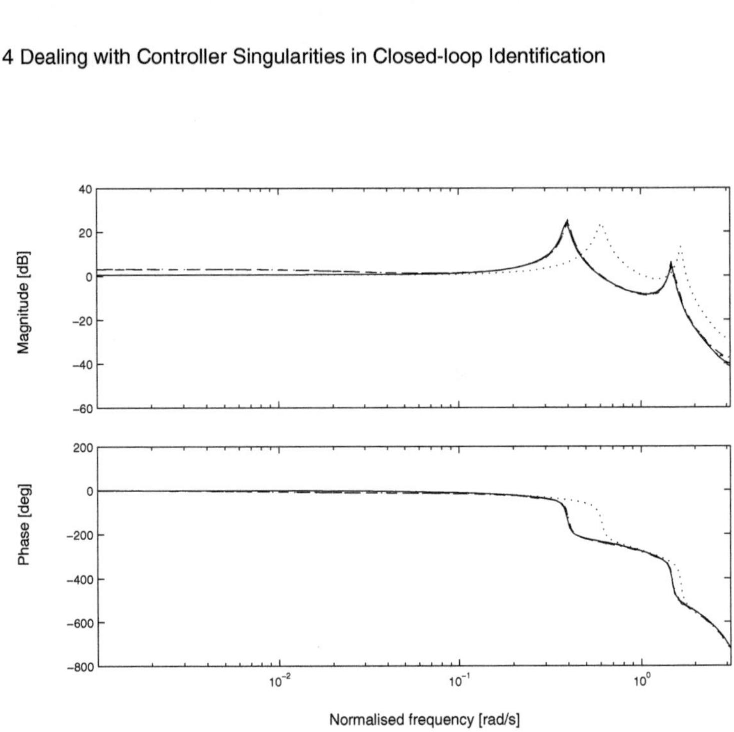

Figure 4.30. Dual Youla parameter identification with $r_1(t)$ and $r_2(t)$: Bode diagrams of G_0 (—), $\hat{G}$ (– –), $\check{G}$ (–·) and G_{aux} (···)

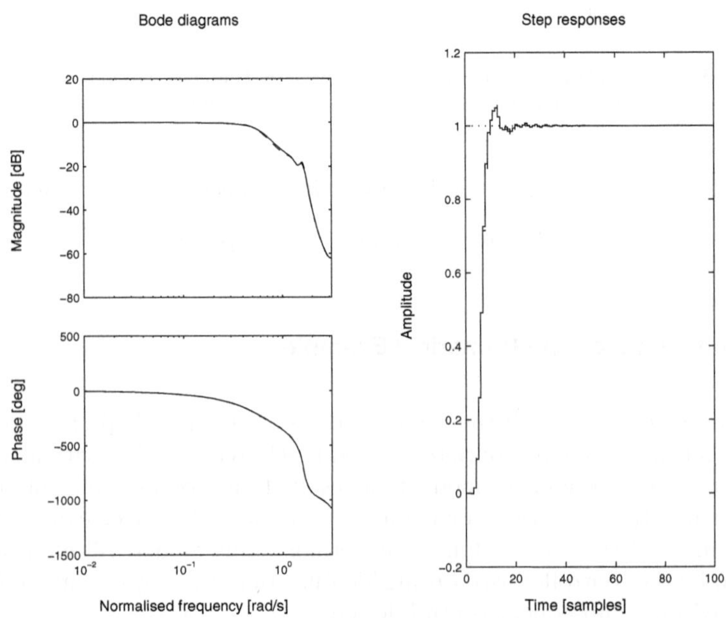

Figure 4.31. Control design *via* dual Youla parameter identification using $r_1(t)$ and $r_2(t)$ followed by model reduction: Bode diagrams (left) and step responses (right) of $\frac{\check{F}\check{G}}{1+\check{K}\check{G}}$ (—) and $\frac{\check{F}G_0}{1+\check{K}G_0}$ (– –)

worked with a strictly unstable controller, just as it did not work with $r_1(t)$ because the controller had a strictly nonminimum phase.

The direct approach delivered very good results; however, nominal stability cannot be guaranteed beforehand and it should be noted that we have used an unbiased model structure for $(\hat{G}, \hat{H})$, which limited the modelling error to a variance error and prevented the occurrence of possible stability issues caused by the bias error.

The dual Youla parametrisation approach delivered models with guaranteed closed-loop stability; this is the main advantage of this method. However, an inappropriate choice for the excitation source may yield a model with a very small generalised stability margin, which is then not suitable for control design. Therefore, the design of the reference signals must be done with regard to the frequency distribution of the magnitudes of the controller coprime factors.

From a control design point of view, the direct approach and the dual Youla parametrisation approach (with $r_1(t)$ and $r_2(t)$) delivered the best models, *i.e.*, the models with generalised stability margins closest to the achieved one, and the smallest ν-gap distances to the true system. This is summarised in Table 4.2.

Table 4.2. Summary of the numerical values attached to the models obtained *via* different closed-loop identification procedures in the numerical example. The last column summarises the performance of the new controllers C (resp. $\check{C}$) computed from the models $\hat{G}$ (resp. $\check{G}$) when they are applied to the true system G_0: unstable closed loop (−); stable closed loop with poor performance (·); stable closed loop with good performance (+).

Method			$\mathbf{b_{\hat{G},K_{id}}}$	$\delta_\nu(\mathbf{G_0}, \hat{\mathbf{G}})$	$\mathbf{sup_K b_{\hat{G},K}}$	$(\mathbf{G_0}, \mathbf{C})$
Indirect	T_{12}		0	1	0.2389	−
	T_{21}		0	1	0.0032	−
Coprime-factor	r_1		0	1	0.0299	−
	r_1	reduc.	0	1	0.0302	−
	r_2		0.2517	0.1881	0.4560	·
	r_2	reduc.	0.2846	0.2016	0.4585	+
Direct	r_1		0.2751	0.0620	0.4760	+
	r_2		0.2725	0.1269	0.4603	+
	r_1 & r_2		0.2776	0.0507	0.4651	+
Dual Youla param.	r_1		0.0612	0.6506	0.5414	−
	r_1	reduc.	0.2795	0.5121	0.5523	·
	r_2		0.1298	0.2366	0.4493	·
	r_2	reduc.	0.2512	0.2224	0.4504	+
	r_1 & r_2		0.2847	0.1581	0.4574	·
	r_1 & r_2	reduc.	0.2761	0.0290	0.4580	+
True system G_0			0.2761		0.4621	

Another important issue confirmed by this numerical illustration is the fact that models produced by two-stage methods may not be ideal for control design because they are generally of much higher order than the true system, the additional modes being nearly nonminimal. It is then necessary to discard these modes by performing a step of model reduction before control design. The order of the true system, to which the high-order model should at least be reduced[7], can easily be determined by observing that the states in excess have very small Hankel singular values attached to them.

4.7 Summary of the Chapter

In the case where an unstable or nonminimum-phase controller is used during closed-loop identification, some of the classical closed-loop identification methods deliver a model that forms an unstable loop with the acting controller. As a result, standard tools used to guarantee robust stability and performance with a newly designed model-based controller become useless. Four closed-loop identification methods have been studied: the indirect, coprime-factor, direct and dual Youla parametrisation methods. For each of them, we have examined what happens when the controller has an unstable pole, a nonminimum-phase zero or both, leading to the conclusion that some combinations of controller singularities and identification method lead to a guaranteed unstable nominal closed-loop system and, hence, fail to deliver a model that is suited to control design. Unit-circle poles or zeroes, even when they do not cause nominal instability, can be an issue as they can impair the SNR of the signals used for identification, thereby leading to important modelling errors in the frequency range that surrounds them.

The main outcome of this chapter is to propose guidelines that the user can follow in order to choose an identification method and the location of the external source of excitation in function of the known controller singularities, in such a way as to guarantee stability of the nominal closed-loop system. These guidelines are summarised in Table 4.1.

We have also seen that two-stage methods like the indirect, coprime-factor and dual Youla parametrisation approaches often lead to high-order models containing nearly cancelling pairs of poles and zeroes, which can hinder the design of a new model-based controller. A step of model reduction is then required to eliminate these quasi-nonminimal modes.

[7]If a model with lower order than the true system is desired, it is better to use a closed-loop reduction method as proposed in Chapter 6.

CHAPTER 5

Model and Controller Validation for Robust Control in a Prediction-error Framework

5.1 Introduction

5.1.1 The Questions of the Cautious Process Control Engineer

Modern optimal control design methods are generally based on models. Most of the time, an identification step will have to be performed on the process in order to compute such a model and we have seen in Chapters 3 and 4 how to realise the identification in order to maximise our chances of finding a *good* model for control design. In this chapter, we elaborate on the question of model validation for control. This question concerns the verification of the quality of a given model; the way this model was obtained is irrelevant, so this verification procedure can be applied not only to models obtained by identification, but also to any (LTI) model that the control designer could have at their disposal. The question under consideration is thus:

> *Can I trustfully use the model that was given to me for control design? Is it close enough to the real process? How can I quantify the error (the distance) between the real process and this model in a way that is representative of the suitability of the model for control design?*

If the practitioner is satisfied with the model, he or she can use it for control design. Most of the time, good *nominal* stability margins and performance will be sought, with the hope that they will remain after implementation of the controller on the actual process. But hope and faith are not sufficient to guarantee closed-loop robustness, and the engineer is faced with a new question:

Can I trustfully implement the controller I designed? Will the actual stability margins and performance of my controller be close to the designed ones?

To answer this question, controller validation tools are proposed.

5.1.2 Some Answers to these Questions

In this chapter, we consider a validation framework that connects prediction-error identification and robustness theory. This framework consists of a method to design an uncertainty region using the tools of prediction-error identification, coupled with robustness tools that are adapted to this uncertainty region. These robustness tools pertain both to the robustness analysis of a specific controller vis-à-vis all models in the uncertainty region (*controller validation*) and to the quality assessment (for model-based control design) of the model uncertainty region (*model validation for control*).

A. Model validation. There are many ways of understanding the concept of model validation and many different frameworks under which model validation questions have been formulated. For the sake of clarity, we first define the terminology adopted in this chapter. We shall say that *model validation* consists of constructing a model set $\mathcal{D}$ that contains the unknown true system G_0, perhaps at a certain probability level. This model set, called *uncertainty set* or *uncertainty region*, is constructed from a combination of data and prior assumptions. A given nominal model G_{nom} will be called validated if it belongs to $\mathcal{D}$. In contrast, we shall use the term *identification* for the estimation of a model $\hat{G}$, as we did in the previous chapters. It turns out that our validation procedure is actually based on a step of identification.

The history of model validation in prediction-error identification is as old as prediction-error identification itself. A reputable engineer should never deliver a product, whether it be a measurement device or a model, without a statement about its quality. However, the information about the quality of a model resulting from prediction-error identification was classically presented *via* a range of model validation tests, such as the whiteness of the residuals, the cross-correlation between inputs and residuals, parameter and transfer function covariance formulae, *etc.* (see Subsection 2.6.6). Thus, despite its enormous practical success, prediction-error identification was not delivering the classical uncertainty descriptions upon which mainstream robustness theory was built throughout the 1980s, namely frequency domain descriptions.

As a consequence, as already stated in the introduction of this book, a huge gap appeared at the end of the 1980s between robustness theory and prediction-error identification (Smith and Dahleh, 1994). This gap drove the control community to develop new techniques, different from prediction-error identification, in order to obtain, from measured data, an uncertainty region containing the true system. Several directions have been pursued:

- *set membership identification* (Giarré *et al.*, 1997; Giarré and Milanese, 1997; Milanese and Taragna, 1999), in which one estimates a model uncertainty set without estimating a nominal model;
- *model invalidation* (Poolla *et al.*, 1994; Kosut, 1995; Chen, 1997; Boulet and Francis, 1998);
- $\mathcal{H}_\infty$ and *worst-case identification* (Helmicki *et al.*, 1991; Gu and Khargonekar, 1992; Mäkilä and Partington, 1999; Mäkilä *et al.*, 1995).

These new techniques aimed to produce one of the standard linear fractional frequency-domain uncertainty regions that are used in mainstream robust control theory (such as additive, multiplicative and coprime-factor uncertainty regions). The drawbacks of these techniques are that they are based on prior assumptions about the unknown system and the noise that may be difficult to ascertain. In addition, because they are based on worst-case assumptions rather than on the idea of averaging out the noise, they typically lead to conservative uncertainty sets (Hjalmarsson, 1994).

Thus, attempts were made to construct frequency domain uncertainty regions around nominal models identified using prediction-error identification methods. In the case where the chosen model structure is able to represent the true system ($\mathcal{S} \in \mathcal{M}$), the only error in the estimated transfer function is the *variance error* (or noise-induced error), for which reliable formulae exist (Ljung, 1999). An interesting attempt to extend these formulae to the case where undermodelling is present was proposed in (Hjalmarsson and Ljung, 1992). The main difficulty in computing the total (mean square) error around a nominal transfer function, estimated by prediction-error identification, has always been the estimation of the *bias error*. A first attempt at computing explicit expressions for the bias error was made in (Goodwin *et al.*, 1992) using the stochastic embedding approach. The basic idea is to treat the model error (*i.e.*, the bias) as a realisation of a stochastic process whose variance is parametrised by a few parameters and then to estimate these parameters from data; this allows one to compute the size of the unmodelled dynamics, in a mean square sense. The only prior knowledge is the parametric structure of the variance of the unmodelled dynamics. The methodology is however limited to models that are linear in the parameters.

Another approach to estimating the modelling error attached to a nominal restricted complexity model in a prediction-error framework is the *model-error model* approach, proposed in (Ljung, 1997, 1998, 2000). The key idea is to estimate an unbiased model of the error ΔG between the nominal model G_{nom} and the true system G_0 by a simple step of prediction-error identification with an unbiased model structure, using validation data. The mean square error on this model error estimate is then a variance error only, for which standard formulae exist, as already stated.

The model validation approach presented in this chapter, initially published in (Gevers *et al.*, 2003*a*, 2003*b*) and references therein, was inspired by the model-error model approach, but it uses the obvious (and much simpler) alternative of identifying a full-order model directly for the real system G_0, rather than for the error $\Delta G = G_0 - G_{nom}$ between the system G_0 and a (possibly low-order) model G_{nom} of it. Hence, the mean square error is again a variance error only, from which an uncertainty set $\mathcal{D}$ can be built and used for validation purposes. Observe that this validation procedure comprises the identification of an unbiased model $\hat{G}$ contained in $\mathcal{D}$ (hence validated) which comes for free in addition to this uncertainty set and can be used as nominal model for control design. Yet, the user is free to use any other nominal model G_{nom} that is validated.

The *mixed stochastic-deterministic* methods for construction of uncertainty regions presented in (Hakvoort and Van den Hof, 1997; de Vries and Van den Hof, 1995; Venkatesh and Dahleh, 1997) are based on a mixture of stochastic assumptions on the noise and deterministic assumptions on the decay rate of the tail of the system's response. Just as for the stochastic embedding approach, the authors show that they can obtain, from data, an estimate of the noise variance and of the prior bounds on this decay rate. Also common with the stochastic embedding approach is a restriction of the method to models that are linear in the parameters, *i.e.*, fixed-denominators models.

All the methods described above, which compute uncertainty sets on the basis of stochastic assumptions on the noise, do of course lead to stochastic descriptions of the uncertainty sets, *i.e.*, the true system belongs to $\mathcal{D}$ with probability p; the level p is entirely the choice of the designer.

B. Controller validation. Once an uncertainty set $\mathcal{D}$ of models has been constructed by a procedure of model validation, one can address controller validation questions. Namely, for such a set $\mathcal{D}$ and for a tentative controller $C(z)$, one can ask the following two questions:

- Does the controller $C(z)$ stabilise all models in the model set $\mathcal{D}$? This is called *controller validation for stability.*
- Does the controller $C(z)$ achieve a given level of closed-loop performance with all models in the model set $\mathcal{D}$? This is called *controller validation for performance.*

In this chapter, we present a solution to both questions in the context of uncertainty sets obtained by prediction-error methods. The controller validation questions are posed with respect to the uncertainty set $\mathcal{D}$ knowing that the true system G_0 belongs to $\mathcal{D}$ with probability p. A more precise and less conservative estimate of the probability that $C(z)$ stabilises the true system can be obtained using the theory of randomised algorithms; see (Campi *et al.*, 2000, 2002).

Note that the terminology 'controller validation' has been used with different senses. Here, we restrict its definition to the operation that consists in verifying whether a candidate controller fulfils a string of conditions with all systems in an uncertainty set containing the true plant, before its implementation on this plant.

Another definition of controller validation concerns the monitoring of the closed-loop performance after the controller has been commissioned. (DeVries and Wu, 1978), (Desborough and Harris, 1992), (Stanfelj *et al.*, 1993), (Kozub and Garcia, 1993) and (Kozub, 1997) have proposed performance indicators based on some comparison of the actual plant output variance σ_y^2 to the one σ_{MV}^2 that would be obtained under minimum-variance control. A typical performance index is

$$\begin{aligned}\eta(k) &= 1 - \frac{\sigma_{MV}^2}{\sigma_y^2} \\ &= 1 - \frac{1 + \tau_1^2 + \ldots + \tau_{k-1}^2}{1 + \tau_1^2 + \ldots + \tau_{k-1}^2 + \tau_k^2 + \ldots}\end{aligned} \tag{5.1}$$

$0 \leq \eta(k) \leq 1$, where the τ_i's are the coefficients of the closed loop impulse response and k is the number of samples covered by the dead time of the process (observe that a minimum-variance controller yields $\tau_i = 0$ for $i \geq k$). $\eta(k)$ and σ_{MV}^2 can be estimated from routine data if the dead time k is known and do not require additional plant excitation. Higher values of $\eta(k)$ indicate performance deterioration with respect to minimum-variance control. $\eta(k)$ allows one not only to assess the controller performance just after its implementation, but also to monitor its performance degradation along time. Because optimal control is rarely minimum-variance control, an extended horizon index $\eta(k + h)$ can be used. It allows the estimation of several other performance indicators like the settling time, *etc.* An appropriate use of $\eta(k)$ and $\eta(k + h)$ makes it possible to decide for plant maintenance and/or for controller retuning. The detection of changes in the performance indicators requires knowledge or estimation of the distribution of their statistics and is based on fault detection theory and likelihood ratio tests: see (Harris *et al.*, 1999) and references therein for details. A very detailed and well structured monograph on these controller performance assessment methods was recently published in Springer's Advances in Industrial Control series; see (Huang and Shah, 1999). Therefore, we shall not go into more detail about this topic in the present book.

A method for *controller invalidation* has also been proposed by (Safonov and Tsao, 1997). It is closely related to the theories of set membership identification and model invalidation. It is based on the identification of control laws that are consistent with performance objectives and past experimental data (possibly before the control laws are ever inserted in the feedback loop), without using a model of the plant or assumptions about the plant, uncertainties, sensors or noise. It complements model-based control design methods by providing a precise characterisation of how the set of suitable controllers shrinks when new experimental data is found to be inconsistent with prior assumptions or earlier data. The method can be used in real time and results in an adaptive switching controller.

C. Model validation for control. The assessment of the quality of a model cannot be decoupled from the purpose for which the model is to be used. The research on identification for control has, in the last 15 years, focused on the design of identification criteria that delivered a *control-oriented nominal model.* Similarly, the validation experiment must be designed in such a way as to deliver uncertainty sets that are tuned for robust control design. Thus, one must think in terms of '*control-oriented validation design*'. This chapter also highlights the connection between uncertainty sets obtained by prediction-error methods and sets of stabilising controllers.

5.2 Model Validation Using Prediction-error Identification

Let us assume that the true process is the SISO LTI system described by (2.79)

$$\mathcal{S}: \begin{cases} y(t) = G_0(z)u(t) + v(t) \\ v(t) = H_0(z)e(t) \end{cases} \tag{5.2}$$

where, as usual, $G_0(z)$ and $H_0(z)$ are unknown discrete-time rational transfer functions, with $G_0(z)$ strictly proper and $H_0(z)$ stable and inversely stable. We shall further assume that $v(t)$ is zero-mean wide-sense stationary noise, $e(t)$ being a white noise signal. Prediction-error identification theory requires no additional assumptions on the noise $v(t)$; in particular $v(t)$ does not need to be Gaussian: see (Ljung, 1999).

A polynomial representation of $G_0(z)$ is given by

$$\begin{aligned} G_0(z) \triangleq G(z, \theta_0) &= \frac{z^{-d}(b_0 + b_1 z^{-1} + \cdots + b_m z^{-m})}{1 + a_1 z^{-1} + \cdots + a_n z^{-n}} \\ &= \frac{Z_N(z)\theta_0}{1 + Z_D(z)\theta_0} \end{aligned} \tag{5.3}$$

where

- d is the dead time of the process;
- $\theta_0 = \begin{bmatrix} a_1 & \ldots & a_n & b_0 & \ldots & b_m \end{bmatrix}^T \in \mathbb{R}^q$ ($q \triangleq (n + m + 1)$) is the vector containing the parameters of the true system;
- $Z_D(z) = \begin{bmatrix} z^{-1} & z^{-2} & \ldots & z^{-n} & 0 & \ldots & 0 \end{bmatrix}$ is a row vector of size q;

- $Z_N(z) = \begin{bmatrix} 0 & \dots & 0 & z^{-1} & z^{-2} & \dots & z^{-m} \end{bmatrix}$ is a row vector of size q.

► **Remark.** If the true system is not LTI, the theoretical derivations of this chapter are no longer valid. Indeed, they are based on the identification of an unbiased estimate of the system using prediction-error identification and on the construction, from the covariance of this estimate, of a frequency-domain uncertainty region containing the true system. The presence of a small amount of nonlinearities will however probably not destroy the ground rules. Also, as prediction-error identification is often applied to nonlinear systems for computing linear estimates that are valid around some operating point, so can our validation procedure be used to build uncertainty regions that are approximately correct around a chosen operating point. Finally, it should be noted that nonlinearities in the controller will not pose any problem here. ◄

We further assume that one has given us an LTI nominal model G_{nom} (possibly H_{nom} also) that has to be validated in a way that is relevant for control design, using a set of validation data collected on the true system (5.2). It must be remarked here that the way this model was obtained is immaterial. In particular, we do not assume that this model was obtained by prediction-error identification, nor that it comes with a confidence region attached to it or that it is unbiased. The validation procedure will consist in performing some experiment on the true system in order to build an uncertainty region in which the model will be embedded.

5.2.1 Model Validation Using Open-loop Data

The validation procedure consists in computing an unbiased estimate $\hat{G}_{ol}(z) \triangleq G(z, \hat{\theta}_{ol})$ of $G_0(z)$ using a set of data

$$Z_{ol}^N = \left\{ u_{ol}(1),\, y_{ol}(1),\, \dots,\, u_{ol}(N),\, y_{ol}(N) \right\} \tag{5.4}$$

collected on the true system in open loop, with $G(z, \theta)$ in some unbiased (*i.e.*, full-order) model set

$$\mathcal{M}_{ol} = \left\{ G(z, \theta) \mid G(z, \theta) = \frac{Z_N(z)\theta}{1 + Z_D(z)\theta},\ \theta \in \mathbb{R}^q \right\} \ni G_0(z) \tag{5.5}$$

In order to guarantee the unbiasedness of the estimate, at least one of the following two conditions must be fulfilled as well:

1. the model structure for the noise model $\hat{H}_{ol}(z)$ is also unbiased;
2. the noise model is independently parametrised (*e.g.*, an OE model structure is used) and the input signal $u_{ol}(t)$ is uncorrelated with the noise $v(t)$ (this is normally the case with open-loop data, and is part of our assumptions about the true system).

When the conditions for the unbiasedness of the estimate are fulfilled, the only error that affects the estimate $\hat{G}_{ol}$ is, by definition, a covariance error due to the noise $v(t)$ corrupting the output $y_{ol}(t)$ of the process. As a result, the mean of the estimated parameters vector $\hat{\theta}_{ol}$ is θ_0. This is formalised in the following proposition.

Proposition 5.1. (Ljung, 1999) *If the prediction-error identification procedure described in this Subsection is performed using the unbiased model tructure (5.5) and if either condition 1 or 2 above is fulfilled, the identified parameter vector $\hat{\theta}_{ol}$ defines an unbiased model $\hat{G}_{ol}(z) \triangleq G(z, \hat{\theta}_{ol}) \in \mathcal{M}_{ol}$ and is asymptotically a random Gaussian vector with mean θ_0 and covariance $P_{\theta_{ol}}$:*

$$\hat{\theta}_{ol} \in AsN(\theta_0, P_{\theta_{ol}}) \tag{5.6}$$

where $P_{\theta_{ol}} \in \mathbb{R}^{q \times q}$ is an unknown symmetric positive definite matrix.

As for general model structures, prediction-error identification with an unbiased model structure also delivers an estimate $\hat{P}_{\theta_{ol}}$ of the covariance matrix $P_{\theta_{ol}}$; see (Ljung, 1999).

Although the Gaussian distribution property of the identified parameter vector is an asymptotic property obtained when $N \to \infty$, we shall use it in the sequel for a finite but sufficiently large number N of data. This widespread approximation in statistics theory has been proved accurate in (Ljung, 1999). Using this approximation, one can define an ellipsoidal uncertainty region in parameter space, centred at the identified parameter vector $\hat{\theta}_{ol}$ and containing the unknown true parameter vector θ_0 with a chosen probability level.

Proposition 5.2. (Ljung, 1999) *Let us consider*

- *the system $G_0 = G(\theta_0)$ satisfying the assumptions made in this Section;*
- *the identified parameter vector $\hat{\theta}_{ol} \in \mathbb{R}^q$ and the estimate $\hat{P}_{\theta_{ol}} \in \mathbb{R}^{q \times q}$ of its covariance matrix, delivered by a prediction-error identification experiment performed on G_0 using a sufficiently large number N of input-output data and an unbiased model structure $\mathcal{M}_{ol}$*

and let us define

$$\gamma \triangleq \chi^2_p(q) \qquad i.e., \qquad \Pr\big(\chi^2(q) < \gamma\big) = p \tag{5.7}$$

where $\chi^2(q)$ is a chi-square-distributed random variable with q degrees of freedom and $0 \leq p \leq 1$.

Then, the ellipsoid $\mathcal{U}_{ol}$ of size γ defined by

$$\mathcal{U}_{ol} = \left\{\theta \in \mathbb{R}^q \mid (\theta - \hat{\theta}_{ol})^T \hat{P}^{-1}_{\theta_{ol}} (\theta - \hat{\theta}_{ol}) < \gamma\right\} \tag{5.8}$$

contains the true parameter vector θ_0 with a probability p:

$$\Pr(\theta_0 \in \mathcal{U}_{ol}) = p \tag{5.9}$$

Practically, $\hat{\theta}_{ol}$ and $\hat{P}_{\theta_{ol}}$ are the result of the identification experiment while p is chosen by the designer as the degree of confidence he or she wants to have and which determines the size γ of the ellipsoid. Together, these three elements define the ellipsoidal uncertainty region $\mathcal{U}_{ol}$. *e.g.*, if the designer chooses a probability level $p = 0.95$, it actually means that there is a probability of 95% that the ellipsoid $\mathcal{U}_{ol}$ (whose shape depends on the covariance of the estimate $\hat{\theta}_{ol}$ due to the noise $v(t)$ and whose size depends on the chosen value of p) contains the true parameter vector θ_0. Choosing a larger probability (or level of confidence), *e.g.*, $p = 0.99$, would yield a larger value of γ and hence a larger ellipsoid.

► **Remark.** The use of a chi-square probability density function with q degrees of freedom to define the probability density linked to $\mathcal{U}_{ol}$ is actually an approximation. Indeed, $\hat{P}_{\theta_{ol}}$ is only an estimate of $P_{\theta_{ol}} = \text{cov}(\hat{\theta}_{ol})$ obtained with N experimental data samples. As a result, the actual probability density function linked to $\mathcal{U}_{ol}$ is a function of the F-distribution $F(q, N-q)$: $\Pr(\theta_0 \in \mathcal{U}_{ol}) = \Pr(F(q, N-q) < \frac{\gamma}{q})$. Nevertheless, for large N there holds $\Pr(F(q, N-q) < \frac{\gamma}{q}) \approx \Pr(\chi^2(q) < \gamma)$. ◄

The parametric uncertainty region $\mathcal{U}_{ol}$ defines a corresponding uncertainty region $\mathcal{D}_{ol}$ in the space of transfer functions:

$$\mathcal{D}_{ol} = \Big\{ G(z, \theta) \mid G(z, \theta) \in \mathcal{M}_{ol},\, \theta \in \mathcal{U}_{ol} \Big\} \tag{5.10}$$

We then have the following property.

Lemma 5.1. *$G_0(z) \in \mathcal{D}_{ol}$ with probability p.*

The importance of this lemma is that the validation procedure has delivered a model set $\mathcal{D}_{ol}$ in which the true system is guaranteed to lie, at some probability level. We can now define the notion of *validation* of a given nominal model G_{nom}.

Definition 5.1. (MODEL VALIDATION) *A model G_{nom} is called validated with the uncertainty region $\mathcal{D}_{ol}$ if $G_{nom}(z) \in \mathcal{D}_{ol}$.*

According to this definition, the model G_{nom} is validated if the noise acting on the process G_0 is such that it is impossible to clearly differentiate them from each other: they both lie in an uncertainty region which is the best description we can build of the process using the available data. In other words, any attempt to find a lower bound on the distance between the process G_0 and its model G_{nom} is pointless, because the (variance only) error around the estimate of this lower bound is larger than the sought distance itself.

▶ **Example 5.1.** Consider the true ARX system previously used in Chapter 3:

$$\mathcal{S} : (1 - 1.4z^{-1} + 0.45z^{-2})y(t) = (1 + 0.25z^{-1})z^{-1}u(t) + e(t)$$

where $e(t)$ is unit-variance white noise. Thus, $G_0(z) = \frac{z^{-1}+0.25z^{-2}}{1-1.4z^{-1}+0.45z^{-2}}$ and the true parameter vector is $\theta_0 = \begin{bmatrix}-1.4 & 0.45 & 1 & 0.25\end{bmatrix}^T$. A nominal ARX model $G_{nom}(z) = \frac{0.989z^{-1}+0.23z^{-2}}{1-1.393z^{-1}+0.4484z^{-2}}$, $H_{nom}(z) = \frac{1}{1-1.393z^{-1}+0.4484z^{-2}}$ is given and must be validated; its parameter vector is $\theta_{nom} = \begin{bmatrix}-1.393 & 0.4484 & 0.989 & 0.23\end{bmatrix}^T$. The validation procedure consists in exciting the plant G_0 with a unit-variance pseudo-random binary input signal u_{ol} of length 4095. The collected input and output data are used for the identification of an unbiased ARX model $G(z,\theta) = \frac{b_1z^{-1}+b_2z^{-2}}{1+a_1z^{-1}+a_2z^{-2}}$, $H(z,\theta) = \frac{1}{1+a_1z^{-1}+a_2z^{-2}}$. The identification delivers the parameter vector

$$\hat{\theta}_{ol} = \begin{bmatrix}\hat{a}_1 & \hat{a}_2 & \hat{b}_1 & \hat{b}_2\end{bmatrix}^T = \begin{bmatrix}-1.3933 & 0.4447 & 1.0359 & 0.2709\end{bmatrix}^T$$

and an estimate of its covariance matrix

$$\hat{P}_{\theta_{ol}} = 10^{-3} \times \begin{bmatrix} 0.1300 & -0.1259 & 0.0015 & 0.1355 \\ -0.1259 & 0.1273 & -0.0016 & -0.1313 \\ 0.0015 & -0.0016 & 0.2543 & 0.0015 \\ 0.1355 & -0.1313 & 0.0015 & 0.3954 \end{bmatrix}$$

Using the results of Proposition 5.2, with $q = 4$ and $p = 0.99$ we find that the 99% confidence ellipsoid in parameter space is given by

$$\mathcal{U}_{ol} = \left\{\theta \in \mathbb{R}^4 \mid (\theta - \hat{\theta}_{ol})^T \hat{P}_{\theta_{ol}}^{-1}(\theta - \hat{\theta}_{ol}) < \chi^2_{0.99}(4) = 13.3\right\}$$

It is easy to check that

$$(\theta_0 - \hat{\theta}_{ol})^T \hat{P}_{\theta_{ol}}^{-1}(\theta_0 - \hat{\theta}_{ol}) = 6.4 < 13.3$$

hence $\theta_0 \in \mathcal{U}_{ol}$ (this was predictable since $\hat{G}_{ol}$ is unbiased), while

$$(\theta_{nom} - \hat{\theta}_{ol})^T \hat{P}_{\theta_{ol}}^{-1}(\theta_{nom} - \hat{\theta}_{ol}) = 18.36 > 13.3$$

hence $\theta_{nom} \notin \mathcal{U}_{ol}$ which means that $G_{nom} \notin \mathcal{D}_{ol}$: the nominal model G_{nom} could not be validated on the basis of the collected data.

Figure 5.1 shows the Nyquist diagrams of G_0 and G_{nom} and the ellipsoids representing $\mathcal{D}_{ol}$ computed at 50 frequencies (the accurate computation of these ellipsoids in frequency domain, *i.e.*, the mapping from $\mathcal{U}_{ol}$ to $\mathcal{D}_{ol}$, involves the resolution of an LMI-based optimisation problem at each frequency). One can see that $G_0(e^{j\omega})$ is contained in these ellipsoids at all frequencies, while $G_{nom}(e^{j\omega})$ is out of the ellipsoids at some frequencies, which confirms the previous invalidation verdict. Note that this has to be checked pointwise: the fact that the Nyquist diagram of G_{nom} crosses the uncertainty ellipsoid that is obtained at some frequency ω_0 does not mean that G_{nom} is validated at this frequency. It is necessary that the portion of the Nyquist diagram of G_{nom} that is inside this ellipsoid contains the point $G_{nom}(e^{j\omega_0})$. ◀

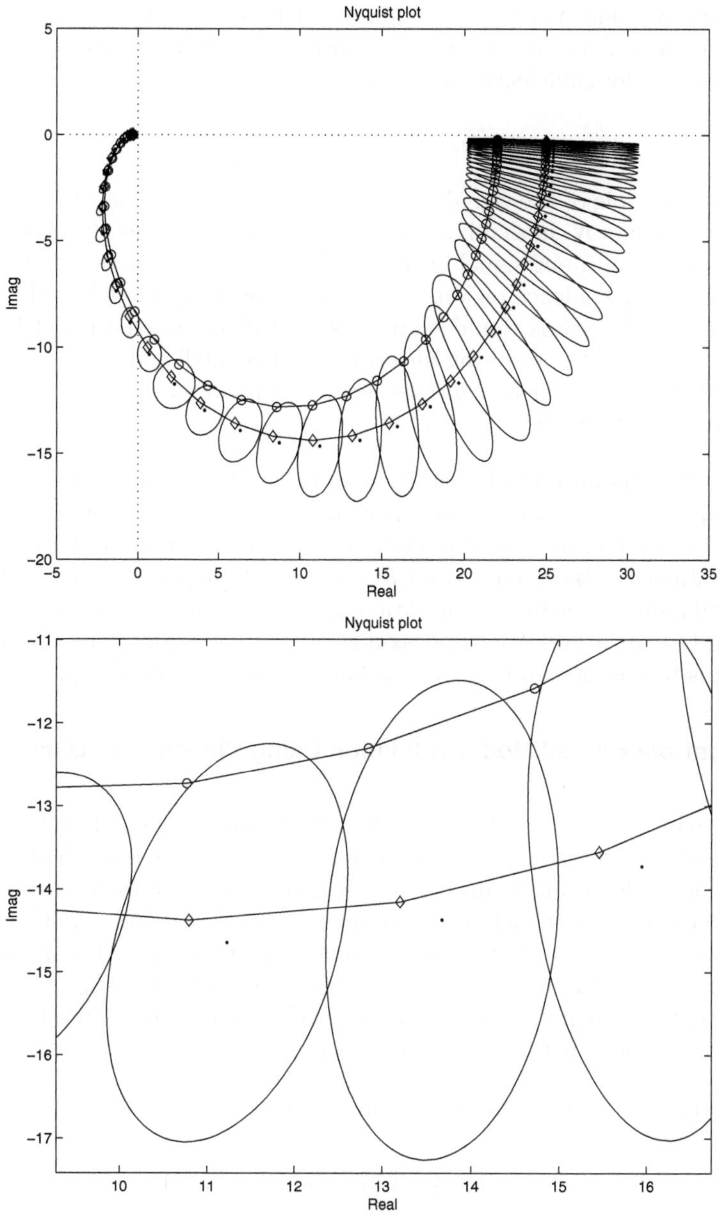

Figure 5.1. Nyquist diagram (bottom picture = zoom). $G_0(e^{j\omega})$ (-◇-), $G_{nom}(e^{j\omega})$ (-o-), $\hat{G}_{ol}(e^{j\omega})$ $(\cdot)$ and $\mathcal{D}_{ol}$ (—)

Observe now that the model set $\mathcal{D}_{ol}$ depends very much on the experimental conditions under which the validation (*i.e.*, the identification of $\hat{G}_{ol}$) is performed. This is perhaps not so apparent in the definition (5.10) of $\mathcal{D}_{ol}$ *via* the parameter covariance matrix $\hat{P}_{\theta_{ol}}$. However, let us recall from (2.130a) that a reasonable approximation for the covariance of the transfer function estimate $\hat{G}_{ol}(z)$ is given, for sufficiently large q and N, by

$$\mathrm{cov}\big(\hat{G}_{ol}(e^{j\omega})\big) \approx \frac{q}{N}\frac{\phi_v(\omega)}{\phi_{u_{ol}}(\omega)} \tag{5.11}$$

This shows the role of the signal spectra $\phi_{u_{ol}}(\omega)$ and $\phi_v(\omega)$, as well as that of the number of data N, in shaping the uncertainty set $\mathcal{D}_{ol}$. Clearly, experiments using a very small input signal energy and/or a small number of data would yield very large confidence regions $\mathcal{D}_{ol}$ (hence almost any model would be validated), but such regions would be useless. A validation experiment is useful for control design if the resulting uncertainty set is small and if its distribution in the frequency domain is 'control-oriented'; this last concept will be made precise in Subsection 5.2.2.

Theoretically, the uncertainty set $\mathcal{D}_{ol}$ can be used for control design even if the model G_{nom} is falsified. However, if the model G_{nom} has failed a range of validation attempts, any sensible designer will want to replace it by a model that is contained in the validated set $\mathcal{D}_{ol}$ (in order to design a robust controller, it is indeed quite reasonable to choose a nominal model in $\mathcal{D}_{ol}$. A possible choice would be $\hat{G}_{ol}$, or a low-order approximation of it in $\mathcal{D}_{ol}$). Thus, the typical situation is where the control design is based on a validated G_{nom}.

5.2.2 Control-oriented Model Validation Using Closed-loop Data

As we have just mentioned, the experimental conditions under which the validation experiment is carried out can have a significant influence on the frequency distribution of the resulting uncertainty region $\mathcal{D}$. Recall also that, as pointed out in Chapter 3, the model G_{nom} should be evaluated by how well it mimics the behaviour of the actual system G_0 when both are connected with the same controller (ideally the optimal, to-be-designed controller) if G_{nom} has to be used for control design. Connecting these two issues leads us naturally to carry out the validation experiment in closed loop.

Suppose that the input signal is determined by a known stabilising controller K_{id} as in Figure 4.1:

$$u_{cl}(t) = r_2(t) - K_{id}(z)\big(y_{cl}(t) - r_1(t)\big) \tag{5.12}$$

For the sake of generality, we make the assumption that $r_1(t)$ and $r_2(t)$ are two possible sources of excitation (reference signals) that are quasi-stationary and uncorrelated with $v(t)$. The case where only one of these two signals is available simply restricts the range of prediction-error identification methods that can be used, as seen in Chapter 4.

Because it is the most straightforward, we shall first study the use of the direct identification method for closed-loop validation.

A. Direct model validation in closed-loop. This closed-loop validation method, which is the most obvious extension of the open-loop validation method described above, uses the direct approach of closed-loop system identification, *i.e.*, the identification of a model $\hat{G}_{cl}(z) \triangleq G(z, \hat{\theta}_{cl})$ from closed-loop input-output data

$$Z_{cl}^N = \Big\{u_{cl}(1),\, y_{cl}(1),\, \ldots,\, u_{cl}(N),\, y_{cl}(N)\Big\} \tag{5.13}$$

As stated in Subsection 5.2.1, the validation procedure that we have adopted in the open-loop case and that we would like to extend here to the closed-loop case can normally be used only if at least one of the following two conditions is fulfilled:

1. the model structure for the noise model $\hat{H}_{cl}(z)$ that is identified during the validation procedure is unbiased;
2. the noise model $\hat{H}_{cl}(z)$ is independently parametrised from $\hat{G}_{cl}(z)$ (*e.g.*, an OE model structure is used) and the input signal $u_{cl}(t)$ is uncorrelated with the noise $v(t)$.

Because the data are generated in closed loop, condition 2 cannot be satisfied, contrary to the open-loop case where it was guaranteed by definition. However, since we identify a strictly causal model, the requirement that $u_{cl}(t)$ an $v(t)$ be uncorrelated can be relaxed and all that is needed is that $u_{cl}(t)$ be uncorrelated with the future noise samples $v(t+1)$, $v(t+2)$, *etc.* Observe now that

a. if $v(t)$ is white, then $u_{cl}(t)$ is only correlated with $v(t)$, $v(t-1)$, *etc.*;
b. otherwise, $v(t)$ is correlated with $v(t+k)$ for some $k > 0$ and therefore $u_{cl}(t)$ is correlated with $v(t+k)$, $v(t+k-1)$, *etc.*

In case a, the choice of an OE model structure would satisfy both conditions 1 and 2 (relaxed).

In case b, it is, however, necessary to find a solution to whiten $v(t)$.

A first possibility is to use a prefilter $L(z) = H_{nom}^{-1}(z)$ which will make $L(z)v(t)$ (almost) white if H_{nom} is a good model of the actual noise process H_0. After prefiltering the input and output data Z_{cl}^N by $L(z)$, any independently parametrised noise model structure $\hat{H}_{cl}(z)$ can be used for the validation experiment and a natural choice would be $\hat{H}_{cl}(z) \equiv 1$, *i.e.*, again an OE model structure. However, the quality of the result will depend very much on the accuracy of the initial noise model H_{nom} which can be hard to assess... before the validation is carried out!

If no model $H_{nom}(z)$ of H_0 is available, or if its accuracy is questionable, it is necessary to use a parametrised noise model during the validation, which amounts to use a parametrised prefilter $L(z, \theta) = H^{-1}(z, \theta)$. The procedure then consists of minimising the experimental variance of the prediction errors, *i.e.*, of computing

$$\hat{\theta}_{cl} = \arg\min_{\theta \in \mathbb{R}^q} \frac{1}{N} \sum_{t=1}^{N} \varepsilon^2(t, \theta) \tag{5.14}$$

where

$$\begin{aligned} \varepsilon(t, \theta) &= H^{-1}(z, \theta)\big(y_{cl}(t) - G(z, \theta)u_{cl}(t)\big) \\ &= H^{-1}(z, \theta)\big(G_0(z) - G(z, \theta)\big)u_{cl}(t) + H^{-1}(z, \theta)H_0(z)e(t) \end{aligned} \tag{5.15}$$

Using results of Section 2.6, we find that $\hat{\theta}_{cl} \to \theta^*$ as $N \to \infty$, where θ^* is given by (2.126). The bias term $B(e^{j\omega}, \theta)$ in (2.126) will be 0 only if the noise model $H(z, \theta)$ is flexible enough to represent the true $H_0(z)$. As a conclusion, this direct approach to closed-loop model validation will deliver an unbiased estimate of the process G_0 only if the model $\big(G(z, \theta), H(z, \theta)\big)$ has an unbiased structure, *i.e.*, if it lies in some model set

$$\mathcal{M}_{cl} = \Big\{\big(G(z, \theta), H(z, \theta)\big) \mid \theta \in \mathbb{R}^q\Big\} \ni \big(G_0(z), H_0(z)\big) \tag{5.16}$$

or if the data is prefiltered by the inverse of the exact noise model H_0, if it is known. The validation procedure then works exactly as the open-loop validation procedure of Section 5.2.1 and delivers uncertainty regions

$$\mathcal{U}_{cl} = \Big\{\theta \in \mathbb{R}^q \mid (\theta - \hat{\theta}_{cl})^T \hat{P}_{\theta_{cl}}^{-1}(\theta - \hat{\theta}_{cl}) < \gamma\Big\} \tag{5.17}$$

and

$$\mathcal{D}_{cl} = \Big\{G(z, \theta) \mid \big(G(z, \theta), H(z, \theta)\big) \in \mathcal{M}_{cl}, \theta \in \mathcal{U}_{cl}\Big\} \tag{5.18}$$

where $\hat{P}_{\theta_{cl}}$ and γ are defined as in Subsection 5.2.1.

▶ **Example 5.2.** Consider again the system G_0 and the model G_{nom} of Example 5.1. This time, the validation experiment is carried out with G_0 in closed loop with a PID controller $K_{id}(z) = \frac{0.09079z^2 - 0.09107z + 0.00874}{z(z-1)}$. The excitation signal is a PRBS with a peak-to-peak amplitude of 38.34, which makes the output variance $\sigma^2_{y_{cl}} = \sigma^2_{y_{ol}} =$ 48.33.

4095 input-output data samples are collected and used for the validation, which consists of the identification of an unbiased ARX model $G(z, \theta) = \frac{b_1 z^{-1} + b_2 z^{-2}}{1 + a_1 z^{-1} + a_2 z^{-2}}$, $H(z, \theta) = \frac{1}{1 + a_1 z^{-1} + a_2 z^{-2}}$. The identification delivers the parameter vector

$$\hat{\theta}_{cl} = \begin{bmatrix}\hat{a}_1 & \hat{a}_2 & \hat{b}_1 & \hat{b}_2\end{bmatrix}^T = \begin{bmatrix}-1.4075 & 0.4581 & 0.9970 & 0.2397\end{bmatrix}^T$$

and an estimate of its covariance matrix

$$\hat{P}_{\theta_{cl}} = 10^{-3} \times \begin{bmatrix} 0.0780 & -0.0766 & -0.0026 & 0.0764 \\ -0.0766 & 0.0809 & -0.0017 & -0.0795 \\ -0.0026 & -0.0017 & 0.0804 & 0.0045 \\ 0.0764 & -0.0795 & 0.0045 & 0.1555 \end{bmatrix}$$

The 99% confidence ellipsoid in parameter space is given by

$$\mathcal{U}_{cl} = \left\{ \theta \in \mathbb{R}^4 \mid (\theta - \hat{\theta}_{cl})^T \hat{P}_{\theta_{cl}}^{-1} (\theta - \hat{\theta}_{cl}) < \chi^2_{0.99}(4) = 13.3 \right\}$$

It is easy to check that

$$(\theta_0 - \hat{\theta}_{cl})^T \hat{P}_{\theta_{cl}}^{-1} (\theta_0 - \hat{\theta}_{cl}) = 0.97 < 13.3$$

and

$$(\theta_{nom} - \hat{\theta}_{cl})^T \hat{P}_{\theta_{cl}}^{-1} (\theta_{nom} - \hat{\theta}_{cl}) = 11.80 < 13.3$$

hence θ_0, $\theta_{nom} \in \mathcal{U}_{cl}$ which means that G_0, $G_{nom} \in \mathcal{D}_{cl}$: the uncertainty region $\mathcal{D}_{cl}$ is centred around an unbiased model $\hat{G}_{cl} \triangleq G(\hat{\theta}_{cl})$ and the nominal model G_{nom} is validated.

Figure 5.2 shows the Nyquist diagrams of G_0 and G_{nom} and the ellipsoids representing $\mathcal{D}_{cl}$ computed at 50 frequencies. One can see that both $G_0(e^{j\omega})$ and $G_{nom}(e^{j\omega})$ are indeed contained in these ellipsoids at all frequencies.

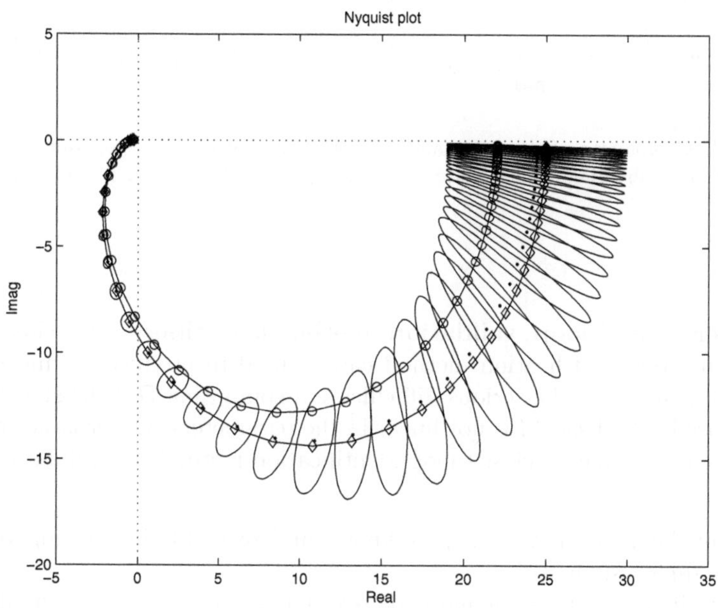

Figure 5.2. Nyquist diagram: $G_0(e^{j\omega})$ (-◇-), $G_{nom}(e^{j\omega})$ (-○-), $\hat{G}_{cl}(e^{j\omega})$ (·) and $\mathcal{D}_{cl}$ (—)

$\mathcal{D}_{ol}$ (from Example 5.1, Figure 5.1) and $\mathcal{D}_{cl}$ (Figure 5.2) may seem very similar at first sight, but a zoom around the cross-over frequency of the open-loop system $G_0 K_{id}$, as made in Figure 5.3, shows that the closed-loop validation procedure has delivered a tighter uncertainty region at frequencies where good precision is usually required for control design, although it is larger at other frequencies (like those where the open-loop validation procedure invalidated the model). ◀

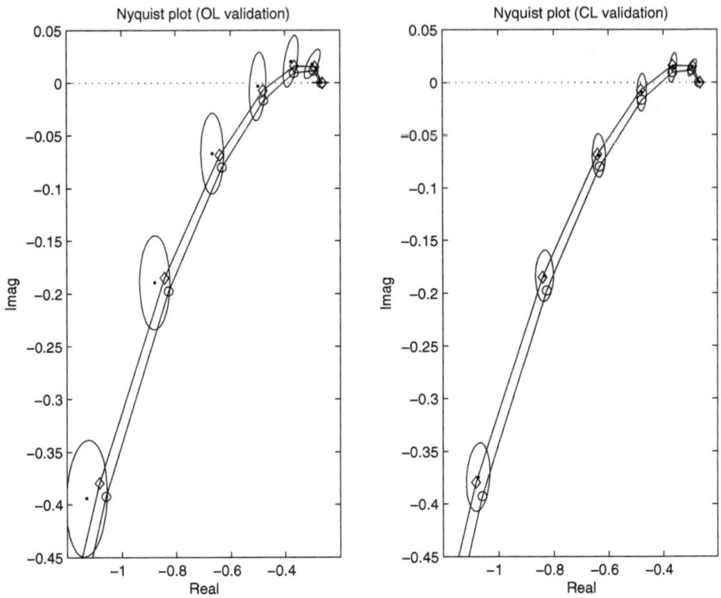

Figure 5.3. Comparison of the uncertainty regions delivered by open-loop validation (left) and closed-loop validation (right). $G_0(e^{j\omega})$ (-◇-), $G_{nom}(e^{j\omega})$ (-○-), $\hat{G}_{ol/cl}(e^{j\omega})$ $(\cdot)$ and $\mathcal{D}_{ol/cl}$ (—)

B. Other closed-loop validation methods. In theory, any closed-loop prediction-error identification method can be used to produce an uncertainty set $\mathcal{D}$. Yet, there may be serious difficulties in computing $\mathcal{D}$ if the structure of the unbiased model set $\mathcal{M}$ is not linear in the parameters. Other issues related to the use of alternative closed-loop identification methods for validation are:

- most of them require that K_{id} be known and/or LTI (which is not required by the direct method);
- the possibility to use some methods is restricted by the type of singularities of K_{id} as is the case for the identification of a nominal model, according to the results of Chapter 4;
- the number of parameters of the identified transfer functions in these methods is often higher than that required for direct estimation of an unbiased model $\hat{G}_{cl}$, which can increase the variance of the estimate and hence, the size of the uncertainty region $\mathcal{D}$.

As an illustration, we give here the mathematical derivations that would lead to such an uncertainty set $\mathcal{D}_{ind}$ using indirect closed-loop identification with the controller K_{id} of (5.12).

The true and the nominal closed-loop transfer matrices are respectively

$$T(G_0, K_{id}) = \begin{bmatrix} \dfrac{G_0 K_{id}}{1 + G_0 K_{id}} & \dfrac{G_0}{1 + G_0 K_{id}} \\ \dfrac{K_{id}}{1 + G_0 K_{id}} & \dfrac{1}{1 + G_0 K_{id}} \end{bmatrix} \triangleq \begin{bmatrix} T_{0,\,11} & T_{0,\,12} \\ T_{0,\,21} & T_{0,\,22} \end{bmatrix} \tag{5.19}$$

and

$$T(G_{nom}, K_{id}) = \begin{bmatrix} \dfrac{G_{nom} K_{id}}{1 + G_{nom} K_{id}} & \dfrac{G_{nom}}{1 + G_{nom} K_{id}} \\ \dfrac{K_{id}}{1 + G_{nom} K_{id}} & \dfrac{1}{1 + G_{nom} K_{id}} \end{bmatrix}$$

$$\triangleq \begin{bmatrix} T_{nom,\,11} & T_{nom,\,12} \\ T_{nom,\,21} & T_{nom,\,22} \end{bmatrix} \tag{5.20}$$

Since r_1 and r_2 are uncorrelated with v, we can apply the open-loop validation procedure described in Subsection 5.2.1 to any of the four entries of the closed-loop transfer matrix (5.19), provided the experiment design guidelines for indirect closed-loop identification given in Chapter 4 are followed.

Thus, an unbiased estimate $T_{ij}(z, \hat{\xi})$ of the closed-loop transfer function $T_{0,\,ij}(z)$ ($i, j \in \{1, 2\}$) is identified using an appropriate set of data (*i.e.*, measurements of either $r_1(t)$ or $r_2(t)$ and of either $y(t)$ or $u(t)$), with $T_{ij}(z, \xi)$ in some unbiased parametrised model set (with parameter vector ξ of length r)

$$\mathcal{M}_{ind} = \left\{ T_{ij}(z, \xi) \mid \xi \in \mathbb{R}^r \right\} \ni T_{0,\,ij}(z) \tag{5.21}$$

The identification delivers $\hat{\xi}$ and an estimate $\hat{P}_\xi$ of $\text{cov}(\hat{\xi})$. The uncertainty region around $\hat{\xi}$ is given by

$$\mathcal{U}_{ind} = \left\{ \xi \in \mathbb{R}^r \mid (\xi - \hat{\xi})^T \hat{P}_\xi^{-1} (\xi - \hat{\xi}) < \chi_p^2(r) \right\} \tag{5.22}$$

where p is the chosen probability level. The true closed-loop transfer function $T_{0,\,ij}$ is then contained (with probability p) in a confidence region

$$\mathcal{E} = \left\{ T_{ij}(z, \xi) \mid T_{ij}(z, \xi) \in \mathcal{M}_{ind},\, \xi \in \mathcal{U}_{ind} \right\} \tag{5.23}$$

In accordance with Subsection 5.2.1, $T_{nom,\,ij}$ is validated if it lies in $\mathcal{E}$. The important observation now is that, if $T_{nom,\,ij}$ is validated with an uncertainty set $\mathcal{E}$, then G_{nom} is validated with an uncertainty set $\mathcal{D}_{ind}$ that can be computed from $\mathcal{U}_{ind}$. Indeed, observe that we can build models $G(z, \xi)$ for the plant using one of the following relations, depending on which closed-loop transfer function has been identified:

$$G(z, \xi) = \frac{T_{11}(z, \xi)}{K_{id}(z) - K_{id}(z) T_{11}(z, \xi)} \tag{5.24a}$$

or

$$G(z, \xi) = \frac{T_{12}(z, \xi)}{1 - K_{id}(z)T_{12}(z, \xi)} \tag{5.24b}$$

or

$$G(z, \xi) = \frac{1}{T_{21}(z, \xi)} - \frac{1}{K_{id}(z)} \tag{5.24c}$$

or

$$G(z, \xi) = \frac{1}{K_{id}(z)} \left(\frac{1}{T_{22}(z, \xi)} - 1 \right) \tag{5.24d}$$

An uncertainty region around $\hat{G}_{ind}(z) \triangleq G(z, \hat{\xi})$, guaranteed to contain G_0 with probability p, is then simply given by

$$\mathcal{D}_{ind} = \Big\{ G(z, \xi) \mid G(z, \xi) \text{ is computed according to (5.24)}, T_{ij}(z, \xi) \in \mathcal{M}_{ind}, \xi \in \mathcal{U}_{ind} \Big\} \tag{5.25}$$

which contains $G_{nom}(z)$ if and only if $T_{nom,\,ij}(z) \in \mathcal{E}$.

A possible way of estimating $\mathcal{D}_{ind}$ would be the following first-order approximation:

$$\mathcal{D}_{ind} \approx \left[\frac{\partial G(z, \xi)}{\partial \xi_1} \quad \cdots \quad \frac{\partial G(z, \xi)}{\partial \xi_r} \right]_{\xi = \hat{\xi}} \times \mathcal{U}_{ind} \times \left[\frac{\partial G(z, \xi)}{\partial \xi_1} \quad \cdots \quad \frac{\partial G(z, \xi)}{\partial \xi_r} \right]^{\star}_{\xi = \hat{\xi}} \tag{5.26}$$

At each frequency ω, this estimate of $\mathcal{D}_{ind}$ is again an ellipsoid (obtained by setting $z = e^{j\omega}$) centred at $\hat{G}_{ind}(e^{j\omega})$. This approximation can however be very imprecise; if it is larger than the true $\mathcal{D}_{ind}$, it may lead to conservative control designs while, on the contrary, it may not contain the true G_0 if it is underestimated. An accurate computation of $\mathcal{D}_{ind}$ requires more complex algorithms involving linear matrix inequalities (LMI's); see (Boyd *et al.*, 1994) for such algorithms.

5.2.3 A Unified Representation of the Uncertainty Zone

It can be shown (Bombois, 2000) that, whatever the method being used (open-loop, direct or indirect closed-loop identification, dual-Youla parametrisation closed-loop identification, or open-loop or closed-loop identification using the model-error model approach of (Ljung and Guo, 1997; Ljung, 2000)), the resulting uncertainty region $\mathcal{D}$ always has the following generic form, which we call *generic prediction-error uncertainty set*:

$$\mathcal{D} = \left\{ G(z,\theta) \mid G(z,\theta) = \frac{\Xi(z) + Z_N(z)\theta}{1 + Z_D(z)\theta} \text{ and } \theta \in \mathcal{U} \subset \mathbb{R}^q \right\} \tag{5.27}$$

where

- $Z_N(z)$ and $Z_D(z)$ are row vectors of size q of known transfer functions;
- $\Xi(z)$ is a known transfer function with a delay equal to that of $G_0(z)$; $\Xi(z) \equiv 0$ in the open-loop case or with the direct and indirect closed-loop methods;
- $\mathcal{U}$ is the uncertainty region in parameter space defined by

$$\mathcal{U} = \left\{ \theta \in \mathbb{R}^q \mid (\theta - \hat{\theta})^T \frac{\hat{P}_\theta^{-1}}{\chi_p^2(q)} (\theta - \hat{\theta}) < 1 \right\} \tag{5.28}$$

where $\hat{P}_\theta$ is the estimate of $\operatorname{cov} \hat{\theta}$ delivered by the prediction-error identification experiment and $\chi_p^2(q)$ is defined as the percentile p ($0 < p < 1$ being the chosen probability level of confidence) of the χ^2-distribution with q degrees of freedom.

5.3 Model Validation for Control and Controller Validation

We have seen in the previous section how a parametric confidence region $\mathcal{D}$, guaranteed to contain the unknown true system G_0 with a chosen probability p, could be built using prediction-error identification. In this section, we describe tools for using the uncertainty region $\mathcal{D}$ in a robust control framework. These tools are, respectively, a control-oriented measure of the size of a generic prediction-error uncertainty set $\mathcal{D}$ of the form (5.27) (Subsection 5.3.1), controller validation tools for stability (Subsection 5.3.2) and controller validation tools for performance (Subsection 5.3.3).

5.3.1 A Control-oriented Measure of the Size of a Prediction-error Uncertainty Set

The uncertainty region $\mathcal{D}$ of (5.27) depends very much on the experimental conditions used for the validation experiment: open-loop or closed-loop set-up, choice of signal spectra, *etc.* Thus, different validation experiments yield different uncertainty regions $\mathcal{D}_i$, each containing the true G_0 with probability p. It is therefore useful to possess a measure of quality of a model set $\mathcal{D}$ that is connected to the size of a controller set that stabilises all models in $\mathcal{D}$. Such a measure is the **worst-case ν-gap** and its frequency-by-frequency counterpart, the **worst-case chordal distance**. They are extensions of Vinnicombe's ν-gap and chordal distance between two transfer functions and are defined as follows.

Definition 5.2. (WORST-CASE CHORDAL DISTANCE) *The worst-case chordal distance at frequency ω between a nominal system G_{nom} and all systems in an uncertainty set $\mathcal{D}$ is defined as*

$$\kappa_{WC}(G_{nom},\mathcal{D},\omega) \triangleq \sup_{G\in\mathcal{D}} \kappa(G_{nom}(e^{j\omega}),G(e^{j\omega})) \tag{5.29}$$

Definition 5.3. (WORST-CASE ν-GAP) *The worst-case ν-gap between a nominal system G_{nom} and all systems in an uncertainty set $\mathcal{D}$ is given by*

$$\begin{aligned}\delta_{WC}(G_{nom},\mathcal{D}) &\triangleq \sup_{G\in\mathcal{D}} \delta_\nu(G_{nom},G)\\ &= \max_\omega \kappa_{WC}(G_{nom},\mathcal{D},\omega)\end{aligned} \tag{5.30}$$

Here $\kappa(G_{nom}(e^{j\omega}),G(e^{j\omega}))$ and $\delta_\nu(G_{nom},G)$ are, respectively, the chordal distance at frequency ω and the ν-gap between G_{nom} and G: see Section 2.5.

It can be shown that $\kappa_{WC}(G_{nom},\mathcal{D},\omega)$ can be obtained as the result of an optimisation problem involving LMI's, as set in the following theorem.

Theorem 5.1. (Bombois, 2000) *Consider a model G_{nom} and an uncertainty region $\mathcal{D}$ given in (5.27). Then, at each frequency ω, the worst-case chordal distance between G_{nom} and all systems in $\mathcal{D}$ can be computed as*

$$\kappa_{WC}(G_{nom},\mathcal{D},\omega) = \sqrt{\rho_{opt}(\omega)} \tag{5.31}$$

where $\rho_{opt}(\omega)$ is the optimal value of ρ in the convex optimisation problem involving LMI constraints, evaluated at frequency ω:

$$\begin{array}{ll} \textit{Minimise} & \rho \\ \textit{over} & \rho,\,\tau \\ \textit{subject to} & \tau \geq 0 \textit{ and} \end{array}$$

$$\begin{bmatrix} Re(a_{11}) & Re(a_{12}) \\ Re(a_{12}^\star) & Re(a_{22}) \end{bmatrix} - \tau \begin{bmatrix} R & -R\hat{\theta} \\ (-R\hat{\theta})^T & \hat{\theta}^T R\hat{\theta} - 1 \end{bmatrix} \tag{5.32}$$

where

$$\begin{aligned} a_{11} =& (Z_N^\star Z_N - Z_N^\star G_{nom} Z_D - Z_D^\star G_{nom}^\star Z_N + Z_D^\star G_{nom}^\star G_{nom} Z_D) \\ & - \rho(Z_N^\star Q Z_N + Z_D^\star Q Z_D) \\ a_{12} =& (Z_N^\star \Xi - Z_N^\star G_{nom} - Z_D^\star \Xi G_{nom}^\star + Z_D^\star G_{nom} G_{nom}^\star) - \rho(Z_N^\star \Xi Q + Z_D^\star Q) \\ a_{22} =& (\Xi\Xi^\star - \Xi^\star G_{nom} - \Xi G_{nom}^\star + G_{nom} G_{nom}^\star) - \rho(\Xi\Xi^\star Q + Q) \\ Q =& 1 + G_{nom}^\star G_{nom} \\ R =& \frac{\hat{P}_\theta^{-1}}{\chi_p^2(q)} \end{aligned}$$

The worst-case ν-gap is then obtained as

$$\delta_{WC}(G_{nom},\mathcal{D}) = \max_\omega \kappa_{WC}(G_{nom},\mathcal{D},\omega) \tag{5.34}$$

The LMI Control Toolbox of MATLAB® contains the tools for solving such a problem. Recall that $\Xi(e^{j\omega}) \equiv 0$ if any of the three methods developed in Section 5.2 is used for building $\mathcal{D}$, which simplifies the expressions of a_{12} and a_{22} above.

► **Example 5.3.** Consider the uncertainty regions $\mathcal{D}_{ol}$ and $\mathcal{D}_{cl}$ and the nominal model $G_{nom}(z)$ of Examples 5.1 and 5.2. Using the result of Theorem 5.1, one can compute the worst-case ν-gaps between $G_{nom}(z)$ and both uncertainty region. Note that $G_{nom}(z)$ does not need to be validated for this computation (recall that this model was not validated by the open-loop experience that produced $\mathcal{D}_{ol}$):

$$\delta_{WC}(G_{nom}, \mathcal{D}_{ol}) = 0.0595$$

$$\delta_{WC}(G_{nom}, \mathcal{D}_{cl}) = 0.0255$$

Observe that $\delta_{WC}(G_{nom}, \mathcal{D}_{cl})$ is a better estimation of the real $\delta_\nu(G_{nom}, G_0) = 0.0115$. ◄

The following theorem gives sufficient conditions for robust stabilisation of all plants in a prediction-error uncertainty set $\mathcal{D}$ by a given controller K (typically, a new controller aimed to replace the possible current controller K_{id}).

Theorem 5.2. *Consider a prediction-error uncertainty set $\mathcal{D}$ containing the true system G_0 with probability p, a nominal model G_{nom} such that $\delta_{WC}(G_{nom}, \mathcal{D}) < 1$ and let K be a controller (typically but not necessarily computed from G_{nom}) that stabilises G_{nom}. Then K stabilises all systems in $\mathcal{D}$ (and it therefore stabilises G_0 with probability p) if*

$$\kappa_{WC}(G_{nom}, \mathcal{D}, \omega) < \kappa\left(G_{nom}(e^{j\omega}), \frac{-1}{K(e^{j\omega})}\right) \quad \forall\omega \tag{5.35}$$

In particular (Min-Max version), K stabilises all models in $\mathcal{D}$ (and it therefore stabilises G_0 with probability p) if

$$\delta_{WC}(G_{nom}, \mathcal{D}) < b_{G_{nom},K} \tag{5.36}$$

where $b_{G_{nom},K}$ is the generalised stability margin of the nominal closed-loop system, defined by (2.11) or (2.12).

Proof. This is an immediate consequence of Propositions 2.8 and 2.9 and of Definitions 5.2 and 5.3. □

► **Example 5.4.** Consider again the uncertainty regions $\mathcal{D}_{ol}$ and $\mathcal{D}_{cl}$, the nominal model $G_{nom}(z)$, and the controller $K_{id}(z)$ of Examples 5.1 and 5.2. $G_{nom}(z)$ is used to compute a new PID controller $K(z) = \frac{0.3469z^2-0.3757z+0.1148}{z(z-1)}$ offering a faster response than K_{id} in closed loop. The various chordal distances and worst-case chordal distances are computed respectively as in Section 2.5 and Theorem 5.1, and plotted in Figure 5.4, where it can be seen that the sufficient conditions for stability of Theorem 5.2 are fulfilled with both $\mathcal{D}_{ol}$ and $\mathcal{D}_{cl}$, *i.e.*, the new controller K stabilises all systems in $\mathcal{D}_{ol}$ and $\mathcal{D}_{cl}$. ◀

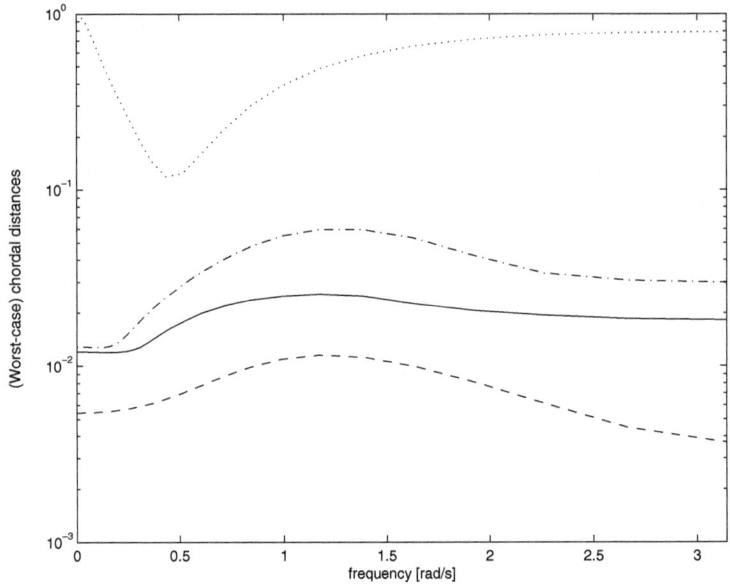

Figure 5.4. $\kappa_{WC}(G_{nom}, \mathcal{D}_{ol}, \omega)$ $(-\cdot)$, $\kappa_{WC}(G_{nom}, \mathcal{D}_{cl}, \omega)$ $(—)$, $\kappa(G_{nom}(e^{j\omega}), G_0(e^{j\omega}))$ $(--)$, $\kappa\left(G_{nom}(e^{j\omega}), \frac{-1}{K(e^{j\omega})}\right)$ $(\cdots)$

The worst-case ν-gap is essentially a control-oriented measure of the size of $\mathcal{D}$. In other words, it is a tool for validating the experimental design under which the validation is carried out. If two prediction-error uncertainty sets $\mathcal{D}_1$ and $\mathcal{D}_2$ obtained from two different validation experiments are such that $\delta_{WC}(G_{nom}, \mathcal{D}_1) < \delta_{WC}(G_{nom}, \mathcal{D}_2)$, then the set of G_{nom}-based controllers that robustly stabilise $\mathcal{D}_2$ is contained within the set of controllers that robustly stabilise $\mathcal{D}_1$. Hence, the smaller $\delta_{WC}(G_{nom}, \mathcal{D})$, the larger the set of controllers that stabilise $\mathcal{D}$. It is therefore important to design the validation experiment in such a way as to obtain a set $\mathcal{D}$ with a worst-case ν-gap (with respect to G_{nom}) that is as small as possible. Observe that, since a major property of the ν-gap is that its resolution is largest around the cross-over frequency of the plant, we expect that closed-loop validation experiments will deliver prediction-error uncertainty sets with smaller worst-case ν-gap than open-loop validation experiments.

5.3.2 Controller Validation for Stability

The stability conditions of Theorem 5.2 are only sufficient and may be quite conservative and lead to reject a controller that, in fact, would stabilise all systems in $\mathcal{D}$. The following theorem (only valid in the SISO case) gives a necessary and sufficient condition for a given controller $K(z)$ to robustly stabilise all models in a prediction-error uncertainty set $\mathcal{D}$.

Theorem 5.3. (Bombois *et al.*, 2001*b*) *Consider a generic prediction-error uncertainty set* $\mathcal{D}$ *of the form (5.27) and a controller* $K(z) = \frac{X(z)}{Y(z)}$ *(where* $X(z)$ *and* $Y(z)$ *represent the numerator and denominator polynomials of* $K(z)$*, respectively) that stabilises the centre of that set,* $G(z, \hat{\theta})$*. Then all models in* $\mathcal{D}$ *are stabilised by* $K(z)$ *if and only if*

$$\max_{\omega} \mu_{\phi}\big(M_{\mathcal{D}}^{K}(e^{j\omega})\big) \leq 1 \tag{5.37}$$

where

- $M_{\mathcal{D}}^{K}(z)$ *is defined as*

$$M_{\mathcal{D}}^{K}(z) = \frac{-\left(Z_D(z) + \dfrac{X(z)\big(Z_N(z) - \Xi(z)Z_D(z)\big)}{Y(z) + \Xi(z)X(z)}\right)T^{-1}}{1 + \left(Z_D(z) + \dfrac{X(z)\big(Z_N(z) - \Xi(z)Z_D(z)\big)}{Y(z) + \Xi(z)X(z)}\right)\hat{\theta}} \tag{5.38}$$

- T *is a square root of the matrix* $R = \frac{\hat{P}_{\theta}^{-1}}{\chi_p^2(q)}$ *defining* $\mathcal{U}$ *in (5.28):* $R = T^T T$*;*
- $\phi = T(\theta - \hat{\theta})$*, whereby* $\theta \in \mathcal{U} \Leftrightarrow \|\phi\|_2 < 1$*;*
- $\mu_{\phi}\big(M_{\mathcal{D}}^{K}(e^{j\omega})\big)$ *is called the* stability radius *of the loop* $(M_{\mathcal{D}}^{K}(z), \phi)$*. For a real vector* ϕ *it is computed as follows:*

$$\mu_{\phi}\big(M_{\mathcal{D}}^{K}(e^{j\omega})\big) = \begin{cases} \sqrt{\|Re(M_{\mathcal{D}}^{K})\|_2^2 - \dfrac{\big(Re(M_{\mathcal{D}}^{K})Im(M_{\mathcal{D}}^{K})^T\big)^2}{\|Im(M_{\mathcal{D}}^{K})\|_2^2}} & \text{if } Im(M_{\mathcal{D}}^{K}) \neq 0 \\ \|M_{\mathcal{D}}^{K}\|_2 & \text{if } Im(M_{\mathcal{D}}^{K}) = 0 \end{cases} \tag{5.39}$$

According to this theorem, the uncertainty region used in Subsection 5.3.1 to assess the quality of the nominal model before control design can also be used afterwards to verify if the designed controller will stabilise the actual process. The condition of this theorem is necessary and sufficient, hence no conservatism is introduced.

► **Example 5.5.** In the continuation of Example 5.4, we now compute the stability radii $\mu_{\phi}\big(M_{\mathcal{D}_{ol}}^{K}(e^{j\omega})\big)$ and $\mu_{\phi}\big(M_{\mathcal{D}_{cl}}^{K}(e^{j\omega})\big)$ associated respectively to the uncertainty regions $\mathcal{D}_{ol}$ and $\mathcal{D}_{cl}$ and the new controller $K(z)$. The condition of Theorem 5.3 is fulfilled in both cases (both stability radii are smaller than 1 at all frequencies, as shown in Figure 5.5), hence K is guaranteed to stabilise all systems in $\mathcal{D}_{ol}$ and $\mathcal{D}_{cl}$. This is of course not a surprise since this guarantee had already been obtained in Example 5.4 *via* a sufficient only (hence more conservative) test.

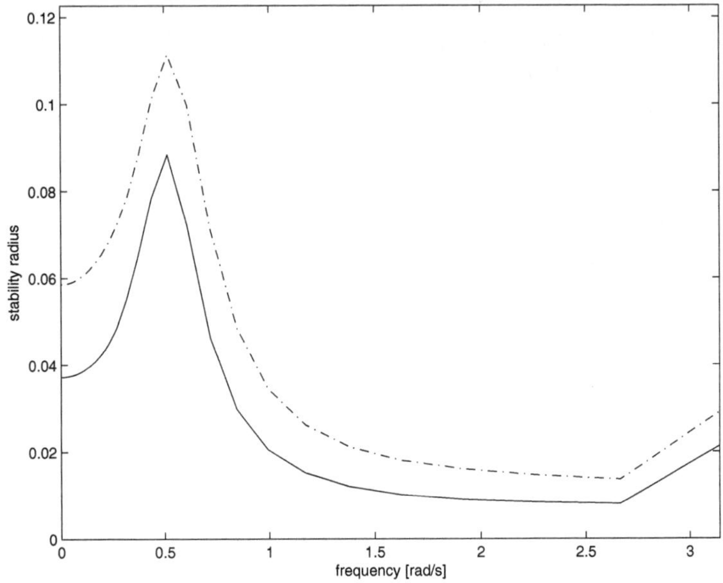

Figure 5.5. $\mu_\phi(M^K_{\mathcal{D}_{ol}}(e^{j\omega}))$ (–·), $\mu_\phi(M^K_{\mathcal{D}_{cl}}(e^{j\omega}))$ (—)

Observe again (Figure 5.5) that $\mathcal{D}_{cl}$ is a less conservative uncertainty region than $\mathcal{D}_{ol}$ since its stability radius is smaller. ◀

5.3.3 Controller Validation for Performance

When designing a controller for a plant G_0, one generally does not only aim to stabilise this plant: there is most often a performance issue (sometimes, the plant is even already stable in open loop). It is thus important to be able to assess the worst-case performance a controller will achieve over a prediction-error uncertainty set, since this is a lower bound of the performance the controller will achieve with the real process. Here, we consider a worst-case performance criterion J_{WC} expressed in the frequency domain by weighting the four entries of the closed-loop transfer matrix individually:

$$J_{WC}(\mathcal{D}, K, W_l, W_r, \omega) = \max_{G\in\mathcal{D}} \bar{\sigma}\left(\overbrace{\begin{bmatrix} W_{l1} & 0 \\ 0 & W_{l2} \end{bmatrix}}^{W_l(e^{j\omega})} \times T(G(e^{j\omega}), K(e^{j\omega})) \times \overbrace{\begin{bmatrix} W_{r1} & 0 \\ 0 & W_{r2} \end{bmatrix}}^{W_r(e^{j\omega})}\right) \quad (5.40)$$

where $W_{l1}(z), W_{l2}(z)$ and $W_{r1}(z), W_{r2}(z)$ are frequency weighting filters that allow one to define specific performance levels and where $\bar{\sigma}(\cdot)$ denotes the largest singular value. The frequency function J_{WC} defines a template. Any function that is derived from J_{WC} can of course also be handled, such as $\|J_{WC}\|_\infty$, for example. J_{WC} can be computed by solving an optimisation problem involving LMI constraints.

Theorem 5.4. (Bombois *et al.*, 2001*b*) *Consider an uncertainty region $\mathcal{D}$ defined in (5.27) and a controller $K(z) = \frac{X(z)}{Y(z)}$ where $X(z)$ and $Y(z)$ represent the numerator and denominator polynomials of $K(z)$, respectively. Then, at each frequency ω, the criterion function (5.40) is obtained as*

$$J_{WC}(\mathcal{D}, K, W_l, W_r, \omega) = \sqrt{\rho_{opt}(\omega)} \tag{5.41}$$

where $\rho_{opt}(\omega)$ is the optimal value of ρ in the convex optimisation problem involving LMI constraints, evaluated at frequency ω:

$$\begin{array}{ll} \textit{Minimise} & \rho \\ \textit{over} & \rho, \tau \\ \textit{subject to} & \tau \geq 0 \textit{ and} \end{array}$$

$$\begin{bmatrix} Re(a_{11}) & Re(a_{12}) \\ Re(a_{12}^\star) & Re(a_{22}) \end{bmatrix} - \tau \begin{bmatrix} R & -R\hat{\theta} \\ (-R\hat{\theta})^T & \hat{\theta}^T R\hat{\theta} - 1 \end{bmatrix} \tag{5.42}$$

where

$$\begin{aligned} a_{11} &= (Z_N^\star W_{l1}^\star W_{l1} Z_N + Z_D^\star W_{l2}^\star W_{l2} Z_D) - \rho(Q\Psi^\star\Psi) \\ a_{12} &= (Z_N^\star W_{l1}^\star W_{l1} \Xi + W_{l2}^\star W_{l2} Z_D^\star) - \rho(Q\Psi^\star(Y + \Xi X)) \\ a_{22} &= (\Xi^\star W_{l1}^\star W_{l1} \Xi + W_{l2}^\star W_{l2}) - \rho(Q(Y + \Xi X)^\star(Y + \Xi X)) \\ Q &= 1/(X^\star W_{r1}^\star W_{r1} X + Y^\star W_{r2}^\star W_{r2} Y) \\ \Psi &= X Z_N + Y_Z D \\ R &= \frac{\hat{P}_\theta^{-1}}{\chi_p^2(q)} \end{aligned}$$

▶ **Remark.** A crucial feature that makes the computation of $J_{WC}(\mathcal{D}, K, W_l, W_r, \omega)$ possible is the rank one property of the matrix $T(G, K)$; this property does not hold, in general, for MIMO systems. ◀

▶ **Example 5.6.** In the continuation of Example 5.5, we now compute the worst-case performance obtained in four different cases:

1. $W_{l1} = W_{r1} = 1$, $W_{l2} = W_{r2} = 0$,
   i.e., $J_{WC}(\mathcal{D}, K, W_l, W_r, \omega) = \max_{G\in\mathcal{D}} \left| T_{11}\left(G(e^{j\omega}), K(e^{j\omega})\right) \right|$,
2. $W_{l1} = W_{r2} = 1$, $W_{l2} = W_{r1} = 0$,
   i.e., $J_{WC}(\mathcal{D}, K, W_l, W_r, \omega) = \max_{G\in\mathcal{D}} \left| T_{12}\left(G(e^{j\omega}), K(e^{j\omega})\right) \right|$,

3. $W_{l2} = W_{r1} = 1,\ W_{l1} = W_{r2} = 0,$
   i.e., $J_{WC}(\mathcal{D},\ K,\ W_l,\ W_r,\ \omega) = \max_{G\in\mathcal{D}}\big|T_{21}\big(G(e^{j\omega}),\ K(e^{j\omega})\big)\big|,$
4. $W_{l2} = W_{r2} = 1,\ W_{l1} = W_{r1} = 0,$
   i.e., $J_{WC}(\mathcal{D},\ K,\ W_l,\ W_r,\ \omega) = \max_{G\in\mathcal{D}}\big|T_{22}\big(G(e^{j\omega}),\ K(e^{j\omega})\big)\big|,$

where the T_{ij}'s are the four entries of the standard closed-loop transfer matrix.

The worst case performance of the new controller $K(z)$ over the uncertainty sets $\mathcal{D}_{ol}$ and $\mathcal{D}_{cl}$ is compared to its actual performance with $G_0(z)$ in the frequency responses magnitude diagrams of Figure 5.6 (exceptionally presented with linear rather than logarithmic axes for clarity). Observe that $J_{WC}(\mathcal{D}_{cl},\ K,\ W_l,\ W_r,\ \omega)$ is always closer to the actual transfer function than $J_{WC}(\mathcal{D}_{ol},\ K,\ W_l,\ W_r,\ \omega)$. More realistic examples will be given in Section 5.5. ◀

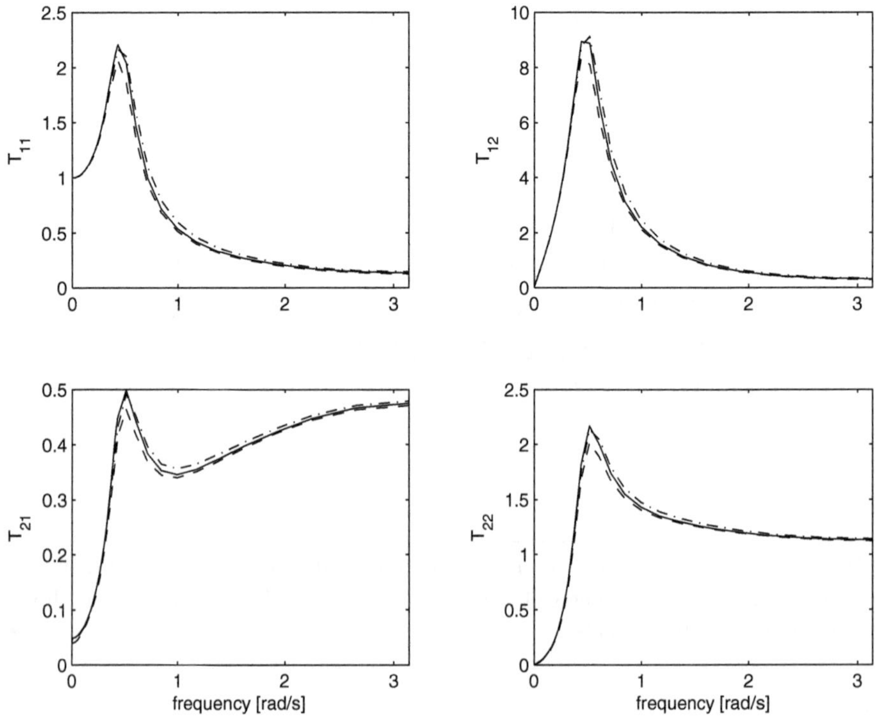

Figure 5.6. Magnitude of the frequency responses of $J_{WC}(\mathcal{D}_{ol},\ K,\ W_l,\ W_r,\ \omega)$ (−·), $J_{WC}(\mathcal{D}_{cl},\ K,\ W_l,\ W_r,\ \omega)$ (—) and $T_{ij}\big(G_0(e^{j\omega}),\ K(e^{j\omega})\big)$ (−−)

5.4 The Effect of Overmodelling on the Variance of Estimated Transfer Functions

A major issue in the model validation procedure proposed in this chapter is the computation of an *unbiased* estimate of the system. The estimated covariance matrix $\hat{P}_\theta$ of the parameters of the obtained model is then used to compute an uncertainty region $\mathcal{D}$ in transfer function space that is guaranteed to contain the true system (at some probability). In this section, we try to give some insight into the choice of the number of parameters and of the model structure, such as to avoid prohibitive covariance increase due to overmodelling.

5.4.1 The Effect of Superfluous Poles and Zeroes

Since bias must absolutely be avoided for the model $\hat{G}$ that is computed during the validation procedure, it may be very appealing to voluntarily choose an overparametrised model structure. The question that arises, in this case, is *'what is the effect of superfluous poles and/or zeroes on the frequency distribution of the uncertainty region?'*.

It is very easy to show that every parameter in excess results in an increased Cramér-Rao lower bound for the estimate of the other parameters. Indeed, let

$$F(\theta_1, \theta_2) = \begin{bmatrix} F_{11} & F_{12} \\ F_{21} & F_{22} \end{bmatrix} = \frac{1}{\lambda_0} \sum_{t=1}^{N} E\psi(t, \theta_0)\psi^T(t, \theta_0) \tag{5.44}$$

denote the Fisher information matrix of the overparametrised model (θ_1 denotes the parameter vector of the minimal unbiased model, which we call the full-order model, while θ_2 represents the extra parameters). The covariance matrix of the overparametrised model parameter vector is thus always lower-bounded as follows:

$$P_{\left[\begin{smallmatrix}\theta_1\\ \theta_2\end{smallmatrix}\right]} \triangleq \begin{bmatrix} P^{over}_{\theta_1} & P^{over}_{\theta_1,\theta_2} \\ P^{over}_{\theta_2,\theta_1} & P^{over}_{\theta_2} \end{bmatrix} \geq F^{-1}(\theta_1, \theta_2) \triangleq \begin{bmatrix} M^{over}_{11} & M^{over}_{12} \\ M^{over}_{21} & M^{over}_{22} \end{bmatrix} \tag{5.45}$$

where

$$M^{over}_{11} = F^{-1}_{11} + F^{-1}_{11}F_{12}\left(F_{22} - F_{21}F^{-1}_{11}F_{12}\right)^{-1}F_{21}F^{-1}_{11} \tag{5.46a}$$

$$M^{over}_{12} = -F^{-1}_{11}F_{12}\left(F_{22} - F_{21}F^{-1}_{11}F_{12}\right)^{-1} \tag{5.46b}$$

$$M^{over}_{21} = -\left(F_{22} - F_{21}F^{-1}_{11}F_{12}\right)^{-1}F_{21}F^{-1}_{11} \tag{5.46c}$$

$$M^{over}_{22} = \left(F_{22} - F_{21}F^{-1}_{11}F_{12}\right)^{-1} \tag{5.46d}$$

while that of the full-order model parameter vector is subject to

$$P^{full}_{\theta_1} \geq F^{-1}_{11} \triangleq M^{full}_{11} \tag{5.47}$$

It follows that the covariance matrix $P^{over}_{\theta_1}$ of the subset θ_1 of parameters, in the case of an overparametrised model, is subject to a higher lower bound than the covariance matrix $P_{\theta_1}^{full}$ of the parameters of a full-order model:

$$P_{\theta_1}^{over} \geq M_{11}^{over} = F_{11}^{-1} + F_{11}^{-1} F_{12} M_{22}^{over} F_{21} F_{11}^{-1} \geq M_{11}^{full} = F_{11}^{-1} \qquad (5.48)$$

This shows that adding superfluous parameters to an unbiased model increases its uncertainty. The asymptotic expressions for the variance in transfer function space (see Subsection 2.6.5, § B.) yield the same conclusion, since the number of parameters n is a factor of the asymptotic covariance:

$$\operatorname{cov}\big(\hat{G}(e^{j\omega})\big) \approx \frac{n}{N}\frac{\phi_v(\omega)}{\phi_u(\omega)} \qquad \text{in open loop and} \qquad (5.49)$$

$$\operatorname{cov}\big(\hat{G}(e^{j\omega})\big) \approx \frac{n}{N}\frac{\phi_v(\omega)}{\phi_u^r(\omega)} \qquad \text{in closed loop.} \qquad (5.50)$$

These expressions mean that the uncertainty on the parametric model is proportional to that on the nonparametric estimate, but that the parametric model has an averaging effect over frequency which results in a reduction of the effect of the noise by a factor $\frac{n}{N}$. However, these results are only valid asymptotically, *i.e.*, for n and N tending to infinity. Although it is very obvious that any additional parameter will increase the variance of the estimate, one can expect that this increase will not be uniform over the frequency range.

Using the results of Subsection 2.6.5, § A., with $(G(z, \theta), H(z, \theta))$ in an unbiased model structure $\mathcal{M}$, we find that the following asymptotic property holds:

$$\hat{\theta} \to \theta^* = \arg\min_{\theta \in D_\theta} \bar{V}(\theta) \qquad \text{w.p. 1 as } N \to \infty \qquad (5.51)$$

where $\bar{V}(\theta) = \bar{E}\left|\varepsilon(t, \theta)\right|^2 = \frac{1}{2\pi}\int_{-\pi}^{\pi} \phi_\varepsilon(\omega)d\omega$, *i.e.*,

$$\bar{V}(\theta) = \frac{1}{2\pi}\int_{-\pi}^{\pi} \left|G_0 - G(\theta)\right|^2 \left|H^{-1}(\theta)\right|^2 \phi_u d\omega + \frac{1}{2\pi}\int_{-\pi}^{\pi} \left|H^{-1}(\theta)\right|^2 \phi_v d\omega \qquad (5.52)$$

in the open-loop case and

$$\bar{V}(\theta) = \frac{1}{2\pi}\int_{-\pi}^{\pi} \left|G_0 - G(\theta)\right|^2 \left|H^{-1}(\theta)\right|^2 \phi_u^r d\omega + \frac{1}{2\pi}\int_{-\pi}^{\pi} \left|1 + G(\theta)K_{id}\right|^2 \left|S(G_0, K_{id})\right|^2 \left|H^{-1}(\theta)\right|^2 \phi_v d\omega \qquad (5.53)$$

with

$$\phi_u^r = |S(G_0, K_{id})|^2 \phi_{r_2} + |S(G_0, K_{id})|^2 |K_{id}|^2 \phi_{r_1} \qquad (5.54)$$

in the closed-loop case.

Imagine for instance that a pole is added to the unbiased structure $\mathcal{M}$. Since $\mathcal{S} \in \mathcal{M}$ without this additional pole, it is clear that this pole is not useful and that its estimate will tend to a value such that the impact on the cost function $\bar{V}(\theta)$ remains small. Hence, this pole will tend towards frequencies where the

SNR is small (open-loop case) or where the sensitivity function $S(G_0, K_{id})$ is small (closed-loop case), *i.e.*, typically outside the frequency band of interest. In the closed-loop case, its impact on the overall model variance will remain small at frequencies where the sensitivity function is large, *i.e.*, around the cross-over frequency and the frequency where the gain margin is determined.

On the contrary, if both a pole and a zero are added, they will tend to cancel. As this cancellation can happen at any frequency, this will probably result in a uniform increase in variance along the frequency axis.

► **Example 5.7.** Let us consider the following ARX 'true' system:

$$G_0(z) = \frac{1+0.5z^{-1}}{1-0.1z^{-1}}z^{-1} \qquad H_0(z) = \frac{1}{1-0.1z^{-1}}$$

This system is connected to the following feedback controller:

$$K_{id}(z) = \frac{0.15z^{-1}}{1-z^{-1}}$$

and excited using Gaussian white noise with zero mean and unit variance for $r_2(t)$ (see Figure 4.1). The white noise $e(t)$ acting on the system has zero mean and variance 100. 1000 input-output data samples are collected and used to identify five unbiased models using a direct closed-loop identification method. These models have the following structures, respectively:

1. ARX[1,2,1], which has the exact structure of (G_0, H_0);
2. ARX[5,2,1], which has four additional poles (total: 4 parameters in excess);
3. ARX[1,6,1], which has four additional zeroes (total: 4 parameters in excess);
4. ARX[3,4,1], which has two additional poles and two additional zeroes (total: 4 parameters in excess);
5. ARX[9,2,1], which has height additional poles (total: 8 parameters in excess).

100 Monte-Carlo runs of this experiment are carried out. During each of these runs, the five models mentioned above are identified, hence 100 realisations of each model are available. These realisations are used to compute the experimental standard deviations of the estimates, which are plotted in Figure 5.7 together with the closed-loop system sensitivity function $S(G_0, K_{id})$.

For the ARX[3,4,1] model, the increase in variance with respect to the full-order model ARX[1,2,1] is, as expected, nearly uniform over frequency due to the pole-zero cancellations. For the two models with additional poles only, ARX[5,2,1] and ARX[9,2,1], the variance remains relatively small where $S(G_0, K_{id})$ is large, but is much larger at low frequencies, where the sensitivity function is small. As a result, pole-zero cancellations will typically lead to more conservative uncertainty regions than additional poles only when the purpose is to make robust control design; however, since pole-zero cancellations are also easier to detect, this should not be a major problem. Additional zeroes, as in the ARX[1,6,1] model, also lead to a quasi-uniform increase in variance, but with a lesser effect than pole-zero cancellations. ◄

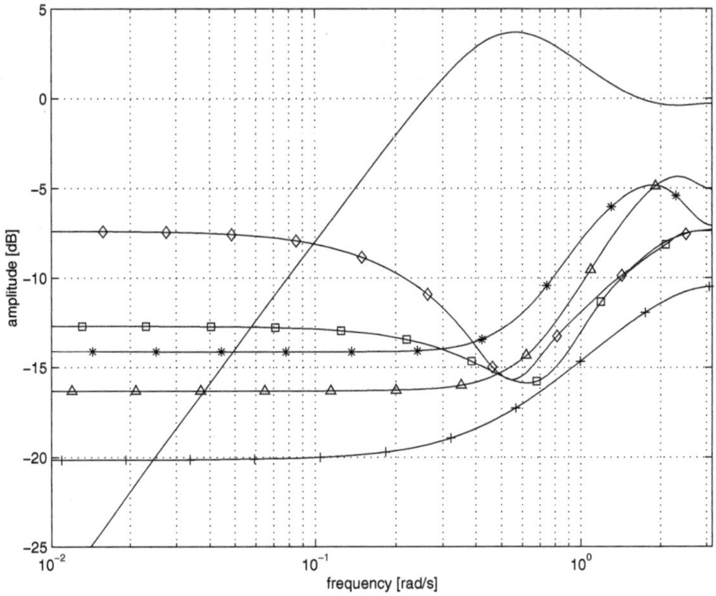

Figure 5.7. Magnitude of the frequency responses of $S(G_0,\ K_{id})$ (—) and of the experimental standard deviations of ARX[1,2,1] (+), ARX[5,2,1] (□), ARX[1,6,1] (△), ARX[3,4,1] (∗) and ARX[9,2,1] (◊)

5.4.2 The Choice of a Model Structure

It has been argued in (Ljung, 1997) that, in the case of model validation by identification of a model-error model, the elements of the cross-correlation function $R_{\varepsilon u}(\tau)$ between the residuals $\varepsilon(t)$ and a finite vector of past inputs $u(t)$ could be taken as an FIR estimate of the model error $\Delta G = G_0 - G_{nom}$.

More generally, any transfer function in $\mathcal{RH}_\infty$ can be approximated arbitrarily well by a FIR model, provided a sufficient number of Markov parameters (impulse response coefficients) are identified. However, this FIR model will generally have a larger number of parameters than an infinite impulse response model, which will generally also result in an increase of the variance.

► **Example 5.8.** Consider the same system $(G_0,\ H_0)$ as in Example 5.7. Figure 5.8 shows that G_0 can be almost perfectly approximated by a FIR model with 4 nonzero parameters and that the noise process $v(t) = H_0(z)e(t)$ can be represented by a moving-average process with 4 nonzero parameters (among which the first one is 1, since H_0 is monic). As a result, the system can be represented by an ARMAX[0,4,3,1] model (*i.e.*, an ARMAX model with denominator polynomial of order 0, input-output numerator polynomial of order 3 with 4 free parameters, noise numerator polynomial of order 3 with 3 free parameters and system delay 1).

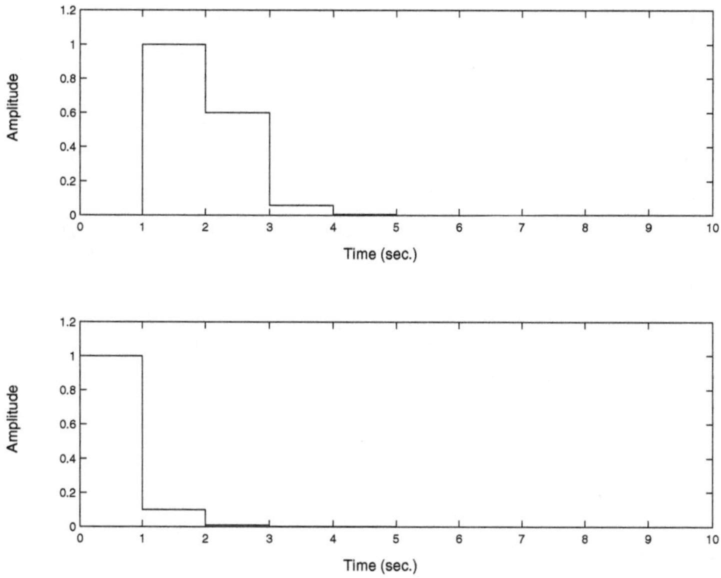

Figure 5.8. Impulse responses of G_0 (top) and H_0 (bottom)

During the Monte-Carlo runs described in Example 5.7, the following models were also identified:

1. ARMAX[0,4,3,1], which is the lowest-order unbiased FIR representation of (G_0, H_0). It has the same number of parameters as the overparametrised ARX[5,2,1] obtained in Example 5.7.
2. ARMAX[0,8,3,1], which has four additional parameters. It has the same number of parameters as the overparametrised ARX[9,2,1] obtained in Example 5.7.

The experimental standard deviations of these models are compared to those of ARX[1,2,1], ARX[5,2,1] and ARX[9,2,1] in Figure 5.9. Clearly, the ARMAX[0,4,3,1] (FIR) model has a larger variance error attached to it than the ARX[1,2,1] model, although both represent the true system without overmodelling. Furthermore, the variance of the ARMAX[0,4,3,1] model has a distribution over frequency that is very different from that of the ARX[5,2,1] model, which has a smaller variance where the closed-loop sensitivityfunction is large. The same conclusion holds when comparing the ARMAX[0,8,3,1] and ARX[9,2,1] models. ◄

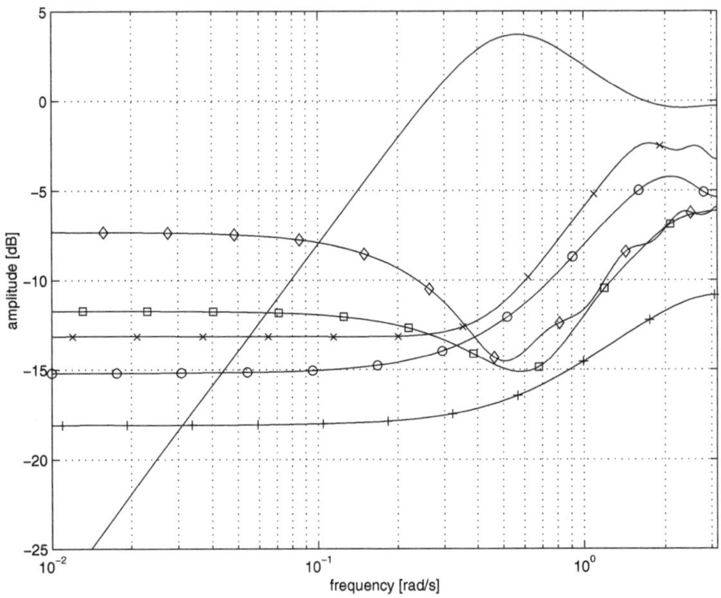

Figure 5.9. Magnitude of the frequency responses of $S(G_0, K_{id})$ (—) and of the experimental standard deviations of ARX[1,2,1] (+), ARX[5,2,1] (□), ARX[9,2,1] (◇), ARMAX[0,4,3,1] (○) and ARMAX[0,8,3,1] (×)

5.5 Case Studies

In order to illustrate further the theory developed in this chapter, we now propose two realistic case studies where the complete validation procedure is applied and where open-loop and closed-loop results are compared.

5.5.1 Case Study I: Flexible Transmission System

Our first illustration is representative of control applications of flexible mechanical structures subject to step disturbances. It is based on a benchmark proposed by I.D. Landau in a special edition of the *European Journal of Control* in 1995 (Landau *et al.*, 1995*b*). We illustrate the fact that, for control-oriented validation, a closed-loop experiment with a flat reference spectrum typically yields a smaller worst-case ν-gap and an uncertainty region that leads to less conservative control designs than an open-loop experiment with a similar flat input spectrum.

A. Problem setting. We consider as 'true system' the half-load model of the flexible transmission system used in the benchmark study (Landau *et al.*,

1995*b*):

$$G_0(z) = \frac{B(z)}{A(z)} z^{-3} = \frac{(0.10276 + 0.18123z^{-1})z^{-3}}{1 - 1.99185z^{-1} + 2.20265z^{-2} - 1.84083z^{-3} + 0.89413z^{-4}} \tag{5.55}$$

The output of the system is subject to step disturbances filtered through

$$H_0(z) = \frac{1}{A(z)} \tag{5.56}$$

One of the major specifications of the benchmark was that the designed controllers should ensure *rejection of step output disturbances filtered by* $\frac{1}{A(z)}$ *within 1.2 s* (the sampling period being 0.05 s). This means that the plant can be seen as a nonstandard ARX system described by

$$A(z)y(t) = B(z)z^{-3}u(t) + p(t) \tag{5.57}$$

where $u(t)$ is the input of the plant, $y(t)$ its output and $p(t)$ a sequence of step disturbances with zero mean and variance σ_p^2. From a statistical point of view, $p(t)$ is equivalent, up to second moments, to white noise filtered through an integrator, *i.e.*, to $\frac{1}{\Delta(z)}e(t)$ where $\Delta(z) = 1 - z^{-1}$ and where $e(t)$ is a sequence of Gaussian white noise with zero mean and appropriate variance. Hence, a standard ARX description of the plant is given by

$$A(z)\underbrace{\Delta(z)y(t)}_{y_f(t)} = B(z)z^{-3}\underbrace{\Delta(z)u(t)}_{u_f(t)} + e(t) \tag{5.58}$$

and the standard prediction-error identification algorithm for ARX models can be used to identify the system, provided the data is prefiltered by $L(z) = \Delta(z)$, *i.e.*, is differentiated.

Our objective is to validate (or invalidate) the following model and controller on the basis of data collected on G_0:

$$G_{nom}(z) = \frac{B_{nom}(z)}{A_{nom}(z)} z^{-3} = \frac{(0.0990 + 0.1822z^{-1})z^{-3}}{1 - 1.9878z^{-1} + 2.2004z^{-2} - 1.8456z^{-3} + 0.9038z^{-4}} \tag{5.59}$$

and

$$K(z) = \frac{0.0355 + 0.0181z^{-1}}{1 - z^{-1}} \times \frac{18.8379 - 43.4538z^{-1} + 26.4126z^{-2}}{1 + 0.6489z^{-1} + 0.1478z^{-2}} \times \frac{0.5626 - 0.7492z^{-1} + 0.3248z^{-2}}{1 - 1.4998z^{-1} + 0.6379z^{-2}} \times \frac{1.3571 - 1.0741z^{-1} + 0.4702z^{-2}}{1 - 0.6308z^{-1} + 0.3840z^{-2}} \times \frac{1.0461 + 0.5633z^{-2}}{1 + 0.4564z^{-1} + 0.1530z^{-2}} \tag{5.60}$$

This controller, obtained in (Nordin and Gutman, 1995), is the one that satisfied the largest number of specification of the benchmark. It also has a very good performance when applied to our nominal model G_{nom}. Of course, for the purpose of illustrating the theory, we pretend not to know if this controller stabilises the true system.

B. Open-loop validation experiment. We first performed an open-loop validation experiment. The input signal $u_{ol}(t)$ applied to the true system $G_0(z)$ was chosen as a PRBS with variance $\sigma^2_{u_{ol}} = 0.1$ and flat spectrum. The output step disturbances $p(t)$ were simulated as a random binary sequence with variance $\sigma_p^2 = 0.01$ and cut-off frequency at 0.05 times the Nyquist frequency; that is, the mean length of the steps of $p(t)$ is about twenty times longer than that of the steps of $u_{ol}(t)$ while the amplitude of the steps of $p(t)$ is $\sqrt{10}$ times smaller than that of $u_{ol}(t)$. The spectra and a realisation of $u_{ol}(t)$ and $p(t)$ are shown in Figure 5.10. $p(t)$ was filtered by $\frac{1}{A(z)}$ and added to the output of the system. 256 data samples $\{u_{ol}(1), y_{ol}(1), \ldots, u_{ol}(256), y_{ol}(256)\}$ were measured and a model $(\hat{G}_{ol}, \hat{H}_{ol})$ with the same ARX[4,2,3] structure as (G_0, H_0) was identified after prefiltering these data by $\Delta(z)$.

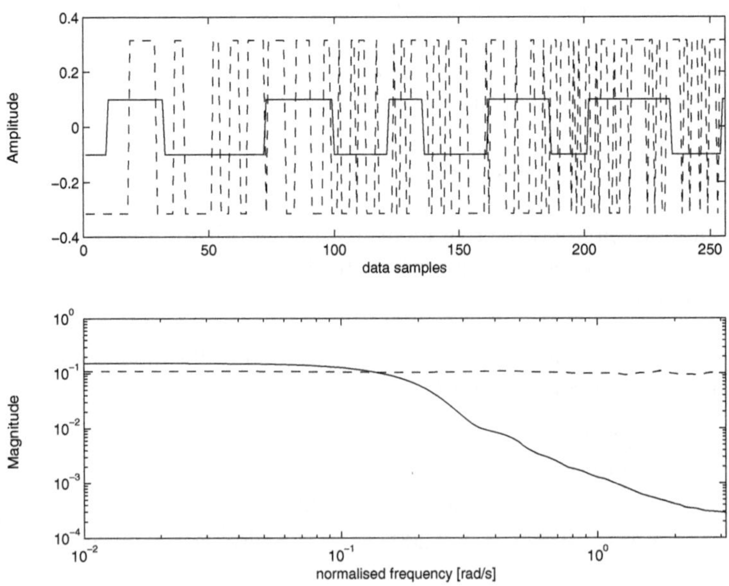

Figure 5.10. Open-loop validation. Top: realisations of $u_{ol}(t)$ (−−) and $p(t)$ (—). Bottom: $\phi_{u_{ol}}(\omega)$ (−−) and $\phi_p(\omega)$ (—)

The obtained model is

$$\hat{G}_{ol}(z) = \frac{(0.1052 + 0.1774z^{-1})z^{-3}}{1 - 1.997z^{-1} + 2.23z^{-2} - 1.876z^{-3} + 0.9039z^{-4}} \tag{5.61}$$

and the estimated covariance matrix of the parameter vector is

$$\hat{P}_{\theta_{ol}} = 0.001 \times \begin{bmatrix} 0.2034 & -0.2970 & 0.2411 & -0.1150 & -0.0139 & -0.0027 \\ -0.2970 & 0.5735 & -0.5136 & 0.2397 & 0.0119 & -0.0064 \\ 0.2411 & -0.5136 & 0.5725 & -0.2962 & -0.0130 & 0.0008 \\ -0.1150 & 0.2397 & -0.2962 & 0.2013 & 0.0094 & 0.0020 \\ -0.0139 & 0.0119 & -0.0130 & 0.0094 & 0.0392 & 0.0126 \\ -0.0027 & -0.0064 & 0.0008 & 0.0020 & 0.0126 & 0.0391 \end{bmatrix} \quad (5.62)$$

The following observations can be made:

$$(\theta_{nom} - \hat{\theta}_{ol})^T \hat{P}_{\theta_{ol}}^{-1} (\theta_{nom} - \hat{\theta}_{ol}) = 9.7168 < \chi^2_{0.95}(6) = 12.6 \quad (5.63)$$

and

$$(\theta_0 - \hat{\theta}_{ol})^T \hat{P}_{\theta_{ol}}^{-1} (\theta_0 - \hat{\theta}_{ol}) = 5.7555 < \chi^2_{0.95}(6) = 12.6 \quad (5.64)$$

where θ_0 and θ_{nom} denote respectively the vectors of coefficients of $G_0(z)$ and $G_{nom}(z)$ in the same order as those of $\hat{G}_{ol}(z)$. This means that both G_0 and G_{nom} lie within the 95% uncertainty region $\mathcal{D}_{ol}$ around $\hat{G}_{ol}$, hence G_{nom} is validated at this probability level.

C. Closed-loop validation experiment. Here, the validation experiment is carried out in closed loop with a suboptimal controller that was obtained using a combined pole placement/sensitivity function shaping method in (Landau *et al.*, 1995*a*). Its feedback part is described by

$$K_{id}(z) = \frac{0.401602 - 1.079378z^{-1} + 0.284895z^{-2} + 1.358224z^{-3}}{1 - 1.031142z^{-1} - 0.995182z^{-2} + 0.752086z^{-3}} \frac{-0.986549z^{-4} - 0.271961z^{-5} + 0.306937z^{-6}}{+0.710744z^{-4} - 0.242297z^{-5} - 0.194209z^{-6}} \quad (5.65)$$

It also has a feed-forward part that we shall not consider here, since we want an excitation signal with a flat spectrum. According to the benchmark specifications, the performance of this controller is not as good as that of the to-be-validated controller K. We use it here only for the validation experiment.

The closed-loop system (G_0, K_{id}) was excited by means of a reference signal $r(t)$ injected at the input of G_0 ($r_2(t)$ in Figure 4.1). In order to establish a fair comparison with the results obtained in open-loop validation, $r(t)$ was chosen as a PRBS with variance $\sigma_r^2 = 0.5541$, while $p(t)$ was the same as in the open-loop validation experiment. With these settings, the total output variance was the same in closed loop as in open loop: $\sigma^2_{y_{cl}} = \sigma^2_{y_{ol}} = 0.8932$.

We used the direct closed-loop identification method[1] to compute a prediction-error uncertainty set $\mathcal{D}_{cl}$. Once again, 256 data samples $\{u_{cl}(1), y_{cl}(1), \ldots, u_{cl}(256), y_{cl}(256)\}$ were measured and a model $(\hat{G}_{cl}, \hat{H}_{cl})$ with the same ARX[4,2,3] structure as (G_0, H_0) was identified after prefiltering these data by $\Delta(z)$.

[1] K_{id} has unstable poles and nonminimum-phase zeroes. Therefore, the indirect closed-loop identification method could not be used for validation, as it would systematically have delivered a model $\hat{G}_{cl}$ that would have been destabilised by K_{id}.

The obtained model is

$$\hat{G}_{cl}(z) = \frac{(0.1016 + 0.1782z^{-1})z^{-3}}{1 - 1.986z^{-1} + 2.187z^{-2} - 1.824z^{-3} + 0.8897z^{-4}} \tag{5.66}$$

and the covariance matrix of its parameter vector is

$$\hat{P}_{\theta_{cl}} = 0.001 \times \begin{bmatrix} 0.0840 & -0.1166 & 0.1024 & -0.0532 & -0.0062 & -0.0027 \\ -0.1166 & 0.2145 & -0.1966 & 0.1009 & 0.0057 & 0.0008 \\ 0.1024 & -0.1966 & 0.2184 & -0.1197 & -0.0074 & -0.0041 \\ -0.0532 & 0.1009 & -0.1197 & 0.0853 & 0.0063 & 0.0037 \\ -0.0062 & 0.0057 & -0.0074 & 0.0063 & 0.0064 & 0.0021 \\ -0.0027 & 0.0008 & -0.0041 & 0.0037 & 0.0021 & 0.0061 \end{bmatrix} \tag{5.67}$$

As in the open-loop case, the nominal model is validated with respect to the 95% uncertainty region $\mathcal{D}_{cl}$ around $\hat{G}_{cl}$, which also contains the true system:

$$(\theta_{nom} - \hat{\theta}_{cl})^T \hat{P}_{\theta_{cl}}^{-1} (\theta_{nom} - \hat{\theta}_{cl}) = 9.7590 < \chi^2_{0.95}(6) = 12.6 \tag{5.68}$$

$$(\theta_0 - \hat{\theta}_{cl})^T \hat{P}_{\theta_{cl}}^{-1} (\theta_0 - \hat{\theta}_{cl}) = 4.7050 < \chi^2_{0.95}(6) = 12.6 \tag{5.69}$$

D. Model set validation for control: comparison of the validated sets. We computed the worst-case chordal distance and the worst-case ν-gap between G_{nom} and all models in $\mathcal{D}_{ol}$ and, respectively, $\mathcal{D}_{cl}$, and we found that

$$\begin{aligned} \delta_{WC}(G_{nom}, \mathcal{D}_{ol}) = 0.3010 > b_{G_{nom},K} &= 0.2381 \\ &> \delta_{WC}(G_{nom}, \mathcal{D}_{cl}) = 0.1423 \end{aligned} \tag{5.70}$$

Hence, according to the Min-Max version of Theorem 5.2, the open-loop validation experiment does not guarantee that K stabilises the true system G_0, while the closed-loop validation experiment does. However, since

$$\kappa_{WC}(G_{nom}, \mathcal{D}_{ol}, \omega) < \kappa\left(G_{nom}(e^{j\omega}), \frac{-1}{K(e^{j\omega})}\right) \tag{5.71}$$

and

$$\kappa_{WC}(G_{nom}, \mathcal{D}_{cl}, \omega) < \kappa\left(G_{nom}(e^{j\omega}), \frac{-1}{K(e^{j\omega})}\right) \tag{5.72}$$

at all frequencies (see Figure 5.11), both validation experiments lead to the conclusion that K stabilises G_0. This illustrates the conservatism of the Min-Max test. Yet, the uncertainty region $\mathcal{D}_{cl}$ delivered by the closed-loop experiment should be preferred to $\mathcal{D}_{ol}$ delivered by the open-loop experiment, since it leads to the largest set of stabilising controllers.

E. Controller validation for stability. Using the open-loop and closed-loop prediction-error uncertainty sets $\mathcal{D}_{ol}$ and $\mathcal{D}_{cl}$, we built the dynamic matrices $M^K_{\mathcal{D}_{ol}}(e^{j\omega})$ and $M^K_{\mathcal{D}_{cl}}(e^{j\omega})$ corresponding to the candidate controller K and we computed their stability radii according to Theorem 5.3. Their respective values are

$$\max_{\omega} \mu_{\phi}(M^K_{\mathcal{D}_{ol}}(e^{j\omega})) = 0.3227 < 1 \tag{5.73}$$

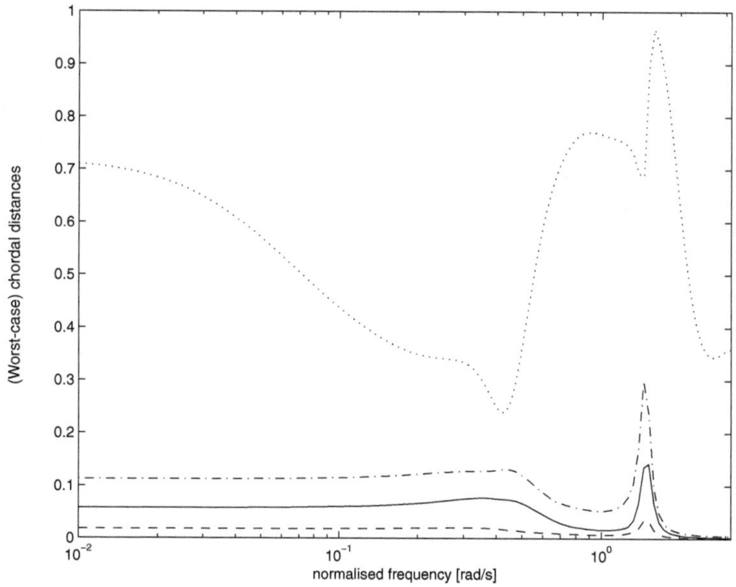

Figure 5.11. $\kappa_{WC}(G_{nom}, \mathcal{D}_{ol}, \omega)$ (–·), $\kappa_{WC}(G_{nom}, \mathcal{D}_{cl}, \omega)$ (—), $\kappa(G_{nom}(e^{j\omega}), G_0(e^{j\omega}))$ (−−), $\kappa\left(G_{nom}(e^{j\omega}), \frac{-1}{K(e^{j\omega})}\right)$ (···)

and

$$\max_{\omega} \mu_{\phi}\big(M^{K}_{\mathcal{D}_{cl}}(e^{j\omega})\big) = 0.2384 < 1 \tag{5.74}$$

According to this test, K stabilises all systems contained in both uncertainty sets $\mathcal{D}_{ol}$ and $\mathcal{D}_{cl}$, hence both the open-loop and the closed-loop approach validate the controller. This was predictable since such a conclusion had already been obtained from the more conservative test based on the worst-case chordal distance.

F. Controller validation for performance. A major specification of the benchmark was that the designed controller should ensure a maximum value smaller than 6 dB for the closed-loop sensitivity function (which is equivalent to a modulus margin greater than 0.5). Our performance criterion is thus

$$J_{WC}(\mathcal{D}, K, \omega) = \max_{G(e^{j\omega}, \theta)\in\mathcal{D}} \left| \frac{1}{1 + G(e^{j\omega}, \theta)K(e^{j\omega})} \right| \tag{5.75}$$

obtained by setting $W_{l1} = W_{r1} = 0$ and $W_{l2} = W_{r2} = 1$ in (5.40) and the controller is validated if $\max_{\omega} J_{WC}(\mathcal{D}, K, \omega) < 6$ dB.

For the open-loop and closed-loop prediction-error uncertainty sets $\mathcal{D}_{ol}$ and $\mathcal{D}_{cl}$, we have respectively

$$\max_{\omega} J_{WC}(\mathcal{D}_{ol}, K, \omega) = 5.97 \text{ dB} \tag{5.76}$$

and

$$\max_{\omega} J_{WC}(\mathcal{D}_{cl}, K, \omega) = 5.00 \text{ dB} \tag{5.77}$$

which means that the open-loop validation procedure nearly leads to rejection of the controller, while the closed-loop validation procedure leads to its acceptance. The Bode diagrams of the worst-case and achieved sensitivity functions are depicted in Figure 5.12.

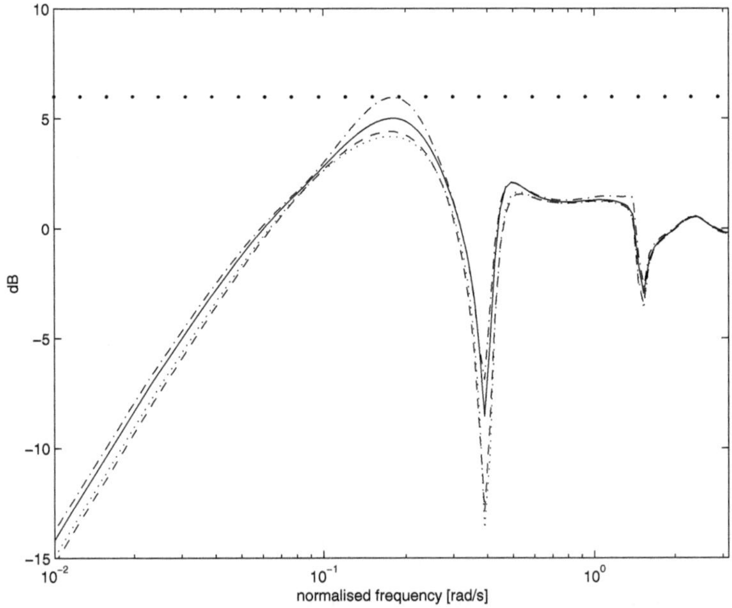

Figure 5.12. $J_{WC}(\mathcal{D}_{ol}, K, \omega)$ (–·), $J_{WC}(\mathcal{D}_{cl}, K, \omega)$ (—), $|S(G_0(e^{j\omega}), K(e^{j\omega}))|$ (– –), $|S(G_{nom}(e^{j\omega}), K(e^{j\omega}))|$ (···), 6 dB limit (·)

5.5.2 Case Study II: Ferrosilicon Production Process

We now present an application that is representative of industrial process control applications, in which the control objective is the rejection of stochastic disturbances.

A. Problem setting. The plant model and the controllers used in this simulation example are taken from (Ingason and Jonsson, 1998). Ferrosilicon is a two-phase mixture of the chemical compound $FeSi_2$ and the element silicon. The balance between silicon and iron is regulated around 76% of the total weight in silicon, 22% in iron and 2% in aluminium by adjusting the input of raw materials to the furnace. Those are charged batch-wise to the top of the furnace, each batch consisting of a fixed amount of quartz (SiO_2) and a variable quantity of coal/coke (C) and iron oxide (Fe_2O_3). The quantity of coal/coke which is burned in the furnace does not influence the silicon ratio in the mixture, hence the control input is the amount of iron oxide.

In (Ingason and Jonsson, 1998), the following ARX model was obtained by identification from data collected on the process:

$$y(t) + ay(t-1) = bu(t-1) + d + e(t) \tag{5.78}$$

where the sampling period is one day, $y(t)$ is the percentage of silicon in the mixture that must be regulated around 76%, $u(t)$ is the quantity of iron oxide in the raw materials (expressed in kilogrammes), d is a constant and $e(t)$ is a stochastic disturbance. The nominal values of the parameters and their standard deviations are:

$$a = -0.44 \qquad b = -0.0028 \qquad d = 46.1 \tag{5.79a}$$

$$\sigma_a = 0.07 \qquad \sigma_b = 0.001 \qquad \sigma_d = 5.6 \tag{5.79b}$$

Here, for the sake of illustrating our theory, we make the assumption that the true system is

$$G_0(z) = \frac{b_0 z^{-1}}{1 + a_0 z^{-1}} = \frac{-0.0032z^{-1}}{1 - 0.34z^{-1}} \tag{5.80a}$$

$$H_0(z) = \frac{1}{1 + a_0 z^{-1}} = \frac{1}{1 - 0.34z^{-1}} \tag{5.80b}$$

for which[2] $a_0 \in [a - 2\sigma_a,\, a + 2\sigma_a]$ and $b_0 \in [b - 2\sigma_b,\, b + 2\sigma_b]$, while the nominal model is the one resulting from the identification experiment:

$$G_{nom}(z) = \frac{bz^{-1}}{1 + az^{-1}} = \frac{-0.0028z^{-1}}{1 - 0.44z^{-1}} \tag{5.81a}$$

$$H_{nom}(z) = \frac{1}{1 + az^{-1}} = \frac{1}{1 - 0.44z^{-1}} \tag{5.81b}$$

This model is used to compute a Generalised Predictive Controller minimising the cost function

$$J_{GPC}(u) = E\left(\sum_{i=1}^{2}\big(y(t+i) - r(t+i)\big)^2 + \sum_{i=1}^{2}\lambda\big(\Delta(z)u(t+i-1)\big)^2\right) \tag{5.82}$$

where $\Delta(z) = 1 - z^{-1}$. The optimal control sequence is given by

$$u(t) = \begin{bmatrix}1 & 0\end{bmatrix}\left(H^T H + F^T \Lambda F\right)^{-1} \times \Big(H^T\left(w(t) - x(t)\right) - F^T \Lambda g(t)\Big) \tag{5.83}$$

[2]Since we have no access to the real plant, we have decided to randomly pick one system in the two-standard-deviation confidence interval around the nominal model and to use it as the true system.

where

$$H = \begin{bmatrix} b & 0 \\ -ab & b \end{bmatrix} \tag{5.84a}$$

$$F = \begin{bmatrix} 1 & 0 \\ -1 & 1 \end{bmatrix} \tag{5.84b}$$

$$x(t) = \begin{bmatrix} -ay(t) + d & a^2y(t) - ad + d \end{bmatrix}^T \tag{5.84c}$$

$$w(t) = \begin{bmatrix} r(t) & r(t+1) \end{bmatrix}^T \tag{5.84d}$$

$$g(t) = \begin{bmatrix} u(t-1) & 0 \end{bmatrix}^T \tag{5.84e}$$

$$\Lambda = \lambda I \tag{5.84f}$$

λ is a tuning parameter. The resulting controller is a two-degree-of-freedom controller $C_\lambda(z)$:

$$\begin{aligned} u(t) &= C_\lambda(z) \begin{bmatrix} r(t) \\ y(t) \end{bmatrix} \\ &= \begin{bmatrix} F_\lambda(z) & K_\lambda(z) \end{bmatrix} \begin{bmatrix} r(t) \\ y(t) \end{bmatrix} \end{aligned} \tag{5.85}$$

where

$$F_\lambda(z) = \frac{b^3 + 2b\lambda - ab\lambda}{\left(b^4 + 3b^2\lambda + a^2b^2\lambda + \lambda^2 - 2ab^2\lambda\right) - \left(b^2\lambda + \lambda^2\right)z^{-1}} \tag{5.86a}$$

and

$$K_\lambda(z) = \frac{ab^3 + ab\lambda - a^2b\lambda + a^3b\lambda}{\left(b^4 + 3b^2\lambda + a^2b^2\lambda + \lambda^2 - 2ab^2\lambda\right) - \left(b^2\lambda + \lambda^2\right)z^{-1}} \tag{5.86b}$$

Experiments carried out on the real process showed that, for a constant reference $r(t) = 76$ and for $\lambda = 0.0007$, a control input variance $\sigma_u^2 = 20$ was obtained (Ingason and Jonsson, 1998). This can only result from a white noise variance $\sigma_e^2 = 0.078$ if we assume that the true system is described by (5.80).

In this example, we shall compare open-loop and closed-loop experiments for the validation of the GPC controller $C_\lambda(z)$ with $\lambda = 0.0007$.

B. Open-loop validation experiment. The 'true plant' model (G_0, H_0) was excited with $u(t)$ chosen as a PRBS with variance $\sigma_{u_{ol}}^2 = 20$, which is the maximum input variance admissible for this process. The disturbance signal $e(t)$ was a Gaussian white noise sequence with variance $\sigma_e^2 = 0.078$, as explained above. The variance of the output was then $\sigma_{y_{ol}}^2 = 0.0884$. 300 input-output data samples were collected, corresponding approximately to one year of measurements. This data was used to identify an ARX model with exact structure

$$G(z,\theta) = \frac{\theta_2 z^{-1}}{1+\theta_1 z^{-1}} \tag{5.87a}$$

$$H(z,\theta) = \frac{1}{1+\theta_1 z^{-1}} \tag{5.87b}$$

We found

$$\hat{\theta}_{ol} = \begin{bmatrix} \hat{\theta}_1 \\ \hat{\theta}_2 \end{bmatrix} = \begin{bmatrix} -0.3763 \\ -0.0073 \end{bmatrix} \tag{5.88}$$

$$\hat{P}_{\theta_{ol}} = \begin{bmatrix} 2.8131 \times 10^{-3} & -1.2784 \times 10^{-5} \\ -1.2784 \times 10^{-5} & 1.4887 \times 10^{-5} \end{bmatrix} \tag{5.89}$$

and the nominal model is validated with respect to the 95% uncertainty region $\mathcal{D}_{ol}$ around $G(z,\hat{\theta}_{ol})$:

$$\left(\begin{bmatrix} a \\ b \end{bmatrix} - \hat{\theta}_{ol}\right)^T \hat{P}_{\theta_{ol}}^{-1} \left(\begin{bmatrix} a \\ b \end{bmatrix} - \hat{\theta}_{ol}\right) = 2.6151 < \chi^2_{0.95}(2) = 5.99 \tag{5.90}$$

Observe that $\mathcal{D}_{ol}$, whose probability of containing the true plant G_0 is $p = 95\%$, does effectively contain it:

$$\left(\begin{bmatrix} a_0 \\ b_0 \end{bmatrix} - \hat{\theta}_{ol}\right)^T \hat{P}_{\theta_{ol}}^{-1} \left(\begin{bmatrix} a_0 \\ b_0 \end{bmatrix} - \hat{\theta}_{ol}\right) = 1.6744 < \chi^2_{0.95}(2) = 5.99 \tag{5.91}$$

C. Closed-loop validation experiment. The closed-loop validation was performed with a suboptimal GPC controller obtained by setting $\lambda = 0.001$ in (5.86). Since this value of λ reduces the input variance (compared to the optimal one $\lambda = 0.0007$), it was possible to add a PRBS signal to the reference $r(t)$ without violating the constraint $\sigma_u^2 \leq 20$. The variance of $r(t)$ was chosen as $\sigma_r^2 = 0.014$, $e(t)$ having the same properties as during the open-loop validation experiment. The input and output variances were then, respectively, $\sigma^2_{u_{cl}} = 20$ and $\sigma^2_{y_{cl}} = 0.0880$. Observe that the input variance is the same as in open loop and that the output variance is very close to that of the open-loop experiment. Again, 300 input-output data samples were collected and used to identify an ARX model with the same structure as in open-loop validation (5.87), using a direct prediction-error method. We found

$$\hat{\theta}_{cl} = \begin{bmatrix} \hat{\theta}_1 \\ \hat{\theta}_2 \end{bmatrix} = \begin{bmatrix} -0.3575 \\ -0.0067 \end{bmatrix} \tag{5.92}$$

$$\hat{P}_{\theta_{cl}} = \begin{bmatrix} 2.8323 \times 10^{-3} & -8.7845 \times 10^{-6} \\ -8.7845 \times 10^{-6} & 6.2416 \times 10^{-6} \end{bmatrix} \tag{5.93}$$

The nominal model is validated with respect to the 95% uncertainty region $\mathcal{D}_{cl}$ around $G(z,\hat{\theta}_{cl})$:

$$\left(\begin{bmatrix} a \\ b \end{bmatrix} - \hat{\theta}_{cl}\right)^T \hat{P}_{\theta_{cl}}^{-1} \left(\begin{bmatrix} a \\ b \end{bmatrix} - \hat{\theta}_{cl}\right) = 4.5575 < \chi^2_{0.95}(2) = 5.99 \tag{5.94}$$

One can also verify that $\mathcal{D}_{cl}$ effectively contains the true G_0:

$$\left(\begin{bmatrix} a_0 \\ b_0 \end{bmatrix} - \hat{\theta}_{cl}\right)^T \hat{P}_{\theta_{cl}}^{-1} \left(\begin{bmatrix} a_0 \\ b_0 \end{bmatrix} - \hat{\theta}_{cl}\right) = 2.1611 < \chi^2_{0.95}(2) = 5.99 \tag{5.95}$$

D. Model set validation for control: comparison of the validated sets. Observe that the nominal model G_{nom} is closer to the centre of the validated set $\mathcal{D}_{ol}$ than to the centre of the validated set $\mathcal{D}_{cl}$, when measured by the normalised distance in *parameter space*: compare (5.90) with (5.94). However, when measured in the control-oriented measures κ_{WC} or δ_{WC}, which are computed in *transfer function space*, the conclusions are reversed. Figure 5.13 shows the worst-case chordal distance between G_{nom} and all models in $\mathcal{D}_{ol}$ and all models in $\mathcal{D}_{cl}$, respectively. Clearly, $\kappa_{WC}(G_{nom}, \mathcal{D}_{cl}, \omega)$ is smaller than $\kappa_{WC}(G_{nom}, \mathcal{D}_{ol}, \omega)$ at all frequencies and

$$\delta_{WC}(G_{nom}, \mathcal{D}_{cl}) = 0.0156 < \delta_{WC}(G_{nom}, \mathcal{D}_{ol}) = 0.0225 \tag{5.96}$$

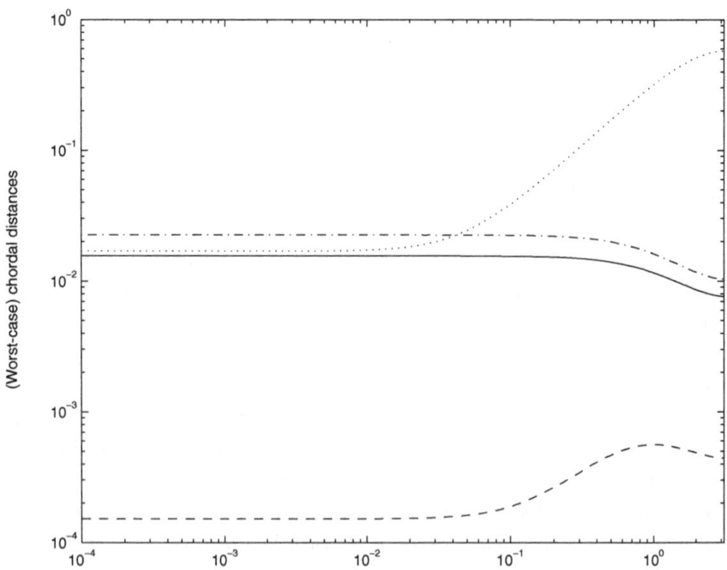

Figure 5.13. $\kappa_{WC}(G_{nom}, \mathcal{D}_{ol}, \omega)$ $(-\cdot)$, $\kappa_{WC}(G_{nom}, \mathcal{D}_{cl}, \omega)$ $(—)$, $\kappa(G_{nom}(e^{j\omega}), G_0(e^{j\omega}))$ $(--)$, $\kappa\left(G_{nom}(e^{j\omega}), \frac{-1}{K_{0.0007}(e^{j\omega})}\right)$ $(\cdots)$

We observe that the open-loop validation procedure does not guarantee that the GPC controller $K_{0.0007}$ will stabilise G_0, since

$$\kappa_{WC}(G_{nom}, \mathcal{D}_{ol}, \omega) > \kappa\left(G_{nom}(e^{j\omega}), \frac{-1}{K_{0.0007}(e^{j\omega})}\right) \tag{5.97}$$

at some frequencies, while the closed-loop validation procedure does guarantee such stability with G_0, since

$$\kappa_{WC}(G_{nom}, \mathcal{D}_{cl}, \omega) < \kappa\left(G_{nom}(e^{j\omega}), \frac{-1}{K_{0.0007}(e^{j\omega})}\right) \tag{5.98}$$

at all frequencies. The Min-Max version of this stability test yields the same conclusion, since

$$\begin{aligned}\delta_{WC}(G_{nom}, \mathcal{D}_{ol}) = 0.0225 > b_{G_{nom},K_{0.0007}} &= 0.0169 \\ > \delta_{WC}(G_{nom}, \mathcal{D}_{cl}) &= 0.0156 \end{aligned} \tag{5.99}$$

We conclude from this analysis that the uncertainty region $\mathcal{D}_{cl}$ delivered by the closed-loop experiment should be preferred to $\mathcal{D}_{ol}$ delivered by the open-loop experiment, since it leads to the largest set of stabilising controllers.

E. Controller validation for stability. Using the open-loop and closed-loop prediction-error uncertainty sets $\mathcal{D}_{ol}$ and $\mathcal{D}_{cl}$, we built the dynamic matrices $M_{\mathcal{D}_{ol}}^{K_{0.0007}}(e^{j\omega})$ and $M_{\mathcal{D}_{cl}}^{K_{0.0007}}(e^{j\omega})$ corresponding to the candidate controller $K_{0.0007}$ and we computed their stability radii. Their respective values are

$$\max_{\omega} \mu_{\phi}\big(M_{\mathcal{D}_{ol}}^{K_{0.0007}}(e^{j\omega})\big) = 0.6572 < 1 \tag{5.100}$$

$$\max_{\omega} \mu_{\phi}\big(M_{\mathcal{D}_{cl}}^{K_{0.0007}}(e^{j\omega})\big) = 0.2111 < 1 \tag{5.101}$$

According to this test, $K_{0.0007}$ stabilises all systems contained in both uncertainty sets $\mathcal{D}_{ol}$ and $\mathcal{D}_{cl}$, hence both the open-loop and the closed-loop approach validate the controller. This shows that the necessary and sufficient condition for closed-loop stability is less conservative than the sufficient condition based on the chordal distance or the ν-gap stability results. Observe that the stability radius obtained with $\mathcal{D}_{cl}$ is smaller than (and thus preferable to) that obtained with $\mathcal{D}_{ol}$.

F. Controller validation for performance. Since the control objective is to reject the noise $v(t) = H_0(z)e(t)$ which is essentially located at low frequencies ($H_0(e^{j\omega})$ is a first-order low-pass filter; see the thick dotted line in Figure 5.14), an appropriate performance specification in the frequency domain is that the sensitivity function $S(G_0, K) = \frac{1}{1+G_0K}$ must be small at low frequencies, since it filters $v(t)$ in closed loop. However, since $S(G_0, K)$ cannot be made arbitrarily small at all frequencies, a tradeoff must be made between reducing the noise at low frequencies and amplifying it at high frequencies. Actually, the ideal situation would be that $|H_0S(G_0, K)|$ be uniform over frequency, *i.e.*, that the noise be whitened by the closed-loop system dynamics.

A possible worst-case performance criterion is thus

$$J_{WC}^{(1)}(\mathcal{D}, K_{0.0007}, \omega) = \max_{G(e^{j\omega},\theta)\in\mathcal{D}} \left|\frac{1}{1 + G(e^{j\omega}, \theta)K_{0.0007}(e^{j\omega})}\right| \tag{5.102}$$

obtained by setting $W_{l1} = W_{r1} = 0$ and $W_{l2} = W_{r2} = 1$ in (5.40). The controller is validated if $J_{WC}^{(1)}$ remains at all frequencies below some template, which is high-pass since we want the worst-case sensitivity to be small in the frequency range where the noise is large. The amplitude Bode diagrams of the worst-case and achieved sensitivity functions are depicted in Figure 5.14.

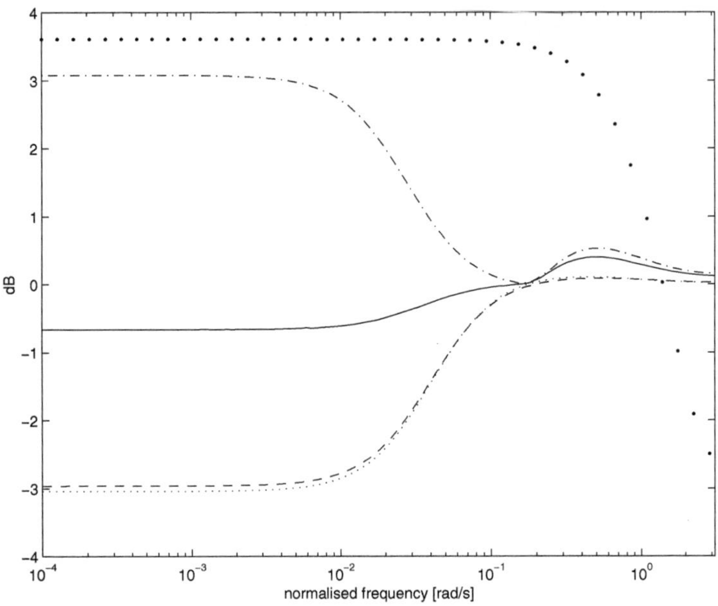

Figure 5.14. $J_{WC}^{(1)}(\mathcal{D}_{ol}, K_{0.0007}, \omega)$ $(-\cdot)$, $J_{WC}^{(1)}(\mathcal{D}_{cl}, K_{0.0007}, \omega)$ $(—)$, $|S(G_0(e^{j\omega}), K_{0.0007}(e^{j\omega}))|$ $(--)$, $|S(G_{nom}(e^{j\omega}), K_{0.0007}(e^{j\omega}))|$ $(\cdots)$, $|H_0(e^{j\omega})|$ $(\cdot)$

Clearly, for most reasonable templates, the controller is invalidated by the open-loop validation experiment yielding $\mathcal{D}_{ol}$, while it can be validated by the closed-loop experiment yielding $\mathcal{D}_{cl}$.

Now, observe that a noise model $H(z, \hat{\theta}) = \frac{1}{1+\hat{\theta}_1 z^{-1}}$, sharing its parameter with $G(z, \hat{\theta})$, is also identified during each validation experiment, due to the use of an ARX model structure. As a result, each θ in $\mathcal{U}_{ol}$ or $\mathcal{U}_{cl}$ defines not only a plant model $G(\theta)$ in $\mathcal{D}_{ol}$ or $\mathcal{D}_{cl}$, but also a noise model $H(\theta)$ associated with this plant model. An alternative performance criterion[3] could thus be

$$J_{WC}^{(2)}(\mathcal{U}, K_{0.0007}, \omega) = \max_{\theta \in \mathcal{U}} \left| \frac{H(e^{j\omega}, \theta)}{1 + G(e^{j\omega}, \theta) K_{0.0007}(e^{j\omega})} \right| \tag{5.103}$$

In this case, the distribution of $J_{WC}^{(2)}$ over frequency should be as close as possible to a uniform distribution and its amplitude should be as small as possible. As shown in Figure 5.15, the worst-case performance attached to the closed-loop uncertainty region is closer to the performance achieved by the

[3]This performance criterion is not one of the form (5.40). However, it only requires a slight modification of the LMI-based algorithm used to compute (5.40).

controller on the actual system, and to the uniform distribution we would like to obtain, than the worst-case performance attached to the open-loop uncertainty region.

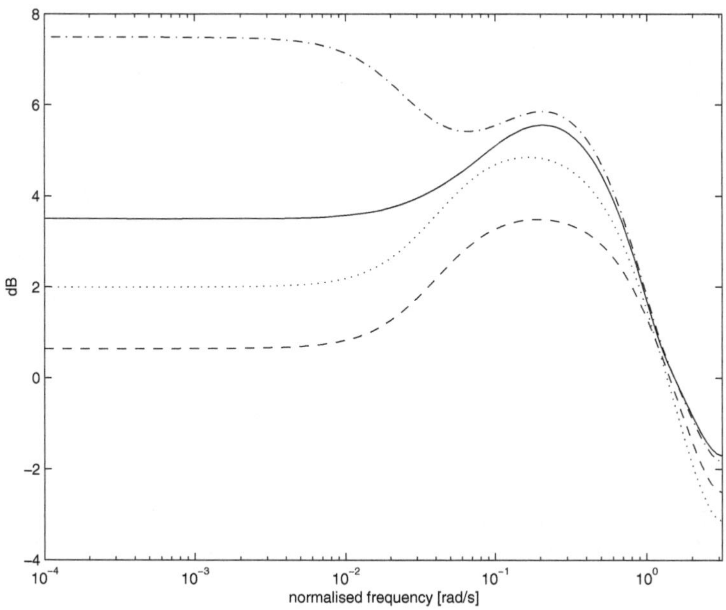

Figure 5.15. $J^{(2)}_{WC}(\mathcal{U}_{ol}, K_{0.0007}, \omega)$ (−·), $J^{(2)}_{WC}(\mathcal{U}_{cl}, K_{0.0007}, \omega)$ (—), $|H_0(e^{j\omega})S(G_0(e^{j\omega}), K_{0.0007}(e^{j\omega}))|$ (−−), $|H_{nom}(e^{j\omega})S(G_{nom}(e^{j\omega}), K_{0.0007}(e^{j\omega}))|$ $(\cdots)$

In practice, it may happen that the noise model is not estimated but given *a priori* and assumed to be exact. If we make the assumption that H_0 is known exactly, a third possible performance criterion is the following:

$$J^{(3)}_{WC}(\mathcal{D}, K_{0.0007}, H_0, \omega) = \max_{G(e^{j\omega},\theta)\in\mathcal{D}} \left| \frac{H_0(e^{j\omega})}{1 + G(e^{j\omega}, \theta)K_{0.0007}(e^{j\omega})} \right| \tag{5.104}$$

obtained by setting $W_{l1} = W_{r1} = 0$, $W_{l2} = H_0$ and $W_{r2} = 1$ in (5.40). The controller is then validated if the amplitude of $J^{(3)}_{WC}$ is smaller than that of H_0 at almost all frequencies, *i.e.*, the worst-case sensitivity reduces the actual noise at almost all frequencies. The results are depicted in Figure 5.16. Once again, the controller is validated when the worst-case performance criterion is computed over $\mathcal{D}_{cl}$, but not if it is computed over $\mathcal{D}_{ol}$.

The results we have obtained here confirm that the prediction-error uncertainty set $\mathcal{D}_{cl}$ obtained *via* a closed-loop experiment is to be preferred to $\mathcal{D}_{ol}$ obtained *via* an open-loop experiment, as the worst-case ν-gap test already indicated, when the purpose is to validate a given candidate controller.

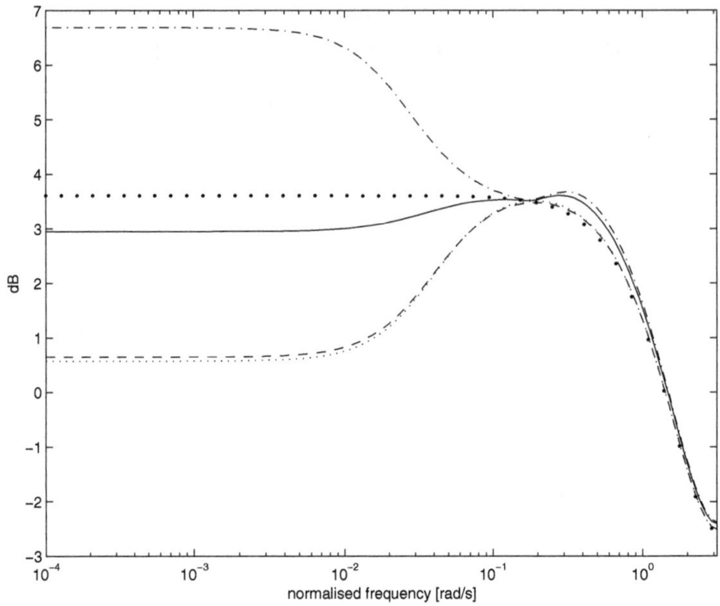

Figure 5.16. $J^{(3)}_{WC}(\mathcal{D}_{ol}, K_{0.0007}, H_0, \omega)$ (–·), $J^{(3)}_{WC}(\mathcal{D}_{cl}, K_{0.0007}, H_0, \omega)$ (—), $\left|H_0(e^{j\omega})S\big(G_0(e^{j\omega}), K_{0.0007}(e^{j\omega})\big)\right|$ (– –), $\left|H_0(e^{j\omega})S\big(G_{nom}(e^{j\omega}), K_{0.0007}(e^{j\omega})\big)\right|$ (···), $\left|H_0(e^{j\omega})\right|$ (·)

5.6 Summary of the Chapter

We have seen in Chapters 3 and 4 how the experiment carried out on the process should be designed to collect the best data for building a good model $\hat{G}$ for control design. Here, we have seen that the same data, collected using the same rules of good practice, can also be used

- before control design, to assess the quality of the identified model $\hat{G}$ (or that of any given model G_{nom}) with respect to the control design objective. This 'quality' is quantified and represented by a single number, the *worst-case* ν*-gap*, or by its frequency-by-frequency counterpart, the *worst-case chordal distance*;
- after control design, but prior to the implementation of the controller, to make sure that no significant loss of stability or performance will happen in comparison with those achieved with the nominal model.

The practical procedure consists of the following steps:

1. Collect data on the process, if possible in closed loop with a controller K_{id} that is not too different from the optimal, yet to be designed, one (hence, if

necessary, finely tune the possible current PID controllers before collecting the data). Use for instance a PRBS as reference signal to excite the system.

2. Use this data to identify an unbiased and, preferably, not overparametrised model $G(\hat{\theta})$ of the process, taking into account the recommendations of Chapter 4. The covariance matrix P_θ of the parameters must be estimated during the identification procedure; this is standard in the MATLAB® Identification Toolbox.
3. Use $\hat{P}_\theta$ to compute an uncertainty region $\mathcal{D}$ and the worst-case ν-gap between $G(\hat{\theta})$ (or any other model G_{nom} one would like to use for control design; in this case, it is necessary to ascertain that $G_{nom} \in \mathcal{D}$) and $\mathcal{D}$. The smaller this number, the better the uncertainty zone, hence it can be used to choose among different uncertainty sets produced by different validation experiments.
4. Design a controller K offering the required level of performance and stability when applied to the nominal model $G(\hat{\theta})$ or G_{nom}.
5. Verify the worst-case stability and performance of K over $\mathcal{D}$. If they are satisfactory, it can be implemented on the real process.

► **Remark.** The unbiasedness of the model $G(\hat{\theta})$ implies that its order is at least equal to that of G_0. If it is used for control design, it may be necessary to reduce its order beforehand (generally for numerical reasons) and/or to reduce the order of the designed controller afterwards (generally for implementation reasons). Techniques for performing these reductions in a way that tunes the introduced bias error in an optimal way are given in Chapter 6. The user can then come back to the present chapter to validate the reduced-order model or controller. ◄

CHAPTER 6

Control-oriented Model Reduction and Controller Reduction

6.1 Introduction

6.1.1 From High to Low Order for Implementation Reasons

In many practical applications, low-order controllers are preferred to high-order ones for numerous reasons (*e.g.*, ease of implementation or maintenance, numerical robustness, *etc.*). However, it often happens that the design has to be carried out on the basis of a high-order model of the plant obtained by methods like physical (white-box) modelling, finite-element modelling, linearisation of a high-order nonlinear simulator, *etc.* In this case, three approaches can be considered for the design of a low-order controller (Obinata and Anderson, 2000), as depicted in Figure 6.1. The direct approach, which would evidently be the most appealing *a priori*, suffers several major problems: it involves very complicated equations with no intuitive content which may present a very great number of solutions; solving these equations is far from straightforward; there is no commercial software available for the direct computation of a model-based low-order controller for a high-order plant. Examples of direct methods can be found in (Bernstein and Hyland, 1984) and (Gangsaas *et al.*, 1986). On the contrary, the two indirect methods, on which this chapter focuses, involve a step of model or controller reduction and a step of control design for which efficient, well understood and very popular tools and software packages are available. Examples of such tools are $\mathcal{H}_\infty$ and LQG for control design, and balanced truncation and Hankel-norm approximation for model or controller order reduction.

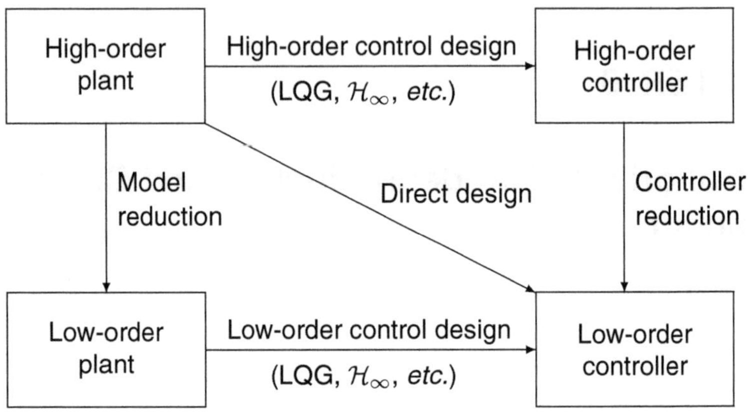

Figure 6.1. Three approaches for the design of a low-order controller for a high-order system

6.1.2 High-order Controllers

In industry, 95% of controllers are probably PID controllers, and the idea of high-order controllers may sound somewhat odd. However, our concern here is the design of high-performance, possibly multivariable controllers, which are obtained by minimising some performance cost function like an LQG or $\mathcal{H}_\infty$ criterion. These optimal design methods generally result in observer-based controllers with static state feedback, *i.e.*, in controllers of the form

$$\begin{aligned} \dot{\hat{x}}(t) &= A\hat{x}(t) + Bu(t) + L\big(\hat{y}(t) - y(t)\big) \\ \hat{y}(t) &= C\hat{x}(t) + Du(t) \\ u(t) &= F\hat{x}(t) \end{aligned} \tag{6.1}$$

Here, $u(t)$ and $y(t)$ are the plant input and output, respectively; $\hat{x}(t)$ and $\hat{y}(t)$ are the observer state and output, respectively; A, B, C and D are the state matrices of a minimal realisation of a model $G(s)$ of the plant and L and F are the observer gain matrix and the state feedback gain matrix, respectively. *grosso modo*, what makes a difference between methods like LQG, $\mathcal{H}_\infty$, *etc.*, is the way L and F are computed. In transfer function form, the controller can be expressed as

$$\begin{aligned} u(t) &= -K(s)y(t) \\ K(s) &= F\big(sI - (A + BF + LC + LDF)\big)^{-1}L \end{aligned} \tag{6.2}$$

As a result, the order of the controller is generally equal to that of the design model $G(s) = C(sI - A)^{-1}B + D$.

If loop-shaping filters $W_l(s)$ and $W_r(s)$ are used, which is common practice, the situation is even worse. Indeed, a controller $\tilde{K}(s)$ is first designed for the augmented model $\tilde{G}(s) = W_l(s)G(s)W_r(s)$ (which makes its order equal to that of $G(s)$ plus those of $W_l(s)$ and $W_r(s)$); then, these loop-shaping filters must be added to the controller, resulting in a controller $K(s) = W_r(s)\tilde{K}(s)W_l(s)$ that can be used with $G(s)$. As a result, the final controller has got the order of $G(s)$ plus twice those of $W_l(s)$ and $W_r(s)$. This shows the importance of using a low-order model for the design and/or to reduce the order of the controller afterwards.

6.1.3 Contents of this Chapter

In this chapter, the heuristic results of Chapter 3, which lead to the use of closed-loop identification to tune the modelling errors in a suitable way for control design, are extended to the case of model and controller reduction. They result in a closed-loop performance-based criterion for the reduction of a model or a controller. This criterion is further motivated by a 'small-gain theorem'-based result. Using the theory of Chapter 2, we show how this criterion leads to frequency-weighted balanced truncation of the plant or controller coprime factors with closed-loop performance induced weightings. Other more classical model and controller reduction techniques are reviewed: unweighted plant or controller normalised coprime-factor balanced truncation, frequency-weighted balanced truncation of the plant or controller coprime factors with closed-loop stability induced weightings, Closed-loop Balanced Truncation, and special methods for observer-based controllers.

Numerical examples and an industrial case study illustrate the theory.

The goal of this chapter is to give an answer to two practical questions:

Question 1 (experiment design): *Does the incorporation of a closed-loop performance objective in the reduction criterion yield significantly better results than open-loop model or controller reduction?*

Question 2 (methodology): *When an order reduction step is required, should it be carried out on the model (i.e., before control design), or on the controller (i.e., after controller design)?*

6.2 A Closed-loop Criterion for Model or Controller Order Reduction

As stated above, there are essentially two ways of designing a low-order controller for a high-order plant. In the first approach, the plant model order is reduced prior to designing the controller while, in the second approach, a high-order controller is first designed, its order being reduced afterwards. It should be obvious that, in both cases, closed-loop considerations must be taken into account in order to carry out the reduction step in the most optimal way.

If the controller order reduction approach is chosen, it is natural to require that the reduced-order controller behaves as much as possible like the original high-order one when connected to the same plant.

If the model order reduction approach is used, the same heuristic reasoning as in Chapter 3 can be made. Recall that it was advocated that performing the identification of a low-order model in closed loop is the best way of tuning the bias error when the final objective is control design, because this error will be small around the cross-over frequency of the present control loop, thereby allowing use of the model for the design of a new controller with a (slightly) increased bandwidth. This heuristic reasoning can be transposed to the case of model reduction, since the problem is the same in terms of modelling error: a bias error is introduced during the order reduction procedure and it is important to tune this error so as to make it small in the frequency range of interest for control design, *i.e.*, around the cross-over frequency of the control loop.

Thus, in both cases, the order reduction criterion should be related to closed-loop performance and/or stability. We suggest using the following performance degradation index:

$$J_{perf}(r) = \left\| T_o(G,\, K) - \hat{T}_o(r) \right\|_\infty \tag{6.3}$$

where $T_o(G, K)$ is defined in (2.3) and $\hat{T}_o(r)$ is the approximation of $T_o(G, K)$ that is obtained by replacing either the plant model G or the controller K by a low-order approximation of order r. ($T_i(G, K)$ could also be used in place of $T_o(G, K)$. This is discussed further in the text.) According to the chosen method, this criterion will thus be developed either as

$$J_{perf}(\hat{G}_r) = \left\| T_o(G_n,\, K) - T_o(\hat{G}_r,\, K) \right\|_\infty \tag{6.4}$$

where G_n is the high-order model (of order n), $\hat{G}_r$ its r-th order approximation and K the current to-be-replaced controller, in case of model order reduction, or as

$$J_{perf}(\hat{K}_r) = \left\| T_o(G_n,\, K_m) - T_o(G_n,\, \hat{K}_r) \right\|_\infty \tag{6.5}$$

where G_n is the high-order model (of order n), K_m the high-order controller (of order m) and $\hat{K}_r$ its r-th order approximation, in case of controller order reduction.

The choice of this criterion is further motivated by the 'small-gain theorem'-based result of Proposition 2.1:

$$J_{perf}(\hat{G}_r) = \left\| T_o(G_n,\, K) - T_o(\hat{G}_r,\, K) \right\|_\infty < \varepsilon \Longrightarrow \left| b_{G_n,K} - b_{\hat{G}_r,K} \right| < \varepsilon \tag{6.6}$$

provided the closed loops (G_n, K) and $(\hat{G}_r, K)$ are both internally stable. This means that the reduction criterion $J_{perf}(\hat{G}_r)$ is an upper bound of the stability robustness degradation measured by the generalised stability margin if the resulting low-order model $\hat{G}_r$ is stabilised by K, which is easy to check (but which may not be guaranteed by the reduction procedure). One might criticise the fact that the minimisation of $J_{perf}(\hat{G}_r)$ in (6.6) does not imply the minimisation of $|b_{G_n,K} - b_{\hat{G}_r,K}|$, hence our approach may seem somewhat conservative. Recall, however, that our primary concern is $J_{perf}(\hat{G}_r)$ and not $|b_{G_n,K} - b_{\hat{G}_r,K}|$; (6.6) only tells us that it is not a bad choice regarding robust stability, provided nominal stability is preserved. Of course, the dual result holds if the reduction is applied to the controller:

$$J_{perf}(\hat{K}_r) = \left\| T_o(G_n, K_m) - T_o(G_n, \hat{K}_r) \right\|_\infty < \varepsilon \implies \left| b_{G_n,K_m} - b_{G_n,\hat{K}_r} \right| < \varepsilon \quad (6.7)$$

provided the closed loops (G_n, K_m) and $(G_n, \hat{K}_r)$ are both internally stable.

Finally, observe that it is easy to put more weight on some inputs and/or outputs of $T_o(G, K)$ in (6.3) and/or over some frequency range, by using input and output static weights or frequency weighting filters W_{in} and W_{out}. The criterion then becomes

$$J_{perf}(r) = \left\| W_{out}\big(T_o(G, K) - \hat{T}_o(r)\big) W_{in} \right\|_\infty \quad (6.8)$$

For instance, a stable filter W_{in} can be used to model the spectral contents of the exogenous signals $r_1(t)$ and $r_2(t)$ entering the closed-loop system, while a static matrix W_{out} can be used to penalise only the degradation of some output signals of interest.

► **Remark.** If a two-degree-of-freedom controller $C = \begin{bmatrix} F & K \end{bmatrix}$ is used, *i.e.*, if the control input signal of the plant is defined as

$$u(t) = Fr_2(t) - Ky(t) \quad (6.9)$$

it is possible to use the same reduction criterion by observing that the closed-loop transfer matrix can be expressed as $T_o\left(\left[\begin{smallmatrix} 0 \\ G \end{smallmatrix}\right], \begin{bmatrix} F & K \end{bmatrix}\right)$ where the number of outputs of the 0 block is equal to the number of inputs of F. The reduction procedure will then deliver an approximation $\left[\begin{smallmatrix} 0 \\ \hat{G}_r \end{smallmatrix}\right]$ of $\left[\begin{smallmatrix} 0 \\ G_n \end{smallmatrix}\right]$ in case of model order reduction, or an approximation $\hat{C}_r = \begin{bmatrix} \hat{F}_r & \hat{K}_r \end{bmatrix}$ of C_m in case of controller order reduction. When one wants to reduce the order of K only, without altering F, which might not have the same state as K (meaning that the order of C is the sum of the respective orders of K and F, which is likely to occur), the standard definition of the closed-loop transfer matrix T_o can be used and F can be used in the weighting filter W_{in} in order to take its shaping effect on the spectral density of $r_2(t)$ into account. See Example 2.3.

In a similar way, if a disturbance model H is given, *i.e.*, if the plant output is defined by

$$y(t) = Gu(t) + Hd(t) \quad (6.10)$$

where $d(t)$ is a vector signal containing all measured and/or unmeasured disturbances, then the transfer matrix can be expressed as $T_o\left(\begin{bmatrix} H & G \end{bmatrix}, \begin{bmatrix} 0 \\ K \end{bmatrix}\right)$, where a number of zero outputs corresponding to the number of entries of $d(t)$ is added to K. The reduction procedure will then deliver an approximation $\hat{R}_r = \begin{bmatrix} \hat{H}_r & \hat{G}_r \end{bmatrix}$ of $R_n = \begin{bmatrix} H & G \end{bmatrix}$ in case of model order reduction, or an approximation $\hat{K}_r$ of K_m in case of controller order reduction. Observe that G and H may not have the same state, meaning that the order of R_n can be significantly higher than that of G alone. In this case, it can be more efficient to reduce the order of G alone. This can be done by using the following closed-loop transfer matrix in the criterion: $T_o\left(\begin{bmatrix} I & G \end{bmatrix}, \begin{bmatrix} 0 \\ K \end{bmatrix}\right)$, and by incorporating H into W_{in}. Indeed, the order of $\begin{bmatrix} I & G \end{bmatrix}$ is equal to that of G and the reduced-oder approximation will then be obtained as $\begin{bmatrix} I & \hat{G}_r \end{bmatrix}$.

A third issue is when some plant outputs are measured but not controlled, *i.e.*, when the controlled outputs are given by $y(t) = Gu(t) + v(t)$ while the measured but uncontrolled outputs are given by $\bar{y}(t) = \bar{G}u(t) + w(t)$. In this case, the closed-loop transfer matrix can be expressed as $T_o\left(\begin{bmatrix} \bar{G} \\ G \end{bmatrix}, \begin{bmatrix} 0 & K \end{bmatrix}\right)$ where the number of inputs of the 0 block is equal to the number of outputs of $\bar{G}$. In general, the order of $\begin{bmatrix} \bar{G} \\ G \end{bmatrix}$ will be the same as that of G since $y(t)$ and $\bar{y}(t)$ are outputs of the same plant. This is typical of an LFT representation of the closed loop (see Section 2.3), where $z(t)$ contains all controlled outputs while $h(t)$ may contain other measured outputs. The reduction procedure will deliver either $\begin{bmatrix} \hat{\bar{G}}_r \\ \hat{G}_r \end{bmatrix}$ or $\begin{bmatrix} 0 & \hat{K}_r \end{bmatrix}$.

These three cases can be combined and they boil down to recasting an LFT-based representation of the closed-loop system into the standard description based on the T_o transfer matrix. See also Section 2.3. ◄

6.3 Choice of the Reduction Method

The reduction method used in this chapter is (frequency-weighted) balanced truncation. The motivation for using balanced truncation is the fact that it is a purely algebraic method, very easy to implement and therefore, very popular in common practice. Furthermore, it allows one to choose the order of the reduced-order transfer function *a priori* by inspection of the Hankel singular values. One might criticise the fact that it does not minimise the approximation error in any norm and that an upper bound on the approximation error is only available in the $\mathcal{H}_\infty$ norm in the continuous-time case. However, it should be noted that the task of minimising $\|P_n - \hat{P}_r\|_\infty$ or $\|P_n - \hat{P}_r\|_2$ over all $\hat{P}_r$ of order r (here, P_n is either the high-order plant model or controller to be approximated) is very complicated. No very reliable solution exists and most algorithms are iterative, time consuming and deliver suboptimal solutions corresponding to local minima. The problem is even worse in case of a multivariable system and/or if frequency weightings are used.

Since this method can only be applied to a stable system, it is necessary to adapt it if P_n is unstable. Two alternatives can be considered:

1. Separate the stable and unstable parts of P_n. Reduce the stable part, keep the antistable part intact.
2. Reduce the *coprime factors* of P_n instead of P_n itself.

The second solution, initially proposed in (Meyer, 1988, 1990), is generally better because all modes of the system are taken into account during the reduction. The application of the first alternative is totally inefficient if P_n has many unstable modes. Furthermore, the reduction criterion (6.3) can easily be expressed in terms of the plant and controller coprime factors, allowing recasting of the *a priori* complicated problem of approximating G_n or K_m in closed loop by minimising (6.3) into a much simpler problem of frequency-weighted balanced truncation applied to the coprime factors of G_n or K_m. We shall also see that other choices of frequency weightings exist, not induced by (6.3), that have different closed-loop interpretations.

6.4 Model Order Reduction

Here, we assume that a high-order LTI plant model G_n is available (the way it was obtained is irrelevant; it could be, for instance, the result of some linearisation step performed on a simulator, or that of a black-box identification experiment). We consider its reduction to order $r < n$. The resulting low-order model is represented by $\hat{G}_r$.

We shall describe five different approaches to model reduction, all based on (frequency-weighted) balanced truncation of the coprime factors of the plant:

- the first approach is simply the unweighted reduction of these coprime factors (Subsection 6.4.1). It is shown that, although the controller is not taken into account, this approach is somewhat closed-loop oriented if the cross-over frequency of the open-loop plant G_n is close to that of the open-loop system with the controller, $G_n K$;
- the second approach is based on our original closed-loop criterion (6.4) and aims to preserve the performance of the closed-loop system (Subsection 6.4.2);
- the third approach also takes the available controller into account, but aims to preserve closed-loop stability (Subsection 6.4.3);
- the fourth and fifth approaches aim to preserve both the performance and the stability of the closed-loop system, but they are efficient only when the controller present in the loop is based on a state observer (Subsection 6.4.4).

6.4.1 Open-loop Plant Coprime-factor Reduction

Let $\begin{bmatrix}\tilde{N}_n & \tilde{M}_n\end{bmatrix}$ be a left coprime factorisation of G_n (see Proposition 2.2 for the construction of such a factorisation). Because $\tilde{N}_n$ and $\tilde{M}_n$ share the same state, $\begin{bmatrix}\tilde{N}_n & \tilde{M}_n\end{bmatrix}$ is of order n like G_n. It is stable by definition, hence it can be reduced by balanced truncation. It is advantageous to choose $\begin{bmatrix}\tilde{N}_n & \tilde{M}_n\end{bmatrix}$ normalised, *i.e.*, such that

$$\tilde{N}_n\tilde{N}_n^\star + \tilde{M}_n\tilde{M}_n^\star = I \tag{6.11}$$

because the $\mathcal{H}_\infty$ norm of the reduction error gives a bound of the distance between G_n and $\hat{G}_r$ in the gap metric and in the ν-gap metric: see (Georgiou and Smith, 1990), (Meyer, 1990), (Vidyasagar, 1984) and below. Such a normalised left coprime factorisation can be obtained following Proposition 2.3.

Let $\begin{bmatrix}\tilde{N}_r & \tilde{M}_r\end{bmatrix}$ denote the r-th order realisation of the reduced coprime factors obtained by balanced truncation of the normalised $\begin{bmatrix}\tilde{N}_n & \tilde{M}_n\end{bmatrix}$ (note that balanced truncation preserves the normalisation, as shown in (Meyer, 1990), *i.e.* $\tilde{N}_r\tilde{N}_r^\star + \tilde{M}_r\tilde{M}_r^\star = I$):

$$\begin{bmatrix}\tilde{N}_r & \tilde{M}_r\end{bmatrix} = \mathrm{bt}\left(\begin{bmatrix}\tilde{N}_n & \tilde{M}_n\end{bmatrix}, r\right) \tag{6.12}$$

The reduction procedure, which considers $\begin{bmatrix}\tilde{N}_n & \tilde{M}_n\end{bmatrix}$ as a single object, ensures that $\tilde{N}_r$ and $\tilde{M}_r$ share a common state realisation. The reconstruction of a reduced-order model $\hat{G}_r^{olcf} = \tilde{M}_r^{-1}\tilde{N}_r$ must then be carried out without duplicating the states, in such a way that $\hat{G}_r^{olcf}$ has order r and not $2r$. This is possible by observing that $\hat{G}_r^{olcf}$ can be obtained by means of a lower LFT:

$$\hat{G}_r^{olcf} = \mathcal{F}_l\left(\begin{bmatrix}I\\I\end{bmatrix}\left(\begin{bmatrix}\tilde{N}_r & \tilde{M}_r\end{bmatrix} - \begin{bmatrix}0 & I\end{bmatrix}\right), -I\right) \tag{6.13}$$

Recall that upper and lower bounds of the reduction error are given by

$$\varsigma_{r+1} \leq \left\|\begin{bmatrix}\Delta_{\tilde{N}} & \Delta_{\tilde{M}}\end{bmatrix}\right\|_\infty \leq 2\sum_{i=r+1}^{n}\varsigma_i \tag{6.14}$$

where the ς_i's are the Hankel singular values of $\begin{bmatrix}\tilde{N}_n & \tilde{M}_n\end{bmatrix}$ in descending order and

$$\begin{bmatrix}\Delta_{\tilde{N}} & \Delta_{\tilde{M}}\end{bmatrix} = \begin{bmatrix}\tilde{N}_n & \tilde{M}_n\end{bmatrix} - \begin{bmatrix}\tilde{N}_r & \tilde{M}_r\end{bmatrix} \tag{6.15}$$

The interpretation of the approximation error in the gap metric and in the ν-gap metric is given by the following relation (Zhou and Doyle, 1998):

$$\begin{aligned}\mathcal{G}_\varepsilon^d = \Big\{G = \left(\tilde{M}_n + \Delta_{\tilde{M}}\right)^{-1}\left(\tilde{N}_n + \Delta_{\tilde{N}}\right) \mid \\ \begin{bmatrix}\Delta_{\tilde{N}} & \Delta_{\tilde{M}}\end{bmatrix} \in \mathcal{H}_\infty, \left\|\begin{bmatrix}\Delta_{\tilde{N}} & \Delta_{\tilde{M}}\end{bmatrix}\right\|_\infty \leq \varepsilon\Big\} \\ = \Big\{G \mid \vec{\delta}_g(G_n, G) \leq \varepsilon\Big\} \subset \mathcal{G}_\varepsilon^\nu = \Big\{G \mid \delta_\nu(G_n, G) \leq \varepsilon\Big\}\end{aligned} \tag{6.16}$$

where $\vec{\delta}_g(\cdot\,,\,\cdot)$ is the directed gap defined in (2.72) and $\delta_\nu(\cdot\,,\,\cdot)$ the ν-gap defined in (2.65). As a result, the following proposition holds, which gives an upper bound of the distance between G_n and $\hat{G}_r^{olcf}$ in the gap metric and in the ν-gap metric.

Proposition 6.1. *Let $G_n = \tilde{M}_n^{-1}\tilde{N}_n$ be a normalised left coprime factorisation of the n-th order system G_n and let $\hat{G}_r^{olcf} = \tilde{M}_r^{-1}\tilde{N}_r$ be a r-th order system $(r < n)$ obtained by balanced truncation of $\begin{bmatrix}\tilde{N}_n & \tilde{M}_n\end{bmatrix}$. Then,*

$$\vec{\delta}_g(G_n, \hat{G}_r^{olcf}) \leq 2 \sum_{i=r+1}^{n} \varsigma_i \tag{6.17}$$

and

$$\delta_\nu(G_n, \hat{G}_r^{olcf}) \leq 2 \sum_{i=r+1}^{n} \varsigma_i \tag{6.18}$$

where the ς_i's are the Hankel singular values of $\begin{bmatrix}\tilde{N}_n & \tilde{M}_n\end{bmatrix}$ in descending order.

According to this result, what we have called the 'open-loop' reduction of $\begin{bmatrix}\tilde{N}_n & \tilde{M}_n\end{bmatrix}$, *i.e.*, its reduction without frequency weighting, can however be considered as 'closed-loop' or 'control'-oriented, since it preserves as much as possible the properties of G_n around its cross-over frequency (recall that the ν-gap is a measure of distance between two transfer functions with maximum resolution at the cross-over frequency). However, from a control design point of view, the focus should be on the cross-over frequency of G_nK, where K is the optimal controller (or, in an iterative framework, a controller that yields a better performance than open-loop operation), rather than on the cross-over frequency of G_n alone. Thus, this method will give an acceptable closed-loop performance only if the bandwidth of G_n is close to that of G_nK.

▶ **Example 6.1.** Consider the following 7th order plant (Safonov and Chiang, 1988):

$$G_7(s) = \frac{0.05(s^7 + 801s^6 + 1024s^5 + 599s^4 + 451s^3 + 119s^2 + 49s + 5.55)}{s^7 + 12.6s^6 + 53.48s^5 + 90.94s^4 + 71.83s^3 + 27.22s^2 + 4.75s + 0.3}$$

We further assume that the system has an ARX model structure, *i.e.*, the disturbance and plant models share the same state and the disturbance model is

$$H_7(s) = \frac{1}{s^7 + 12.6s^6 + 53.48s^5 + 90.94s^4 + 71.83s^3 + 27.22s^2 + 4.75s + 0.3}$$

This system is stabilised by the following simple controller:

$$K(s) = \frac{0.1}{s}$$

In a first step, the normalised plant coprime factors are computed following Proposition 2.3 and a balanced realisation is obtained using the derivations of Subsection 2.7.2:

$$\begin{bmatrix}\tilde{N}_7 & \tilde{M}_7\end{bmatrix} = \left[\begin{array}{c|c}\breve{A} & \breve{B} \\ \hline \breve{C} & \breve{D}\end{array}\right]$$

where

$$\breve{A} = \begin{bmatrix} -0.03637 & -6.935 & -0.6719 & 0.1728 & 0.02596 & -0.004674 & -0.000207 \\ 5.877 & -39.74 & -7.939 & 4.573 & 1.124 & -0.1718 & -0.007594 \\ 0.5113 & -7.129 & -1.482 & 0.9448 & 0.2648 & -0.03838 & -0.001696 \\ -0.1191 & 3.719 & 0.8556 & -0.7759 & -0.5603 & 0.04887 & 0.002151 \\ 0.01775 & -0.9066 & -0.2379 & 0.5558 & -0.05438 & 0.01818 & 0.0008133 \\ -0.003187 & 0.1382 & 0.0344 & -0.04836 & 0.01814 & -0.1422 & -0.01258 \\ -0.0001411 & 0.006112 & 0.00152 & -0.002129 & 0.0008113 & -0.01258 & -1 \end{bmatrix}$$

$$\breve{B} = \begin{bmatrix} 0.1655 & 0.1607 \\ -5.317 & 4.348 \\ -1.001 & 0.5597 \\ 0.4644 & -0.09168 \\ -0.08776 & -0.001837 \\ 0.01481 & -0.0007292 \\ 0.0006556 & -3.278 \times 10^{-5} \end{bmatrix}$$

$$\breve{C} = \begin{bmatrix} -0.2307 & -6.869 & -1.147 & 0.4734 & 0.08778 & -0.01483 & -0.0006563 \end{bmatrix}$$

$$\breve{D} = \begin{bmatrix} 0.04994 & 0.9988 \end{bmatrix}$$

The balanced Gramians are

$$\mathbf{P} = \mathbf{Q} = \Sigma = \text{diag}(0.7314, 0.5936, 0.4438, 0.1444, 0.0708, 7.74 \times 10^{-4}, 2.15 \times 10^{-7})$$

The Hankel singular values decrease very slowly, showing that it could be difficult to discard some states without introducing an important approximation error, especially for $r \leq 4$. Three low-order approximations $\begin{bmatrix} \tilde{N}_r & \tilde{M}_r \end{bmatrix}$ of orders 5, 4 and 3, respectively, of $\begin{bmatrix} \tilde{N}_7 & \tilde{M}_7 \end{bmatrix}$ are produced by discarding the states corresponding to the smallest Hankel singular values and are used to obtain low-order approximations of G_7 using (6.13):

$$\hat{G}_5^{olcf} = \frac{0.05s^5 + 39.99s^4 + 5.459s^3 + 17.96s^2 + 1.241s + 1.986}{s^5 + 11.46s^4 + 40.23s^3 + 43.31s^2 + 16.51s + 2.141}$$

$$\hat{G}_4^{olcf} = \frac{0.05s^4 + 40s^3 + 1.664s^2 + 9.522s + 2.443}{s^4 + 11.39s^3 + 38.53s^2 + 39.03s + 2.093}$$

$$\hat{G}_3^{olcf} = \frac{0.05s^3 + 39.72s^2 - 12.84s + 10.61}{s^3 + 10.69s^2 + 37.35s + 15.69}$$

The Bode diagrams of these approximations are shown in Figure 6.2, where it can be observed that the reduction error is large at frequencies smaller than 3 rad/s. The open-loop step responses of the reduced-order models (Figure 6.3, top) exhibit an important steady-state error while, in closed loop with controller K (Figure 6.3, bottom), an important dynamic error happens during the transient of the step response.

These full-order and reduced-order models are now used to compute new controllers. The requirement is to reduce the closed-loop rise time by two (we want a 90% rise time of 10 seconds at most) without letting the overshoot become greater than 20%. The chosen method is the $\mathcal{H}_\infty$ loop-shaping design method of (Skogestad and Postlethwaite, 1996), which delivers the central controller of (McFarlane and Glover, 1990). The frequency weighting filter is $W(s) = \frac{1}{s(s+2)}$ and the specified $\mathcal{H}_\infty$ bound is $\gamma = 1.01\gamma_{min}$ where γ_{min} is the smallest achievable bound, *i.e.*, the optimal $\mathcal{H}_\infty$

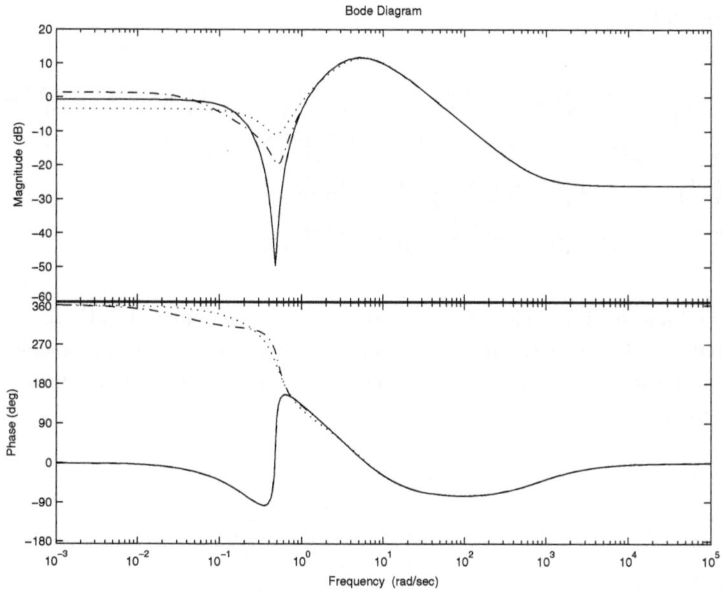

Figure 6.2. Bode diagrams of G_7 (—), $\hat{G}_5^{olcf}$ (−−) (indistinguishable from that of G_7), $\hat{G}_4^{olcf}$ (−·) and $\hat{G}_3^{olcf}$ (···)

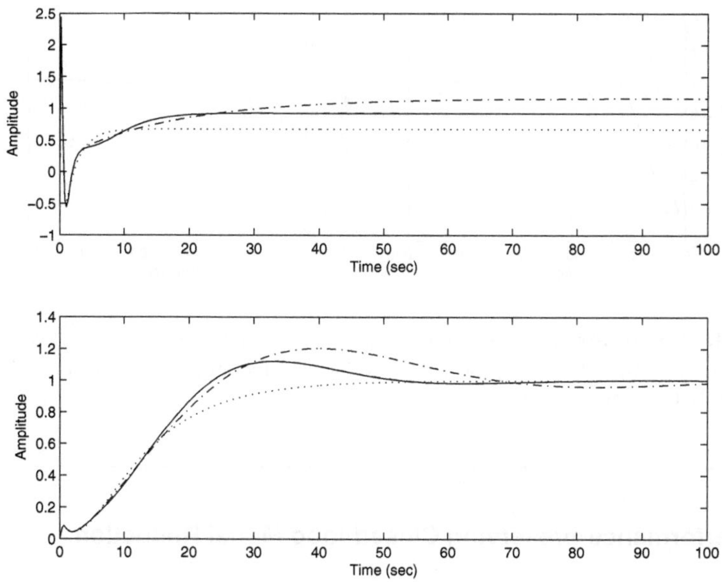

Figure 6.3. Step responses of G_7 (—), $\hat{G}_5^{olcf}$ (−−) (indistinguishable from that of G_7), $\hat{G}_4^{olcf}$ (−·) and $\hat{G}_3^{olcf}$ (···). Top: in open loop; bottom: in closed loop with controller K.

bound. We find:

$$K_{\mathcal{H}_\infty}(G_7) =$$

$$\frac{18.77s^8+274.5s^7+1483s^6+3749s^5+4910s^4+3428s^3+1253s^2+229.4s+15.31}{s^{11}+27.34s^{10}+293.2s^9+1627s^8+4191s^7+6418s^6+7520s^5+5593s^4+2233s^3+630.2s^2+57.02s}$$

$$K_{\mathcal{H}_\infty}(\hat{G}_5^{olcf}) = \frac{18.77s^6+253s^5+1191s^4+2350s^3+2051s^2+744.1s+109.6}{s^9+26.2s^8+263s^7+1322s^6+2641s^5+3208s^4+3473s^3+1163s^2+408s}$$

$$K_{\mathcal{H}_\infty}(\hat{G}_4^{olcf}) = \frac{17.07s^5+228.8s^4+1050s^3+1997s^2+1470s+133.3}{s^8+26.11s^7+259.8s^6+1285s^5+2495s^4+2779s^3+2699s^2+452s}$$

$$K_{\mathcal{H}_\infty}(\hat{G}_3^{olcf}) = \frac{16.15s^4+205s^3+948.2s^2+1460s+564.5}{s^7+24.91s^6+241.3s^5+1168s^4+2017s^3+1881s^2+1916s}$$

The designed and achieved closed-loop step responses are depicted in Figure 6.4. One can see that the designed (nominal) responses always fulfill the requirements on the rise time and the overshoot, while the achieved responses exhibit an unacceptable overshoot for $r \leq 4$. ◀

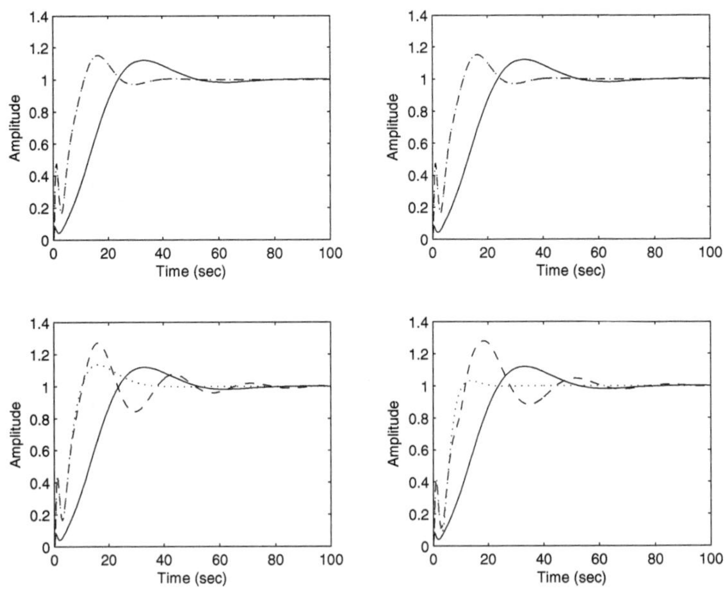

Figure 6.4. Step responses of $T_{11}(G_7, K)$ (—), $T_{11}(G_7, K_{\mathcal{H}_\infty}(\hat{G}_r^{olcf}))$ (– –) and $T_{11}(\hat{G}_r^{olcf}, K_{\mathcal{H}_\infty}(\hat{G}_r^{olcf}))$ (· · ·). Top left: $r = 7$ (no reduction), top right: $r = 5$, bottom left: $r = 4$, bottom right $r = 3$.

6.4.2 Performance-preserving Closed-loop Model Reduction

We shall now see how the closed-loop criterion (6.4) can be used to obtain a low-order model $\hat{G}_r$ preserving those properties of G_n that are important in closed loop with some given controller K. Because this controller is explicitly taken into account in the reduction criterion, a better performance can be expected than with the previous method.

As already mentioned, the computation of $\hat{G}_r$ in (6.4) is not an easy problem and is practically intractable. Therefore, we would like to replace this criterion by another one that could be approximately minimised by (frequency-weighted) balanced truncation of the coprime factors $\begin{bmatrix}\tilde{N}_n & \tilde{M}_n\end{bmatrix}$ of G_n. Let us therefore also consider a right coprime factorisation $\left[\begin{smallmatrix}U\\V\end{smallmatrix}\right]$ of the controller K. Following the derivations of Subsection 2.4.2, the transfer matrix of the closed-loop system can be rewritten as

$$T_o(G_n, K) = \begin{bmatrix}U\\V\end{bmatrix} \Phi^{-1} \begin{bmatrix}\tilde{N}_n & \tilde{M}_n\end{bmatrix} \tag{6.19}$$

where

$$\Phi = \begin{bmatrix}\tilde{N}_n & \tilde{M}_n\end{bmatrix} \begin{bmatrix}U\\V\end{bmatrix} \tag{6.20}$$

is inversely stable.

The criterion (6.4) can then be expressed as

$$J_{perf}(\hat{G}_r) = \left\| \begin{bmatrix}U\\V\end{bmatrix} \Phi^{-1} \begin{bmatrix}\tilde{N}_n & \tilde{M}_n\end{bmatrix} - \begin{bmatrix}U\\V\end{bmatrix} \hat{\Phi}^{-1} \begin{bmatrix}\tilde{N}_r & \tilde{M}_r\end{bmatrix} \right\|_\infty \tag{6.21}$$

where

$$\hat{\Phi} = \begin{bmatrix}\tilde{N}_r & \tilde{M}_r\end{bmatrix} \begin{bmatrix}U\\V\end{bmatrix} \tag{6.22}$$

To make the reduction problem solvable by frequency-weighted balanced truncation, it is necessary to have a criterion of the form

$$\left\| W_l \left(\begin{bmatrix}\tilde{N}_n & \tilde{M}_n\end{bmatrix} - \begin{bmatrix}\tilde{N}_r & \tilde{M}_r\end{bmatrix} \right) W_r \right\|_\infty \tag{6.23}$$

where W_l and W_r are frequency-weighting filters that do not depend on $\begin{bmatrix}\tilde{N}_r & \tilde{M}_r\end{bmatrix}$. Such a criterion can be obtained by first-order approximation of $J_{perf}(\hat{G}_r)$:

$$\begin{aligned} J^{(1)}_{perf}(\hat{G}_r) = \Bigg\| \begin{bmatrix}U\\V\end{bmatrix} & \Phi^{-1}\left(\begin{bmatrix}\tilde{N}_n & \tilde{M}_n\end{bmatrix} - \begin{bmatrix}\tilde{N}_r & \tilde{M}_r\end{bmatrix} \right) \\ & \times \left(I - \begin{bmatrix}U\\V\end{bmatrix} \Phi^{-1} \begin{bmatrix}\tilde{N}_n & \tilde{M}_n\end{bmatrix} \right) \Bigg\|_\infty \end{aligned} \tag{6.24}$$

which has the desired form with frequency-weighting filters

$$W_l = \begin{bmatrix}U\\V\end{bmatrix} \Phi^{-1} \qquad \text{and} \qquad W_r = I - \begin{bmatrix}U\\V\end{bmatrix} \Phi^{-1} \begin{bmatrix}\tilde{N}_n & \tilde{M}_n\end{bmatrix} \tag{6.25}$$

$J^{(1)}_{perf}(\hat{G}_r)$ is obtained by observing that

$$\begin{aligned} &T_o(G_n, K) - T_o(\hat{G}_r, K) \\ &= \left(I - \begin{bmatrix} U \\ V \end{bmatrix} \Phi^{-1} \begin{bmatrix} \Delta_{\tilde{N}} & \Delta_{\tilde{M}} \end{bmatrix}\right)^{-1} \begin{bmatrix} U \\ V \end{bmatrix} \Phi^{-1} \begin{bmatrix} \Delta_{\tilde{N}} & \Delta_{\tilde{M}} \end{bmatrix} (I - T_o(G_n, K)) \\ &\quad = \begin{bmatrix} U \\ V \end{bmatrix} \Phi^{-1} \begin{bmatrix} \Delta_{\tilde{N}} & \Delta_{\tilde{M}} \end{bmatrix} (I - T_o(G_n, K)) + \text{higher order terms} \end{aligned} \tag{6.26}$$

where $\begin{bmatrix} \Delta_{\tilde{N}} & \Delta_{\tilde{M}} \end{bmatrix} = \begin{bmatrix} \tilde{N}_n & \tilde{M}_n \end{bmatrix} - \begin{bmatrix} \tilde{N}_r & \tilde{M}_r \end{bmatrix}$ and where the second equality results from a first-order approximation[1].

Another possibility is to use the following zeroth-order approximation of $J_{perf}(\hat{G}_r)$:

$$J^{(0)}_{perf}(\hat{G}_r) = \left\| \begin{bmatrix} U \\ V \end{bmatrix} \Phi^{-1} \left(\begin{bmatrix} \tilde{N}_n & \tilde{M}_n \end{bmatrix} - \begin{bmatrix} \tilde{N}_r & \tilde{M}_r \end{bmatrix} \right) \right\|_\infty \tag{6.27}$$

obtained by replacing $\hat{\Phi}$ by Φ in $J_{perf}(\hat{G}_r)$. It also has the desired form, but with frequency-weighting filters

$$W_l = \begin{bmatrix} U \\ V \end{bmatrix} \Phi^{-1} \qquad \text{and} \qquad W_r = I \tag{6.28}$$

It will lead to less precise results, but it can help avoid numerical problems caused by the possibly high order of W_r in $J^{(1)}_{perf}(\hat{G}_r)$.

Both criteria can be generalised by adding input and output weighting filters or matrices W_{in} and W_{out} if one wants to give more importance to some frequency range and/or to some inputs and outputs of $T_o(G_n, K)$:

$$\begin{aligned} J^{(1)}_{perf}(\hat{G}_r) = \Bigg\| &\underbrace{W_{out} \begin{bmatrix} U \\ V \end{bmatrix} \Phi^{-1}}_{W_l} \left(\begin{bmatrix} \tilde{N}_n & \tilde{M}_n \end{bmatrix} - \begin{bmatrix} \tilde{N}_r & \tilde{M}_r \end{bmatrix} \right) \\ &\times \underbrace{\left(I - \begin{bmatrix} U \\ V \end{bmatrix} \Phi^{-1} \begin{bmatrix} \tilde{N}_n & \tilde{M}_n \end{bmatrix} \right) W_{in}}_{W_r} \Bigg\|_\infty \end{aligned} \tag{6.29}$$

[1] To ascertain the first equality, just express $T_o(G_n, K)$ and $T_o(\hat{G}_r, K)$ in function of $\begin{bmatrix} \tilde{N}_n & \tilde{M}_n \end{bmatrix}$, $\begin{bmatrix} \tilde{N}_r & \tilde{M}_r \end{bmatrix}$ and $\begin{bmatrix} U \\ V \end{bmatrix}$, then left-multiply both sides of the equality by $I - \begin{bmatrix} U \\ V \end{bmatrix} \Phi^{-1} \begin{bmatrix} \Delta_{\tilde{N}} & \Delta_{\tilde{M}} \end{bmatrix}$ and develop the products.

or

$$J^{(0)}_{perf}(\hat{G}_r) = \left\| \underbrace{W_{out} \begin{bmatrix} U \\ V \end{bmatrix} \Phi^{-1}}_{W_l} \left(\begin{bmatrix} \tilde{N}_n & \tilde{M}_n \end{bmatrix} - \begin{bmatrix} \tilde{N}_r & \tilde{M}_r \end{bmatrix} \right) \underbrace{W_{in}}_{W_r} \right\|_\infty \tag{6.30}$$

A suboptimal solution to the minimisation of these criteria can be obtained by frequency-weighted balanced truncation:

$$\begin{bmatrix} \tilde{N}_r & \tilde{M}_r \end{bmatrix} = \text{fwbt}\left(\begin{bmatrix} \tilde{N}_n & \tilde{M}_n \end{bmatrix}, W_l, W_r, r \right) \tag{6.31a}$$

$$\hat{G}_r^{perf} = \tilde{M}_r^{-1} \tilde{N}_r \tag{6.31b}$$

Note that it is useful to choose $\begin{bmatrix} \tilde{N}_n & \tilde{M}_n \end{bmatrix}$ normalised for the same reasons as given in Subsection 6.4.1.

Finally, remember that the stability of $\begin{bmatrix} \tilde{N}_r & \tilde{M}_r \end{bmatrix}$ cannot be guaranteed in the case of bilateral frequency-weighted balanced truncation, so $\begin{bmatrix} \tilde{N}_r & \tilde{M}_r \end{bmatrix}$ may not be a coprime factorisation of $\hat{G}_r^{perf}$ in the strict sense of the word if $J^{(1)}_{perf}(\hat{G}_r)$ (rather than $J^{(0)}_{perf}(\hat{G}_r)$) and/or an additional weighting filter W_{in} is used.

▶ **Example 6.2.** Consider again the plant model G_7, the disturbance model H_7 and the controller K of Example 6.1. In order to take the disturbances acting on the process into account during the reduction step, we shall express the closed-loop transfer matrix as

$$T_o\left(\begin{bmatrix} H_7 & G_7 \end{bmatrix}, \begin{bmatrix} 0 \\ K \end{bmatrix} \right) = \begin{bmatrix} 0 & 0 & 0 \\ \dfrac{KH_7}{1+G_7K} & \dfrac{KG_7}{1+G_7K} & \dfrac{K}{1+G_7K} \\ \dfrac{H_7}{1+G_7K} & \dfrac{G_7}{1+G_7K} & \dfrac{1}{1+G_7K} \end{bmatrix}$$

with input vector $\begin{bmatrix} e(t) & r_2(t) & -r_1(t) \end{bmatrix}^T$ and output vector $\begin{bmatrix} 0 & f(t) & g(t) \end{bmatrix}^T$. These signals are defined in Figure 2.1 and $e(t)$ is the white noise driving the disturbance system: $v(t) = H_7(s)e(t)$.

A normalised left coprime factorisation $\begin{bmatrix} \tilde{N}_7 & \tilde{M}_7 \end{bmatrix}$ of $\begin{bmatrix} H_7 & G_7 \end{bmatrix}$ and a right coprime factorisation $\left[\begin{smallmatrix} U \\ V \end{smallmatrix}\right]$ of $\left[\begin{smallmatrix} 0 \\ K \end{smallmatrix}\right]$ are computed following Propositions 2.3 and 2.4. The following additional weightings are used:

$$W_{in} = \text{diag}(1, 0, 1) \qquad W_{out} = \text{diag}(0, 1, 1)$$

W_{in} reflects the fact that we are only interested in the responses to the reference input $r_1(t)$ and to the disturbance $e(t)$. The closed-loop performance is determined by the response of the output signal $y(t)$ or, equivalently, of the control error signal $g(t)$ to these inputs, which is why a penalty of 1 is put on $g(t)$ in W_{out}. The penalty on the control signal $f(t)$ was obtained by trial and error. Better results were obtained by setting it to 1 rather than to 0, which reflects the fact that the closed-loop properties of a reduced-order model $\hat{G}_r$ are determined not only by how well its output mimics that of the full-order model G_n, but also by the internal signals of the loop. These weightings are introduced in (6.29) and (6.31) and frequency-weighted balancing is applied to $\begin{bmatrix}\tilde{N}_7 & \tilde{M}_7\end{bmatrix}$.

The frequency-weighted balanced Gramians are

$$\mathbf{P}_{22} = \mathbf{Q}_{22} = \mathrm{diag}(3.292,\ 1.119,\ 1.038,\ 0.1347,\ 0.0372,\ 0.0273,\ 1.19 \times 10^{-3})$$

This time, the Hankel singular values decrease faster, meaning that better results can be expected at low order.

The frequency-weighted balanced realisation of $\begin{bmatrix}\tilde{N}_7 & \tilde{M}_7\end{bmatrix}$ is reduced to produce the following low-order estimates of the system (low-order estimates of the noise model are obtained simultaneously):

$$\hat{G}_3^{perf}(s) = \frac{0.05s^3 + 0.06948s^2 + 0.01207s + 3.607 \times 10^{-5}}{s^3 + 0.1573s^2 + 0.0166s - 4.775 \times 10^{-5}}$$
$$\hat{G}_2^{perf}(s) = \frac{0.05s^2 + 0.06895s + 0.0114}{s^2 + 0.1611s + 0.01368}$$
$$\hat{G}_1^{perf}(s) = \frac{0.05s + 0.04616}{s + 0.02858}$$

The Bode diagrams of Figure 6.5 show that $\hat{G}_3^{perf}$ and $\hat{G}_2^{perf}$ are very precise around 0.1 rad/s, which is the cut-off frequency of G_7K, contrary to $\hat{G}_3^{olcf}$ obtained in Example 6.1, which has its highest precision in a wide range located around the cut-off frequency of G_7 at 4 rad/s. The closed-loop step responses of $\hat{G}_3^{perf}$ and $\hat{G}_2^{perf}$ are also much closer to that of G_7 than that of $\hat{G}_3^{olcf}$, especially during the transient,

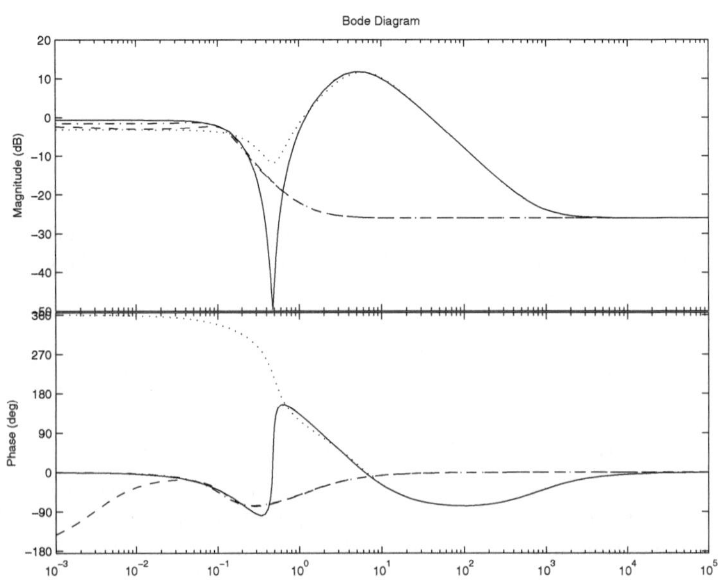

Figure 6.5. Bode diagrams of G_7 (—), $\hat{G}_3^{perf}$ (−−), $\hat{G}_2^{perf}$ (−·) and $\hat{G}_3^{olcf}$ (···)

as can be seen on Figure 6.6. Furthermore, the 1st-order estimate $\hat{G}_1^{perf}(s)$ is still stabilised by K, while closed-loop stability was lost for orders lower than 3 with the unweighted reduction procedure.

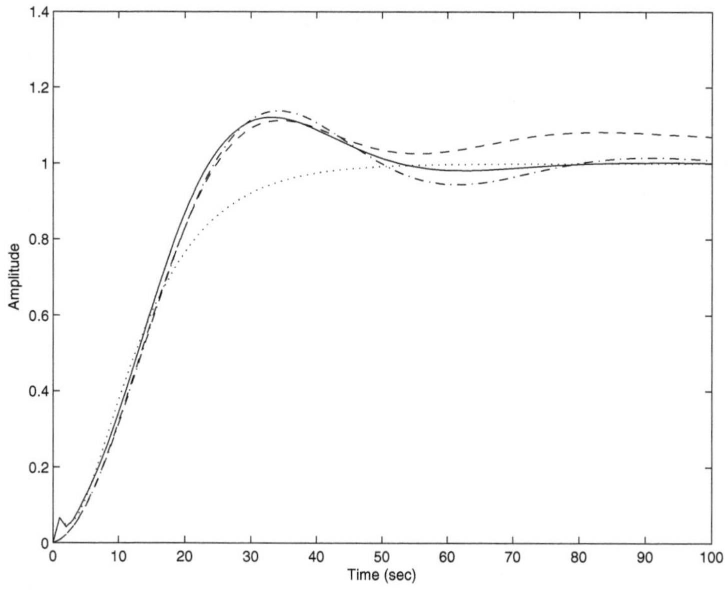

Figure 6.6. Closed-loop step responses of G_7 (—), $\hat{G}_3^{perf}$ (– –), $\hat{G}_2^{perf}$ (–·) and $\hat{G}_3^{olcf}$ (···) with controller K

Following the same procedure as in Example 6.1, these models are used to compute $\mathcal{H}_\infty$ controllers:

$$K_{\mathcal{H}_\infty}(\hat{G}_3^{perf}) = \frac{18.6s^4 + 41.05s^3 + 8.065s^2 + 0.7295s + 0.002062}{s^7 + 13.95s^6 + 49.16s^5 + 61.5s^4 + 23.34s^3 + 2.754s^2 + 0.008162s}$$
$$K_{\mathcal{H}_\infty}(\hat{G}_2^{perf}) = \frac{18.74s^3 + 41.45s^2 + 8.266s + 0.6539}{s^6 + 13.93s^5 + 49.07s^4 + 61.33s^3 + 23.15s^2 + 2.612s}$$
$$K_{\mathcal{H}_\infty}(\hat{G}_1^{perf}) = \frac{16.94s^2 + 35.23s + 2.678}{s^5 + 12.64s^4 + 41.56s^3 + 46.05s^2 + 10.95s}$$

$K_{\mathcal{H}_\infty}(\hat{G}_3^{perf})$ and $K_{\mathcal{H}_\infty}(\hat{G}_2^{perf})$ satisfy all requirements on the rise time and on the overshoot and have performance very close to that of the high-order controller $K_{\mathcal{H}_\infty}(G_7)$ that would be obtained without model reduction, as illustrated in Figure 6.7. $K_{\mathcal{H}_\infty}(\hat{G}_1^{perf})$ yields more oscillating closed-loop responses (not shown) and do not satisfy the requirement on the overshoot, although its behaviour is better than that of $K_{\mathcal{H}_\infty}(\hat{G}_3^{olcf})$ obtained in Example 6.1. The improvement between $K_{\mathcal{H}_\infty}(\hat{G}_3^{olcf})$ and $K_{\mathcal{H}_\infty}(\hat{G}_3^{perf})$ (and even $K_{\mathcal{H}_\infty}(\hat{G}_2^{perf})$) is significant, which shows the advantage of using an explicit closed-loop criterion for the reduction. ◀

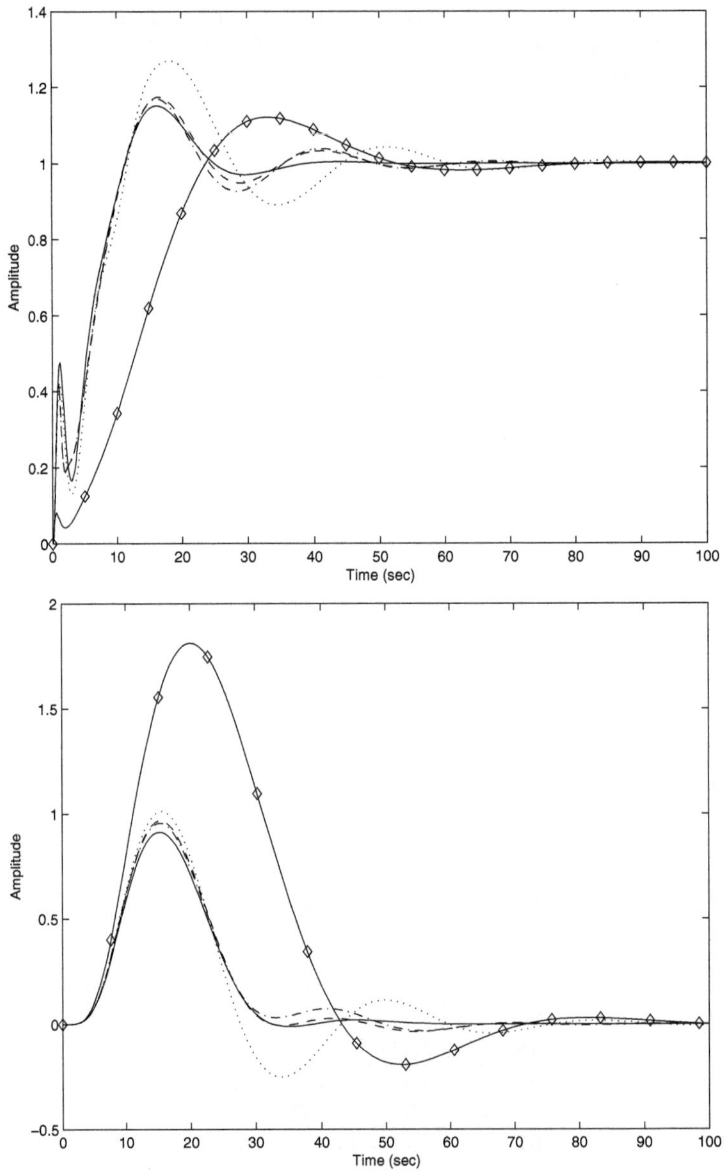

Figure 6.7. Closed-loop step responses from $r_1(t)$ (top) and $e(t)$ (bottom) of (G_7, H_7) with controllers K (-◇-), $K_{\mathcal{H}_\infty}(G_7)$ (—), $K_{\mathcal{H}_\infty}(\hat{G}_3^{perf})$ (−−), $K_{\mathcal{H}_\infty}(\hat{G}_2^{perf})$ (−·) and $K_{\mathcal{H}_\infty}(\hat{G}_3^{olcf})$ (⋯)

► **Remark.** Similar derivations can be made if one decides to use the $T_i(G_n, K)$ closed-loop transfer matrix instead of $T_o(G_n, K)$. We chose T_o because it yields a formulation of the criterion in terms of the plant left coprime factors $\begin{bmatrix}\tilde{N}_n & \tilde{M}_n\end{bmatrix}$ and of the controller right coprime factors $\left[\begin{smallmatrix}U\\V\end{smallmatrix}\right]$, which seems to be of common use in the literature and therefore, also in the other sections of this chapter. With $T_i(G_n, K)$, the criterion would be formulated in terms of the plant right coprime factors $\left[\begin{smallmatrix}N_n\\M_n\end{smallmatrix}\right]$ and of the controller left coprime factors $\begin{bmatrix}\tilde{V} & \tilde{U}\end{bmatrix}$. On the other hand, this transfer function connects the exogenous signals $r_1(t)$ and $r_2(t)$ to $u(t)$ and $y(t)$, which may be easier or more intuitive to use as performance indicators than $f(t)$ and $g(t)$. It is up to the user to choose between T_o and T_i and there is no theoretical reason for preferring one to the other. ◀

6.4.3 Stability-preserving Closed-loop Model Reduction

Instead of focusing on the preservation of the closed-loop performance during the model reduction procedure, one can formulate the reduction criterion in terms of closed-loop stability (recall that it is important to have a nominal model that is stabilised by the current to-be-replaced controller: see Section 4.2). Consider the following lemma.

Lemma 6.1. *Let $G_n = \tilde{M}_n^{-1}\tilde{N}_n$ and $K = UV^{-1}$ be two coprime factorisations satisfying the Bezout identity (2.61). Then, K stabilises any system $G = (\tilde{M}_n + \Delta_{\tilde{M}})^{-1}(\tilde{N}_n + \Delta_{\tilde{N}})$ such that*

$$\left\| \begin{bmatrix}\Delta_{\tilde{N}} & \Delta_{\tilde{M}}\end{bmatrix} \begin{bmatrix}U\\V\end{bmatrix} \right\|_\infty < 1 \tag{6.32}$$

Proof. This lemma is the dual of Lemma 6.2 for stability-based reduction of controller coprime factors (see page 193), for which a proof is given in (Anderson and Liu, 1989). □

It leads very naturally to the definition of the following model reduction criterion:

$$J_{stab}(\hat{G}_r) = \left\| \left(\begin{bmatrix}\tilde{N}_n & \tilde{M}_n\end{bmatrix} - \begin{bmatrix}\tilde{N}_r & \tilde{M}_r\end{bmatrix}\right) \begin{bmatrix}U\\V\end{bmatrix} \right\|_\infty \tag{6.33}$$

where

- $\begin{bmatrix}\tilde{N}_n & \tilde{M}_n\end{bmatrix}$ is any left coprime factorisation of G_n (it can be useful to choose it normalised for the same reasons as given in Subsection 6.4.1);
- $\left[\begin{smallmatrix}U\\V\end{smallmatrix}\right]$ is a right coprime factorisation of K satisfying the Bezout identity (2.61).

When K is not an observer-based controller for G_n, there is no systematic construction method for $\left[\begin{smallmatrix}U\\V\end{smallmatrix}\right]$ satisfying (2.61), so the simplest way is to use any right coprime factorisation $K = UV^{-1}$, to compute $\Phi = \begin{bmatrix}\tilde{N}_n & \tilde{M}_n\end{bmatrix}\left[\begin{smallmatrix}U\\V\end{smallmatrix}\right]$

and to replace the criterion by

$$J_{stab}(\hat{G}_r) = \left\| \left(\begin{bmatrix} \tilde{N}_n & \tilde{M}_n \end{bmatrix} - \begin{bmatrix} \tilde{N}_r & \tilde{M}_r \end{bmatrix} \right) \begin{bmatrix} U \\ V \end{bmatrix} \Phi^{-1} \right\|_\infty \tag{6.34}$$

$\left[\begin{smallmatrix} U \\ V \end{smallmatrix}\right] \Phi^{-1}$ being a coprime factorisation of K satisfying the Bezout identity with $\begin{bmatrix} \tilde{N}_n & \tilde{M}_n \end{bmatrix}$ (see the proof of Lemma 2.2, page 18).

Thus, this criterion penalises the degradation of the Bezout equality during the reduction of $\begin{bmatrix} \tilde{N}_n & \tilde{M}_n \end{bmatrix}$. Since this equality means that

- the factors $\tilde{N}_n$ and $\tilde{M}_n$ are left coprime,
- the factors U and V are right coprime and
- the closed-loop (G_n, K) is stable,

this criterion aims, indeed, to preserve closed-loop stability. The resulting $\hat{G}_r$ may however be very different from G_n regarding closed-loop performance.

As before, frequency-weighted balanced truncation can be used to compute a suboptimal $\hat{G}_r^{stab}$:

$$\begin{bmatrix} \tilde{N}_r & \tilde{M}_r \end{bmatrix} = \text{fwbt}\left(\begin{bmatrix} \tilde{N}_n & \tilde{M}_n \end{bmatrix}, I, \begin{bmatrix} U \\ V \end{bmatrix} \Phi^{-1}, r \right) \tag{6.35a}$$

$$\hat{G}_r^{stab} = \tilde{M}_r^{-1} \tilde{N}_r \tag{6.35b}$$

▶ **Example 6.3.** Consider again the plant model G_7, the disturbance model H_7 and the controller K of Examples 6.1 and 6.2.

A normalised left coprime factorisation $\begin{bmatrix} \tilde{N}_7 & \tilde{M}_7 \end{bmatrix}$ of G_7 and a right coprime factorisation $\left[\begin{smallmatrix} U \\ V \end{smallmatrix}\right]$ of K are computed following Propositions 2.3 and 2.4. Using $\left[\begin{smallmatrix} U \\ V \end{smallmatrix}\right] \Phi^{-1}$ as right-hand-side frequency weighting, a frequency-weighted balanced realisation of $\begin{bmatrix} \tilde{N}_7 & \tilde{M}_7 \end{bmatrix}$ is obtained following the derivations of Subsection 2.7.5 and is then truncated to produce a 3rd-order estimate of the system:

$$\hat{G}_3^{stab}(s) = \frac{0.05s^3 + 39.54s^2 - 15.04s + 8.84}{s^3 + 10.53s^2 + 32.78s + 9.442}$$

As shown in Figure 6.8, it captures the closed-loop dynamic behaviour of G_7 better than $\hat{G}_3^{olcf}$, but not as well as $\hat{G}_2^{perf}$.

Figure 6.9 shows that the estimation error of $\hat{G}_3^{stab}$ is, *grosso modo*, distributed similarly to that of $\hat{G}_3^{olcf}$ and that its precision is poor around 0.1 rad/s (the cross-over frequency of G_7K), contrary to what was obtained by performance-based reduction.

Following the same procedure as in Examples 6.1 and 6.2, this model is used to compute an $\mathcal{H}_\infty$ controller:

$$K_{\mathcal{H}_\infty}(\hat{G}_3^{stab}) = \frac{21.85s^4 + 274.3s^3 + 1181s^2 + 1659s + 486.8}{s^7 + 25.62s^6 + 249.1s^5 + 1189s^4 + 1833s^3 + 1565s^2 + 2080s}$$

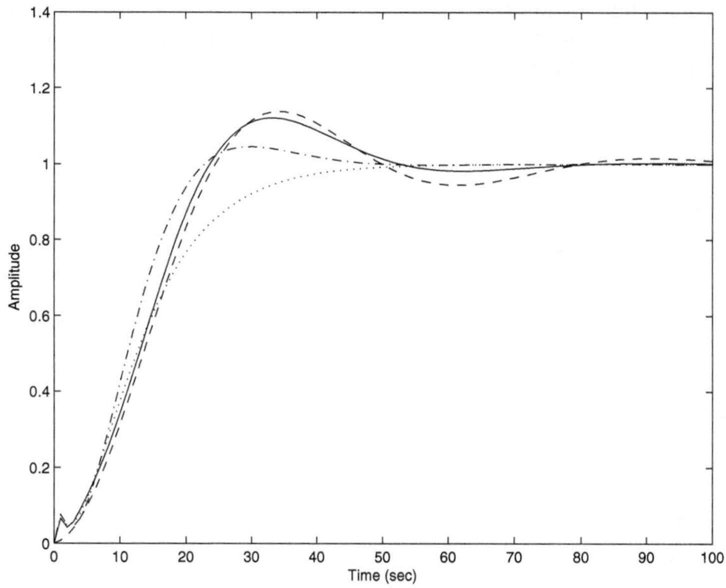

Figure 6.8. Closed-loop step responses of G_7 (—), $\hat{G}_2^{perf}$ (-- -), $\hat{G}_3^{stab}$ (-·) and $\hat{G}_3^{olcf}$ (···) with controller K

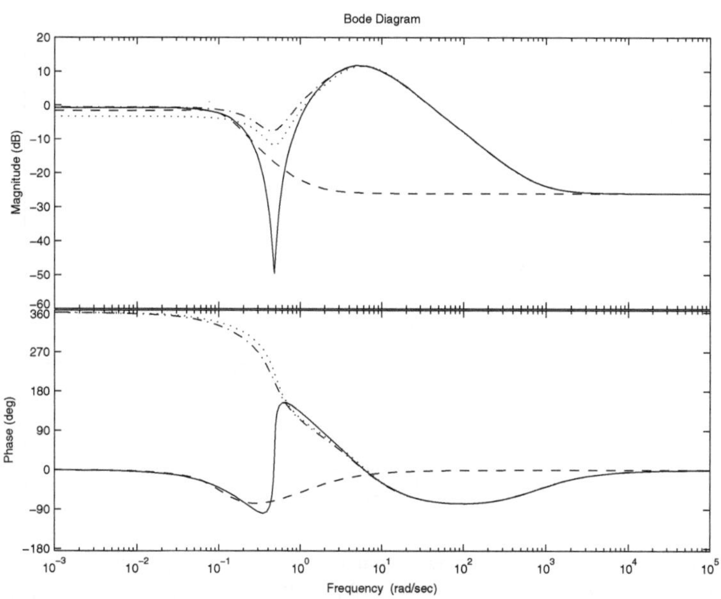

Figure 6.9. Bode diagrams of G_7 (—), $\hat{G}_2^{perf}$ (-- -), $\hat{G}_3^{stab}$ (-·) and $\hat{G}_3^{olcf}$ (···)

It satisfies all requirements on the rise time and on the overshoot, but its more oscillatory behaviour and its poorer ability to reject disturbances make it less suitable than $K_{\mathcal{H}_\infty}(\hat{G}_2^{perf})$, as illustrated in Figure 6.10. ◀

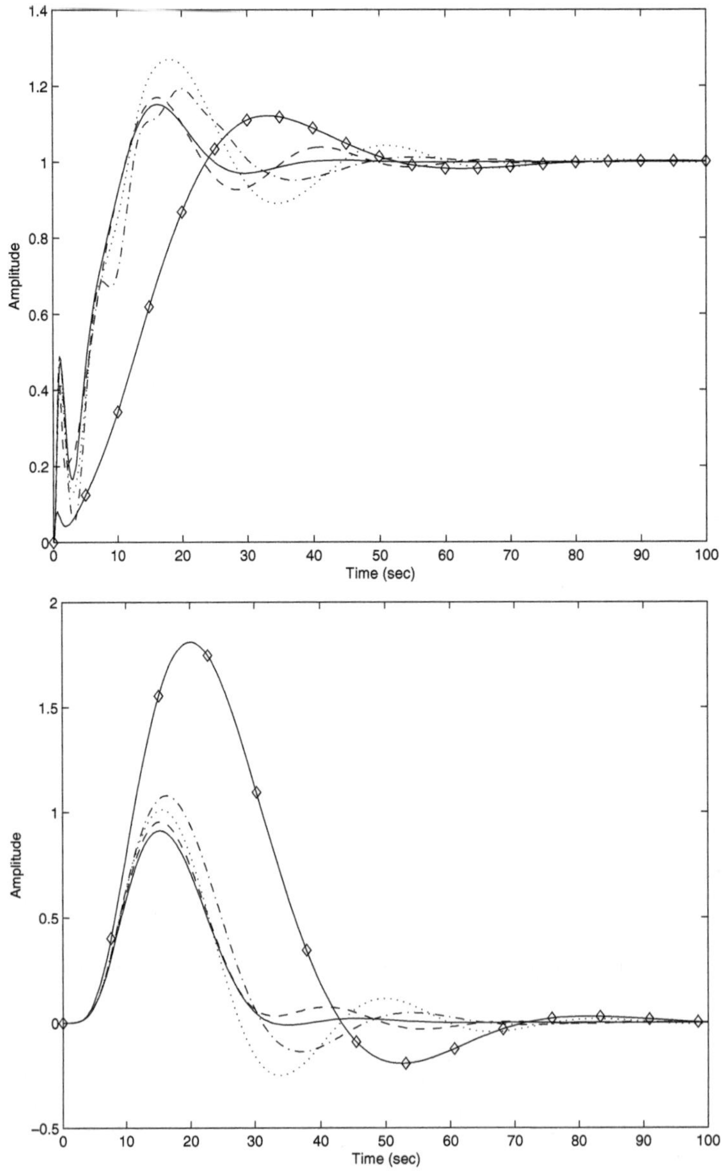

Figure 6.10. Closed-loop step responses from $r_1(t)$ (top) and $e(t)$ (bottom) of (G_7, H_7) with controllers K (-◇-), $K_{\mathcal{H}_\infty}(G_7)$ (—), $K_{\mathcal{H}_\infty}(\hat{G}_2^{perf})$ (−−), $K_{\mathcal{H}_\infty}(\hat{G}_3^{stab})$ (−·) and $K_{\mathcal{H}_\infty}(\hat{G}_3^{olcf})$ (···)

6.4.4 Preservation of Stability and Performance by Closed-loop Model Reduction

In Subsections 6.4.2 and 6.4.3, we have presented closed-loop coprime-factor reduction methods based, respectively, on the preservation of the closed-loop performance and on the preservation of the closed-loop stability. A question that naturally arises is: 'Would it be possible to combine both approaches?'. The answer, in case of coprime-factor reduction, is generally 'no'. However, in case of an observer-based controller, it is possible to use two different methods, as we now show.

A. Unweighted balanced truncation of coprime factors satisfying the Bezout identity. (Anderson and Moore, 1990) proposed this method for the case of an observer-based (LQG) controller. It later received an alternative justification, not based on LQG-performance considerations, in (Zhou and Chen, 1995)[2].

Let $K = UV^{-1}$ and $G_n = \tilde{M}_n^{-1}\tilde{N}_n$ be coprime factorisations satisfying (2.60) and (2.61). Let $\hat{G}_r = \tilde{M}_r^{-1}\tilde{N}_r$ define a coprime factorisation of a reduced-order model $\hat{G}_r$. Then, if

$$\left\| \begin{bmatrix} \tilde{N}_n & \tilde{M}_n \end{bmatrix} - \begin{bmatrix} \tilde{N}_r & \tilde{M}_r \end{bmatrix} \right\|_\infty < \varepsilon < 1 \tag{6.36}$$

it is possible to show that $\hat{G}_r$ is stabilised by K and that

$$\left\| T_o(G_n, K) - T_o(\hat{G}_r, K) \right\|_\infty \le \|T_o(G_n, K)\|_\infty \frac{\varepsilon}{1-\varepsilon} \tag{6.37}$$

Thus, the unweighted reduction of system coprime factors satisfying the Bezout identity (when the controller coprime factors are normalised) makes it possible to guarantee stability and to overbound the performance degradation. However, the order of $\begin{bmatrix} \tilde{N}_n & \tilde{M}_n \end{bmatrix}$ will generally be twice that of G_n plus that of K if K is not an observer-based controller, which makes the method useless except in that particular case. Indeed, let $\begin{bmatrix} \bar{N}_n & \bar{M}_n \end{bmatrix}$ be any n-th order left coprime factorisation of G_n. Then,

$$\begin{bmatrix} \tilde{N}_n & \tilde{M}_n \end{bmatrix} = \begin{bmatrix} \bar{N}_n & \bar{M}_n \end{bmatrix} \left(\bar{N}_n U + \bar{M}_n V \right)^{-1} \tag{6.38}$$

is a left coprime factorisation of G_n that satisfies the Bezout identity (2.61) (see the proof of Lemma 2.2 and the subsequent discussion, page 18). It has a minimal realisation of order at most n if and only if $\left[\begin{smallmatrix} U \\ V \end{smallmatrix}\right]$ is of the form

$$\begin{bmatrix} U(s) \\ V(s) \end{bmatrix} = \begin{bmatrix} F(sI - A - BF)^{-1}L \\ I - C(sI - A - BF)^{-1}L \end{bmatrix} \tag{6.39}$$

where (A, B, C) is a minimal state-space realisation of G_n (assumed here to be strictly proper, *i.e.*, $D = 0$) and F and L are such that the eigenvalues of $A + BF$ and $A + LC$ lie in the open left half plane. This corresponds typically

[2] Actually, in (Anderson and Moore, 1990) and (Zhou and Chen, 1995), controller reduction is considered. Here, we present a dual version of this method, adapted to model reduction.

to an observer-based controller where F is the stabilising feedback gain and L is the observer gain (the Kalman filter gain in the case of LQG control).

B. Multiplicative coprime-factor approximation. The method presented here follows an idea of (Gu, 1995). Let $K = \tilde{V}^{-1}\tilde{U} = UV^{-1}$ and $G_n = \tilde{M}_n^{-1}\tilde{N}_n = N_n M_n^{-1}$ be coprime factorisations satisfying the following double Bezout identity:

$$\begin{bmatrix} \tilde{M}_n & -\tilde{N}_n \\ \tilde{U} & \tilde{V} \end{bmatrix} \begin{bmatrix} V & N_n \\ -U & M_n \end{bmatrix} = I \tag{6.40}$$

Once again, $\left[\begin{smallmatrix} V & N_n \\ -U & M_n \end{smallmatrix}\right]$ has order at most n only if it can be put in the form

$$\begin{bmatrix} V(s) & N_n(s) \\ -U(s) & M_n(s) \end{bmatrix} = \begin{bmatrix} I & 0 \\ 0 & I \end{bmatrix} + \begin{bmatrix} C \\ F \end{bmatrix} (sI - A - BF)^{-1} \begin{bmatrix} -L & B \end{bmatrix} \tag{6.41}$$

where $(A,\, B,\, C)$ is a minimal state-space realisation of G_n and F and L are such that the eigenvalues of $A+BF$ and $A+LC$ lie in the open left half plane.

We can seek U_r, V_r, N_r and M_r of order r so that the $\mathcal{H}_\infty$ norm of $\Delta \in \mathcal{RH}_\infty$ in

$$\begin{bmatrix} V_r & N_r \\ -U_r & M_r \end{bmatrix} = (I - \Delta) \begin{bmatrix} V & N_n \\ -U & M_n \end{bmatrix} \tag{6.42}$$

is minimised. This amounts to find U_r, V_r, N_r and M_r so that

$$\|\Delta\|_\infty = \left\| \left(\begin{bmatrix} V & N_n \\ -U & M_n \end{bmatrix} - \begin{bmatrix} V_r & N_r \\ -U_r & M_r \end{bmatrix} \right) \begin{bmatrix} V & N_n \\ -U & M_n \end{bmatrix}^{-1} \right\|_\infty \tag{6.43}$$

is minimised. $\hat{G}_r$ is then given by $\hat{G}_r = N_r M_r^{-1}$ and the closed loop $(\hat{G}_r,\, K)$ is internally stable if

$$\|\Delta\|_\infty \|T_o(G_n,\, K)\|_\infty < 1 \tag{6.44}$$

Furthermore, there holds

$$\left\| T_o(G_n,\, K) - T_o(\hat{G}_r,\, K) \right\|_\infty \leq \frac{\|T_o(G_n,\, K)\|_\infty^2 \|\Delta\|_\infty}{1 - \|T_o(G_n,\, K)\|_\infty \|\Delta\|_\infty} \tag{6.45}$$

Clearly, the minimisation of $\|\Delta\|_\infty$ aims to preserve closed-loop stability and performance, but the method is generally inefficient if K is not an observer-based controller.

6.5 Controller Order Reduction

In this section, we describe the second method for obtaining a low-order controller from a high-order plant model, namely controller reduction. We assume here that the nominal model G_n has been used to design a high-order controller K_m and our objective is to reduce the latter to a lower order r. We denote this low-order controller by $\hat{K}_r$. As in the case of model reduction, it is not possible to carry out the reduction of K_m in a straightforward way if it is unstable, which is likely since most controllers contain an integrator to ensure zero steady-state error. Therefore, we shall also consider the reduction of the coprime factors of K_m, either in 'open loop', *i.e.* without frequency weightings, or in 'closed loop', *i.e.* with frequency weightings that take the interconnection with the process (represented by its model G_n) into account.

6.5.1 Open-loop Controller Coprime-factor Reduction

As it was the case for model reduction, it is strongly recommended to apply the reduction to *normalised* coprime factors $\begin{bmatrix} U_m \\ V_m \end{bmatrix}$ of the controller K_m. Such normalised factors can be constructed as explained in Proposition 2.5. The reduced-order controller can then be obtained by balanced truncation of $\begin{bmatrix} U_m \\ V_m \end{bmatrix}$:

$$\begin{bmatrix} U_r \\ V_r \end{bmatrix} = \mathrm{bt}\left(\begin{bmatrix} U_m \\ V_m \end{bmatrix}, r\right) \tag{6.46a}$$

$$\hat{K}_r^{olcf} = U_r V_r^{-1} \tag{6.46b}$$

The motivation for using normalised coprime factors comes from the following theorem.

Theorem 6.1. (Vinnicombe, 1993*a*, 1993*b*) *Let G_n be a nominal plant and let K_m be a stabilising controller for this plant. Let $K_m = U_m V_m^{-1}$ define a normalised coprime factorisation of K_m and let U_r, $V_r \in \mathcal{RH}_\infty$ be such that*

$$\left\| \begin{bmatrix} U_m \\ V_m \end{bmatrix} - \begin{bmatrix} U_r \\ V_r \end{bmatrix} \right\|_\infty \leq \varepsilon \tag{6.47}$$

Then $\hat{K}_r \triangleq U_r V_r^{-1}$ stabilises G_n if $\varepsilon < b_{G_n,K_m}$.

Furthermore, there holds

$$\arcsin b_{G,\hat{K}_r} \geq \arcsin b_{G_n,K_m} - \arcsin \varepsilon - \arcsin \beta \quad \forall G : \delta_\nu(G, G_n) \leq \beta \tag{6.48}$$

The first part of this theorem concerns the stabilisation of the nominal model G_n by the reduced-order controller $\hat{K}_r$. It says that finding the best possible approximation of the normalised coprime factors of the optimal controller in the $\mathcal{H}_\infty$ norm is appropriate (although not necessarily optimal, since the condition of the theorem is only sufficient) regarding this stability issue (of course, balanced truncation only delivers a suboptimal approximation). The second part of the theorem concerns a robustness issue: if the nominal model G_n has an uncertainty region of size β attached to it (*i.e.*, if the ν-gap between the underlying unknown true system G_0 and the nominal model G_n is known to be at most equal to β), a lower bound of the generalised stability margin that $\hat{K}_r$ will achieve with any plant in this uncertainty region is given. Observe that this theorem does not make any assumption about the computation of $\begin{bmatrix} U_r \\ V_r \end{bmatrix}$, which means that it can be used to check closed-loop stability with a reduced-order controller $\hat{K}_r$ computed by any method, even if the criterion it induces is unweighted normalised-coprime-factor reduction. Note also that this theorem can be very conservative.

As was the case for the 'open-loop' model reduction procedure, this approach is somewhat closed-loop oriented, since it is based on a sufficient condition for preserving closed-loop stability. However, it does not take the nominal model G_n explicitly into account.

▶ **Example 6.4.** Consider the plant model G_7, the disturbance model H_7 and the controller $K_{\mathcal{H}_\infty}(G_7)$ of Example 6.1. We shall rename this controller K_{11} to emphasise its 11th order and simplify the notation.

The normalised right coprime factors of $K_{11} = U_{11}V_{11}^{-1}$ are computed following Proposition 2.5 and a balanced realisation is obtained using the derivations of Subsection 2.7.2. The balanced Gramians are

$$\mathbf{P} = \mathbf{Q} = \Sigma = \mathrm{diag}(0.7885,\ 0.4072,\ 0.2478,\ 0.1109,\ 0.0592,\ 0.0074,\ 0.0017,\\ 2.25\times 10^{-4},\ 6.33\times 10^{-5},\ 5.74\times 10^{-5},\ 5.28\times 10^{-9})$$

The slow decrease of the Hankel singular values in Σ shows that poor results are expectable for low values of the reduced order r. This is confirmed by the closed-loop responses of Figure 6.11. Three reduced-order controllers were produced:

$$\hat{K}_6^{olcf} = \frac{-0.03674s^5 + 1.044s^4 + 3.172s^3 + 3.846s^2 + 1.401s + 0.2244}{s^6 + 3.728s^5 + 3.874s^4 + 6.458s^3 + 2.084s^2 + 0.8208s + 0.0006637}$$
$$\hat{K}_4^{olcf} = \frac{0.02953s^3 + 0.7731s^2 + 0.9893s + 0.0244}{s^4 + 0.518s^3 + 1.648s^2 + 0.212s - 0.003201}$$
$$\hat{K}_2^{olcf} = \frac{-0.03784s + 1.112}{s^2 + 0.3301s + 0.7487}$$

They all stabilise the plant G_7, but they exhibit an increasing static error as r decreases. Finally, note that

$$\left\|\begin{bmatrix}U_{11}\\V_{11}\end{bmatrix} - \begin{bmatrix}U_6\\V_6\end{bmatrix}\right\|_\infty = 3.50\times 10^{-3} < \left\|\begin{bmatrix}U_{11}\\V_{11}\end{bmatrix} - \begin{bmatrix}U_4\\V_4\end{bmatrix}\right\|_\infty = 1.30\times 10^{-1}$$
$$< b_{G_7,K_{11}} = 2.05\times 10^{-1} < \left\|\begin{bmatrix}U_{11}\\V_{11}\end{bmatrix} - \begin{bmatrix}U_2\\V_2\end{bmatrix}\right\|_\infty = 5.84\times 10^{-1}$$

meaning that the sufficient condition for stability of Theorem 6.1 is only fulfilled for $\hat{K}_6^{olcf}$ and $\hat{K}_4^{olcf}$. ◀

6.5.2 Performance-preserving Closed-loop Controller Reduction

A. Performance-oriented controller reduction using coprime factorisations. We shall now see how the closed-loop criterion (6.5) can be used to obtain a low-order controller $\hat{K}_r^{perf}$ preserving those properties of K_m that are important when it is connected to the plant model G_n. Because this model is

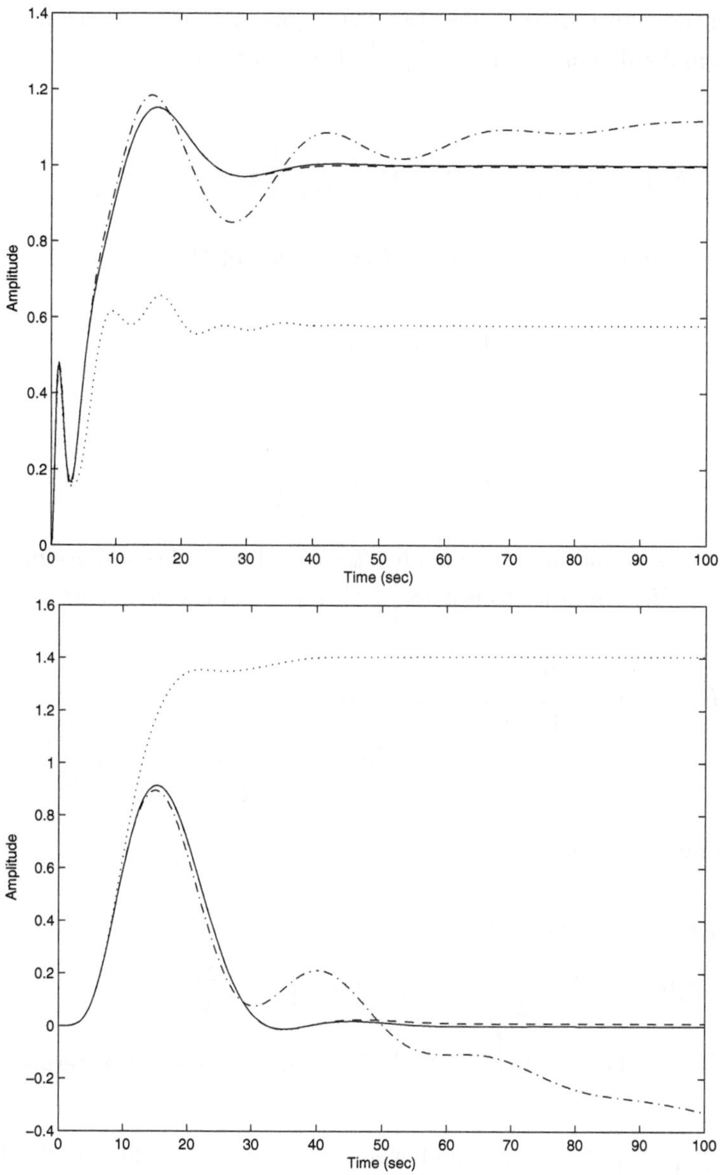

Figure 6.11. Closed-loop step responses from $r_1(t)$ (top) and $e(t)$ (bottom) of $(G_7,\, H_7)$ with controllers K_{11} (—), $\hat{K}_6^{olcf}$ (– –), $\hat{K}_4^{olcf}$ (–·) and $\hat{K}_2^{olcf}$ (···)

explicitly taken into account in the reduction criterion, a better performance can be expected than with the previous method.

Following the derivations of Subsection 2.4.2, (6.5) can be expressed in function of the normalised coprime factors $\begin{bmatrix} U_m \\ V_m \end{bmatrix}$ of K_m as follows:

$$J_{perf}(\hat{K}_r) = \left\| \begin{bmatrix} U_m \\ V_m \end{bmatrix} \Phi^{-1} \begin{bmatrix} \tilde{N}_n & \tilde{M}_n \end{bmatrix} - \begin{bmatrix} U_r \\ V_r \end{bmatrix} \hat{\Phi}^{-1} \begin{bmatrix} \tilde{N}_n & \tilde{M}_n \end{bmatrix} \right\|_\infty \tag{6.49}$$

where $\begin{bmatrix} \tilde{N}_n & \tilde{M}_n \end{bmatrix}$ is any left coprime factorisation of G_n,

$$\Phi = \begin{bmatrix} \tilde{N}_n & \tilde{M}_n \end{bmatrix} \begin{bmatrix} U_m \\ V_m \end{bmatrix} \tag{6.50}$$

and

$$\hat{\Phi} = \begin{bmatrix} \tilde{N}_n & \tilde{M}_n \end{bmatrix} \begin{bmatrix} U_r \\ V_r \end{bmatrix} \tag{6.51}$$

In practice, as we did in Subsection 6.4.2, we shall use a first-order approximation of $J_{perf}(\hat{K}_r)$ in order to isolate $\begin{bmatrix} U_m \\ V_m \end{bmatrix} - \begin{bmatrix} U_r \\ V_r \end{bmatrix}$ in the criterion:

$$\begin{aligned} J^{(1)}_{perf}(\hat{K}_r) = \Bigg\| \left(I - \begin{bmatrix} U_m \\ V_m \end{bmatrix} \Phi^{-1} \begin{bmatrix} \tilde{N}_n & \tilde{M}_n \end{bmatrix} \right) & \\ \times \left(\begin{bmatrix} U_m \\ V_m \end{bmatrix} - \begin{bmatrix} U_r \\ V_r \end{bmatrix} \right) \Phi^{-1} \begin{bmatrix} \tilde{N}_n & \tilde{M}_n \end{bmatrix} \Bigg\|_\infty & \end{aligned} \tag{6.52}$$

obtained by observing that

$$\begin{aligned} & T_o(G_n, K_m) - T_o(G_n, \hat{K}_r) \\ & = (I - T_o(G_n, K_m)) \begin{bmatrix} \Delta_U \\ \Delta_V \end{bmatrix} \Phi^{-1} \begin{bmatrix} \tilde{N}_n & \tilde{M}_n \end{bmatrix} \left(I - \begin{bmatrix} \Delta_U \\ \Delta_V \end{bmatrix} \Phi^{-1} \begin{bmatrix} \tilde{N}_n & \tilde{M}_n \end{bmatrix} \right)^{-1} \\ & \quad = (I - T_o(G_n, K_m)) \begin{bmatrix} \Delta_U \\ \Delta_V \end{bmatrix} \Phi^{-1} \begin{bmatrix} \tilde{N}_n & \tilde{M}_n \end{bmatrix} + \text{higher order terms} \end{aligned} \tag{6.53}$$

where $\begin{bmatrix} \Delta_U \\ \Delta_V \end{bmatrix} = \begin{bmatrix} U_m \\ V_m \end{bmatrix} - \begin{bmatrix} U_r \\ V_r \end{bmatrix}$ and where the second equality results from a first-order approximation.

In case of numerical issues due to the high order of the transfer matrices appearing in this criterion, a zeroth-order approximation can be used instead:

$$J^{(0)}_{perf}(\hat{K}_r) = \left\| \left(\begin{bmatrix} U_m \\ V_m \end{bmatrix} - \begin{bmatrix} U_r \\ V_r \end{bmatrix} \right) \Phi^{-1} \begin{bmatrix} \tilde{N}_n & \tilde{M}_n \end{bmatrix} \right\|_\infty \tag{6.54}$$

obtained by replacing $\hat{\Phi}$ by Φ in $J_{perf}(\hat{K}_r)$.

As in Subsection 6.4.2, these criteria can be generalised by adding input and output weighting filters W_{in} and W_{out} if one wants to filter or give more importance to some inputs and outputs of $T_o(G_n, K_m)$:

$$J^{(1)}_{perf}(\hat{K}_r) = \left\| \underbrace{W_{out}\left(I - \begin{bmatrix} U_m \\ V_m \end{bmatrix} \Phi^{-1} \begin{bmatrix} \tilde{N}_n & \tilde{M}_n \end{bmatrix}\right)}_{W_l} \times \left(\begin{bmatrix} U_m \\ V_m \end{bmatrix} - \begin{bmatrix} U_r \\ V_r \end{bmatrix}\right) \underbrace{\Phi^{-1} \begin{bmatrix} \tilde{N}_n & \tilde{M}_n \end{bmatrix} W_{in}}_{W_r} \right\|_\infty \tag{6.55}$$

$$J^{(0)}_{perf}(\hat{K}_r) = \left\| \underbrace{W_{out}}_{W_l} \left(\begin{bmatrix} U_m \\ V_m \end{bmatrix} - \begin{bmatrix} U_r \\ V_r \end{bmatrix}\right) \underbrace{\Phi^{-1} \begin{bmatrix} \tilde{N}_n & \tilde{M}_n \end{bmatrix} W_{in}}_{W_r} \right\|_\infty \tag{6.56}$$

A suboptimal $\hat{K}_r^{perf}$ is then obtained by frequency-weighted balanced truncation:

$$\begin{bmatrix} U_r \\ V_r \end{bmatrix} = \text{fwbt}\left(\begin{bmatrix} U_m \\ V_m \end{bmatrix}, W_l, W_r, r\right) \tag{6.57a}$$

$$\hat{K}_r^{perf} = U_r V_r^{-1} \tag{6.57b}$$

► **Example 6.5.** Consider again the plant model G_7, the disturbance model H_7 and the controller K_{11} of Example 6.4. In order to take the disturbances acting on the process into account during the reduction step, we shall express the closed-loop transfer matrix as in Example 6.2:

$$T_o\left(\begin{bmatrix} H_7 & G_7 \end{bmatrix}, \begin{bmatrix} 0 \\ K_{11} \end{bmatrix}\right) = \begin{bmatrix} 0 & 0 & 0 \\ \dfrac{K_{11}H_7}{1+G_7K_{11}} & \dfrac{K_{11}G_7}{1+G_7K_{11}} & \dfrac{K_{11}}{1+G_7K_{11}} \\ \dfrac{H_7}{1+G_7K_{11}} & \dfrac{G_7}{1+G_7K_{11}} & \dfrac{1}{1+G_7K_{11}} \end{bmatrix}$$

with input vector $\begin{bmatrix} e(t) & r_2(t) & -r_1(t) \end{bmatrix}^T$ and output vector $\begin{bmatrix} 0 & f(t) & g(t) \end{bmatrix}^T$. These signals are defined in Figure 2.1 and $e(t)$ is the white noise driving the disturbance system: $v(t) = H_7(s)e(t)$.

A left coprime factorisation $\begin{bmatrix} \tilde{N}_7 & \tilde{M}_7 \end{bmatrix}$ of $\begin{bmatrix} H_7 & G_7 \end{bmatrix}$ and a normalised right coprime factorisation $\begin{bmatrix} U_{11} \\ V_{11} \end{bmatrix}$ of $\begin{bmatrix} 0 \\ K_{11} \end{bmatrix}$ are computed following Propositions 2.2 and 2.5. The following additional weightings are used:

$$W_{in} = \text{diag}(1, 0, 1), \quad W_{out} = \text{diag}\left(0, 0, \frac{1}{s+0.001}\right)$$

The filter $\frac{1}{s+0.001}$ in W_{out}, applied to the control error $g(t)$, aims to preserve the low-frequency properties of the closed-loop response, thus a zero static error. W_{in}

shows that the same importance is put on disturbance rejection as on reference tracking. Putting these weightings in (6.55) and (6.57), a frequency-weighted balanced realisation of $\begin{bmatrix} U_{11} \\ V_{11} \end{bmatrix}$ is obtained following the derivations of Subsection 2.7.5. The frequency-weighted Gramians are

$$\mathbf{P}_{22} = \mathbf{Q}_{22} = \text{diag}(77.98,\ 1.326,\ 0.505,\ 0.223,\ 0.145,\ 6.62\times10^{-3},\ 3.26\times10^{-3},\\ 7.09\times10^{-4},\ 9.23\times10^{-5},\ 1.88\times10^{-5},\ 1.56\times10^{-7})$$

The faster decrease of the Hankel singular values allows us to predict better results at low orders.

The frequency-weighted balanced realisation of $\begin{bmatrix} U_{11} \\ V_{11} \end{bmatrix}$ is truncated to produce the following low-order approximations of K_{11}:

$$\hat{K}_4^{perf} = \frac{0.01649s^3 + 0.9199s^2 + 0.2113s + 0.01004}{s^4 + 0.585s^3 + 0.6574s^2 + 0.0371s + 9.819\times10^{-7}}$$

$$\hat{K}_3^{perf} = \frac{0.1805s^2 + 0.6163s + 0.1256}{s^3 + 0.3305s^2 + 0.4709s - 9.2\times10^{-6}}$$

$$\hat{K}_2^{perf} = \frac{0.734s + 0.08115}{s^2 + 0.3204s - 2.33\times10^{-5}}$$

$$\hat{K}_1^{perf} = \frac{0.3097}{s - 2.132\times10^{-5}}$$

$\hat{K}_4^{perf}$ and $\hat{K}_3^{perf}$ satisfy all requirements on the rise time and on the overshoot (defined in Example 6.1) and their responses (see Figure 6.12) are very close to those of the optimal controller K_{11} and better than those obtained by a step of performance-oriented closed-loop model reduction (see Example 6.2, Figure 6.7). $\hat{K}_2^{perf}$ exhibits slightly more oscillating responses with an overshoot of 21%, *i.e.*, 1% more than allowed. $\hat{K}_1^{perf}$ stabilises the system G_7 but has an unacceptable performance. ◀

B. Performance-oriented controller reduction using LFT. A very nice and intuitive performance-based method, called *Closed-loop Balanced Truncation*, has been proposed in (Ceton *et al.*, 1993) and (Wortelboer, 1994). Its principle is the following. Consider a closed-loop system represented by an LFT as in Figure 2.2, where $\Gamma(s)$ is a generalised plant that can be represented by

$$\Gamma(s) = \begin{bmatrix} \Gamma_{11}(s) & \Gamma_{12}(s) \\ \Gamma_{21}(s) & \Gamma_{22}(s) \end{bmatrix} = \left[\begin{array}{c|cc} A & B_1 & B_2 \\ \hline C_1 & D_{11} & D_{12} \\ C_2 & D_{21} & D_{22} \end{array}\right] \tag{6.58}$$

and where $Q(s)$ is a stabilising high-order, one or two-degree-of-freedom, generalised controller (see Section 2.3) with state-space representation

$$Q(s) = \left[\begin{array}{c|c} A_Q & B_Q \\ \hline C_Q & D_Q \end{array}\right] \tag{6.59}$$

Without loss of generality, we make the assumption that $D_{22} = 0$, since if $\Gamma_{22}(s)$ is not strictly proper, the problem can always be transformed into another one where $\Gamma_{22}(s)$ is strictly proper by means of a loop transformation: see (Glover and Doyle, 1988).

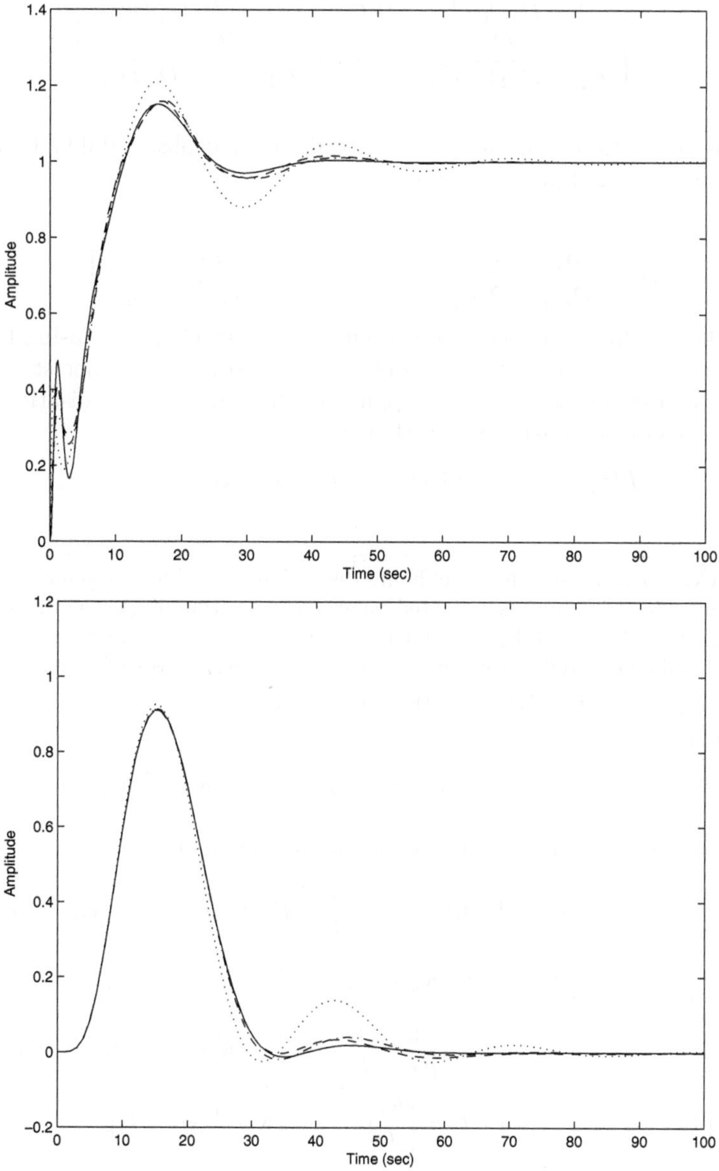

Figure 6.12. Closed-loop step responses from $r_1(t)$ (top) and $e(t)$ (bottom) of (G_7, H_7) with controllers K_{11} (—), $\hat{K}_4^{perf}$ (−−), $\hat{K}_3^{perf}$ (−·) and $\hat{K}_2^{perf}$ (···)

The closed-loop transfer matrix is then given by

$$T_{zw}(s) = \Gamma_{11}(s) + \Gamma_{12}(s)Q(s)\big(I - \Gamma_{22}(s)Q(s)\big)^{-1}\Gamma_{21}(s)$$

$$= \left[\begin{array}{cc|c} A + B_2D_QC_2 & B_2C_Q & B_1 + B_2D_QD_{21} \\ B_QC_2 & A_Q & B_QD_{21} \\ \hline C_1 + D_{12}D_QC_2 & D_{12}C_Q & D_{11} + D_{12}D_QD_{21} \end{array}\right] \tag{6.60}$$

Since this realisation is stable, its controllability and observability Gramians can be computed and partitioned as

$$\mathbf{P} = \begin{bmatrix} \mathbf{P}_\Gamma & \mathbf{P}_{\Gamma Q} \\ \mathbf{P}_{\Gamma Q}^T & \mathbf{P}_Q \end{bmatrix} \quad \text{and} \quad \mathbf{Q} = \begin{bmatrix} \mathbf{Q}_\Gamma & \mathbf{Q}_{\Gamma Q} \\ \mathbf{Q}_{\Gamma Q}^T & \mathbf{Q}_Q \end{bmatrix} \tag{6.61}$$

We can then perform balanced truncation on the controller $Q(s)$ using $\mathbf{P}_Q$ and $\mathbf{Q}_Q$, which are the sub-Gramians corresponding to the controller states: the balancing transformation T that is applied to the state-space realisation of the controller $Q(s)$ is a transformation that makes

$$T\mathbf{P}_QT^T = T^{-T}\mathbf{Q}_QT^{-1} = \Sigma = \text{diag}(\varsigma_1, \ldots, \varsigma_m) \tag{6.62}$$

▶ **Remark.** It can be shown (Schelfhout, 1996) that Closed-loop Balanced Truncation is equivalent to frequency-weighted balanced truncation of $Q(s)$ with frequency weightings $W_l = \Gamma_{12}(I - Q\Gamma_{22})^{-1}$ and $W_r = (I - \Gamma_{22}Q)^{-1}\Gamma_{21}$. In case of a single-degree-of-freedom controller, this amounts to reducing $K_m(s)$ with frequency weightings $W_l = \left[\begin{smallmatrix} I \\ -G_n \end{smallmatrix}\right](I+K_mG_n)^{-1}$ and $W_r = (I+G_nK_m)^{-1}\left[G_n \;\; I\right]$, *i.e.*, the reduction aims to minimise

$$J_{clbt}(\hat{K}_r) = \left\| \begin{bmatrix} I \\ -G_n \end{bmatrix} (I + K_mG_n)^{-1}(K_m - \hat{K}_r)(I + G_nK_m)^{-1} \begin{bmatrix} G_n & I \end{bmatrix} \right\|_\infty \tag{6.63}$$

Now, observe that $J_{clbt}(\hat{K}_r)$ is a first-order approximation of

$$\left\| \begin{bmatrix} I \\ -G_n \end{bmatrix} K_m(I + G_nK_m)^{-1} \begin{bmatrix} G_n & I \end{bmatrix} - \begin{bmatrix} I \\ -G_n \end{bmatrix} \hat{K}_r(I + G_n\hat{K}_r)^{-1} \begin{bmatrix} G_n & I \end{bmatrix} \right\|_\infty$$

$$= \left\| \begin{bmatrix} 0 & 0 \\ G_n & I \end{bmatrix} + \begin{bmatrix} I \\ -G_n \end{bmatrix} K_m(I + G_nK_m)^{-1} \begin{bmatrix} G_n & I \end{bmatrix} \right.$$
$$\left. - \begin{bmatrix} 0 & 0 \\ G_n & I \end{bmatrix} - \begin{bmatrix} I \\ -G_n \end{bmatrix} \hat{K}_r(I + G_n\hat{K}_r)^{-1} \begin{bmatrix} G_n & I \end{bmatrix} \right\|_\infty$$

$$= \left\| \begin{bmatrix} K_m \\ I \end{bmatrix} (I + G_nK_m)^{-1} \begin{bmatrix} G_n & I \end{bmatrix} - \begin{bmatrix} \hat{K}_r \\ I \end{bmatrix} (I + G_n\hat{K}_r)^{-1} \begin{bmatrix} G_n & I \end{bmatrix} \right\|_\infty$$

$$= \left\| T_o(G_n, K_n) - T_o(G_n, \hat{K}_r) \right\|_\infty \tag{6.64}$$

which is precisely the criterion $J_{perf}(\hat{K}_r)$ of (6.5) used in Paragraph A. Hence, there is no theoretical reason to prefer the coprime-factor-based controller reduction procedure of Paragraph A to Closed-loop Balanced Truncation, or vice-versa, since both aim to minimise the same criterion (6.5), involve a first-order approximation of it to make the problem tractable and accept additional input and/or output frequency weightings (see, *e.g.*, (Wortelboer and Bosgra, 1994) for a discussion on such weightings). However, because they are very different from a computational point of view, the first-order approximation of the criterion is not introduced the same way and the object that is approximated is not the same (the controller in one case, its coprime factors in the other), they do not deliver identical results. Our experience showed us that, depending on the case, one method would outperform the other, especially at very low orders, and it is therefore advisable to test both methods whenever possible. ◀

▶ **Example 6.6.** We consider the same problem as in Example 6.5, but this time Closed-loop Balanced Truncation is used to reduce K_{11}. Using the notations of (6.58), the closed-loop representation of Figure 2.2 and the same frequency weightings as in Example 6.5, we obtain:

$$\begin{aligned} w(t) &= \begin{bmatrix} e(t) & r_2(t) & -r_1(t) \end{bmatrix}^T \\ z(t) &= W_{out}(s) \begin{bmatrix} 0 & f(t) & g(t) \end{bmatrix}^T \\ h(t) &= g(t) \\ Q(s) &= K_{11}(s) \end{aligned}$$

$$\Gamma_{11} = W_{out} \begin{bmatrix} 0 & 0 & 0 \\ 0 & 0 & 0 \\ H_7 & G_7 & 1 \end{bmatrix} W_{in} \qquad \Gamma_{12} = W_{out} \begin{bmatrix} 0 \\ 1 \\ -G_7 \end{bmatrix}$$

$$\Gamma_{21} = \begin{bmatrix} H_7 & G_7 & 1 \end{bmatrix} W_{in} \qquad \Gamma_{22} = -G_7$$

which make

$$T_{zw}(s) = \mathcal{F}_l(\Gamma, Q) = W_{out}(s) T_o\left(\begin{bmatrix} H_7 & G_7 \end{bmatrix}, \begin{bmatrix} 0 \\ K_{11} \end{bmatrix} \right) W_{in}(s)$$

Closed-loop Balanced Truncation is used to produce the following low-order approximations of K_{11}:

$$\begin{aligned} \hat{K}_3^{clbt} &= \frac{0.42s^2 + 0.2959s + 0.05917}{s^3 + 0.2348s^2 + 0.2203s + 1.58 \times 10^{-7}} \\ \hat{K}_2^{clbt} &= \frac{0.9749s + 0.1433}{s^2 + 0.564s - 2.355 \times 10^{-5}} \\ \hat{K}_1^{clbt} &= \frac{0.282}{s - 3.085 \times 10^{-5}} \end{aligned}$$

They all stabilise G_7 and have slightly better performance than those obtained by performance-based coprime-factor reduction in Example 6.5: see Figure 6.13. $\hat{K}_1^{clbt}$ (not illustrated), however, still has an unacceptable performance. ◀

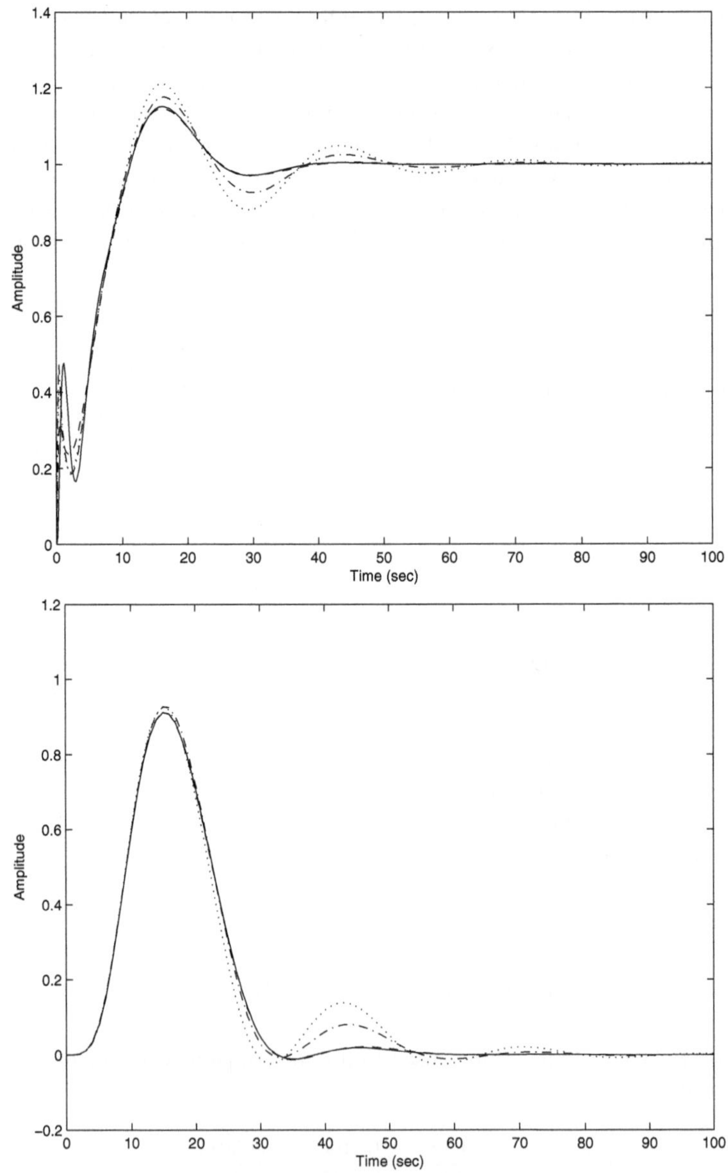

Figure 6.13. Closed-loop step responses from $r_1(t)$ (top) and $e(t)$ (bottom) of (G_7, H_7) with controllers K_{11} (—), $\hat{K}_3^{clbt}$ (−−), $\hat{K}_2^{clbt}$ (−·) and $\hat{K}_2^{perf}$ (···)

6.5.3 Stability-preserving Closed-loop Controller Reduction

A controller reduction criterion frequently addressed in the literature (see, *e.g.*, (Anderson and Liu, 1989), (Liu *et al.*, 1990), (Obinata and Anderson, 2000), (Zhou, 1995) and (Zhou and Chen, 1995)) is based on the following lemma.

Lemma 6.2. (Anderson and Liu, 1989) *Let* $G_n = \tilde{M}_n^{-1}\tilde{N}_n$ *and* $K_m = U_m V_m^{-1}$ *be two coprime factorisations satisfying the Bezout identity (2.61). Then, every* $\hat{K}_r = U_r V_r^{-1}$ *such that*

$$\left\| \begin{bmatrix} \tilde{N}_n & \tilde{M}_n \end{bmatrix} \left(\begin{bmatrix} U_m \\ V_m \end{bmatrix} - \begin{bmatrix} U_r \\ V_r \end{bmatrix} \right) \right\|_\infty < 1 \tag{6.65}$$

stabilises G_n.

This lemma induces a reduction method where the reduced-order controller $\hat{K}_r = U_r V_r^{-1}$ is obtained by minimising (6.65) over all $\left[\begin{smallmatrix} U_r \\ V_r \end{smallmatrix}\right]$ of order r. This method aims to preserve satisfaction of the Bezout identity (2.61), *i.e.*, closed-loop stability with G_n. It says however nothing about the closed-loop performance of $\hat{K}_r$.

Once again, the optimisation problem involved in this method is intractable in practice, but frequency-weighted balanced truncation can be used to compute a suboptimal $\hat{K}_r^{stab}$:

$$\begin{bmatrix} U_r \\ V_r \end{bmatrix} = \text{fwbt}\left(\begin{bmatrix} U_m \\ V_m \end{bmatrix}, \begin{bmatrix} \tilde{N}_n & \tilde{M}_n \end{bmatrix}, I, r \right) \tag{6.66a}$$

or, if $\begin{bmatrix} \tilde{N}_n & \tilde{M}_n \end{bmatrix}$ and $\left[\begin{smallmatrix} U_m \\ V_m \end{smallmatrix}\right]$ do not satisfy the Bezout identity (2.61),

$$\begin{bmatrix} U_r \\ V_r \end{bmatrix} = \text{fwbt}\left(\begin{bmatrix} U_m \\ V_m \end{bmatrix}, \Phi^{-1}\begin{bmatrix} \tilde{N}_n & \tilde{M}_n \end{bmatrix}, I, r \right) \tag{6.66b}$$

where $\Phi = \begin{bmatrix} \tilde{N}_n & \tilde{M}_n \end{bmatrix} \left[\begin{smallmatrix} U_m \\ V_m \end{smallmatrix}\right]$ and

$$\hat{K}_r^{stab} = U_r V_r^{-1} \tag{6.66c}$$

▶ **Example 6.7.** Consider again the plant model G_7, the disturbance model H_7 and the controller K_{11} of Example 6.4. A left coprime factorisation $\begin{bmatrix} \tilde{N}_7 & \tilde{M}_7 \end{bmatrix}$ of G_7 and a normalised right coprime factorisation $\left[\begin{smallmatrix} U_{11} \\ V_{11} \end{smallmatrix}\right]$ of K_{11} are computed following Propositions 2.3 and 2.4. The following reduced-order controllers are then computed using (6.66):

$$\hat{K}_6^{stab} = \frac{-0.045s^5 + 1.103s^4 + 3.31s^3 + 4.277s^2 + 1.502s + 0.2482}{s^6 + 4.03s^5 + 4.06s^4 + 7.063s^3 + 2.287s^2 + 0.8848s + 0.00256}$$

$$\hat{K}_4^{stab} = \frac{0.01007s^3 + 0.9464s^2 + 0.6974s + 0.001524}{s^4 + 0.5849s^3 + 1.306s^2 + 0.1947s - 0.001516}$$

$$\hat{K}_2^{stab} = \frac{0.01042s + 0.9637}{s^2 + 0.5565s + 0.3022}$$

They all stabilise G_7, but their performances are unacceptable and not better than those obtained by unweighted normalised coprime-factor reduction in Example 6.4, as can be seen in Figure 6.14. ◀

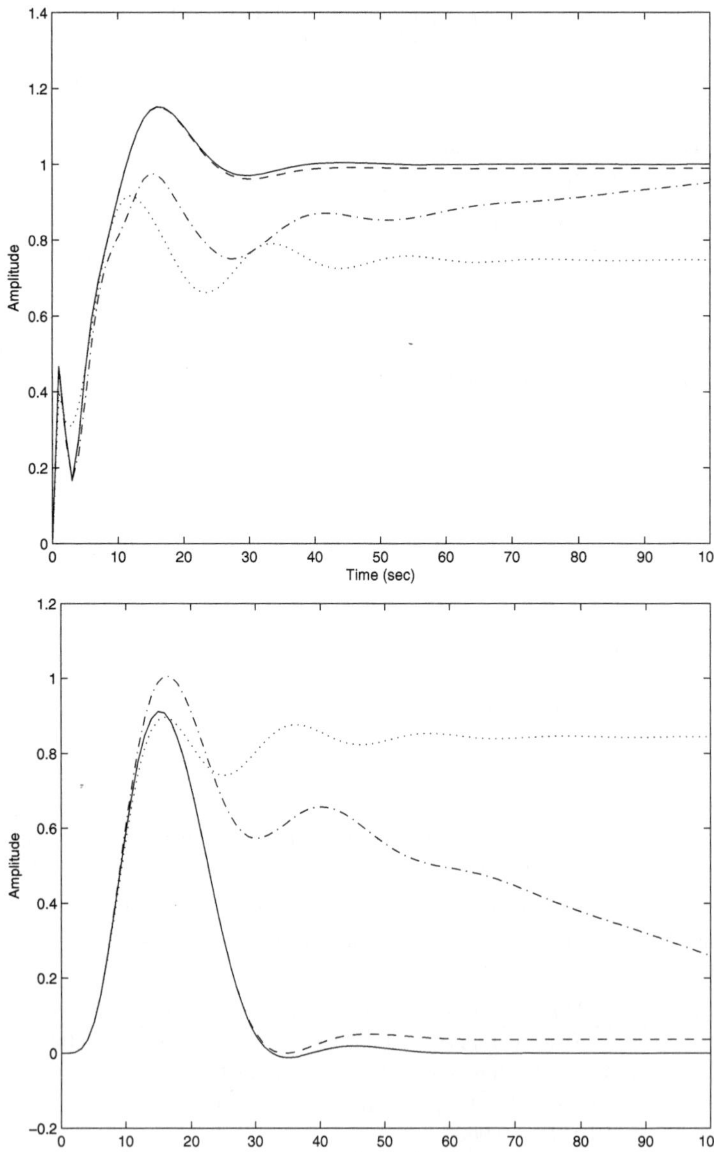

Figure 6.14. Closed-loop step responses from $r_1(t)$ (top) and $e(t)$ (bottom) of (G_7, H_7) with controllers K_{11} (—), $\hat{K}_6^{stab}$ (−−), $\hat{K}_4^{stab}$ (−·) and $\hat{K}_2^{stab}$ (···)

6.5.4 Other Closed-loop Controller Reduction Methods

Here, we briefly describe a few alternative approaches to closed-loop controller reduction. Other methods can be found in, *e.g.*, (Obinata and Anderson, 2000).

A. Performance-oriented approaches based on frequency-weighted balanced truncation. A performance-based, frequency-weighted approach to controller reduction has been proposed in (Goddard and Glover, 1993). It is based on balanced truncation or Hankel-norm approximation of the stable part of the high-order controller, the antistable part remaining unchanged, which impairs the efficiency of the method in case of an unstable controller.

The same authors have proposed a coprime-factorisation approach for performance-preserving frequency-weighted controller approximation (Goddard and Glover, 1994). Although the ultimate goal of this method is essentially the same as in our method of Subsection 6.5.2, § A., the technical aspects are much more involved.

B. Performance and stability-preserving methods that can be used with observer-based controllers. When the controller to be reduced is an observer-based controller, two methods, dual of those presented in Subsection 6.4.4, can be used for preserving closed-loop performance and stability. Recall that, in this case, the order of the controller is that of the system G_n and that it is possible to build plant and controller coprime factorisations $G_n = \tilde{M}_n^{-1}\tilde{N}_n = N_n M_n^{-1}$ and $K_n = \tilde{V}_n^{-1}\tilde{U}_n = U_n V_n^{-1}$ satisfying the following double Bezout identity:

$$\begin{bmatrix} \tilde{M}_n & -\tilde{N}_n \\ \tilde{U}_n & \tilde{V}_n \end{bmatrix} \begin{bmatrix} V_n & N_n \\ -U_n & M_n \end{bmatrix} = I \tag{6.67}$$

where both $\left[\begin{smallmatrix} \tilde{M}_n & -\tilde{N}_n \\ \tilde{U}_n & \tilde{V}_n \end{smallmatrix}\right]$ and $\left[\begin{smallmatrix} V_n & N_n \\ -U_n & M_n \end{smallmatrix}\right]$ are of order at most n.

The first method, proposed by (Anderson and Moore, 1990) and (Zhou and Chen, 1995), is based on the unweighted reduction of the right coprime factors $\left[\begin{smallmatrix} U_n \\ V_n \end{smallmatrix}\right]$ of the controller. These factors are chosen in such a way that the Bezout identity (2.61) holds with normalised left coprime factors of the system G_n. Then, if

$$\left\| \begin{bmatrix} U_n \\ V_n \end{bmatrix} - \begin{bmatrix} U_r \\ V_r \end{bmatrix} \right\|_\infty < \varepsilon < 1 \tag{6.68}$$

it is possible to show that $\hat{K}_r = U_r V_r^{-1}$ stabilises G_n and that

$$\left\| T_o(G_n, K_n) - T_o(G_n, \hat{K}_r) \right\|_\infty \leq \|T_o(G_n, K_n)\|_\infty \frac{\varepsilon}{1-\varepsilon} \tag{6.69}$$

The second method, initially proposed by (Gu, 1995), uses multiplicative error reduction of $\left[\begin{smallmatrix} V_n & N_n \\ -U_n & M_n \end{smallmatrix}\right]$ to compute the reduced-order transfer matrix $\left[\begin{smallmatrix} V_r & N_r \\ -U_r & M_r \end{smallmatrix}\right]$

as explained in Subsection 6.4.4, § B. This time, however, U_r and V_r are used to compute $\hat{K}_r = U_r V_r^{-1}$ and the closed loop $(G_n, \hat{K}_r)$ is internally stable if

$$\|\Delta\|_\infty \|T_o(G_n, K_n)\|_\infty < 1 \tag{6.70}$$

Furthermore, there holds

$$\left\| T_o(G_n, K_n) - T_o(G_n, \hat{K}_r) \right\|_\infty \leq \frac{\|T_o(G_n, K_n)\|_\infty^2 \|\Delta\|_\infty}{1 - \|T_o(G_n, K_n)\|_\infty \|\Delta\|_\infty} \tag{6.71}$$

6.6 Case Study: Design of a Low-order Controller for a PWR Nuclear Power Plant Model

We now present a detailed and realistic case study, realised in the 1990s in collaboration with ÉLECTRICITÉ DE FRANCE (EDF) in the framework of research works aimed at improving the operational flexibility of nuclear power plants by the use of advanced control techniques.

▶ **Remark.** For reasons of confidentiality, upon EDF's request, we are not allowed to publish the transfer functions of the models and controllers used in this application. ◀

6.6.1 Description of the System

A realistic nonlinear simulator, based on a first-principles model describing both primary and secondary circuits of a pressurised water reactor (PWR) nuclear power plant (see Figure 6.15), has been developed at EDF. It includes all local controllers of both primary and secondary circuits.

We focus on the behaviour of the plant around a fixed operating point corresponding to 95% of its maximum load. The linearisation of the simulator around this point results in a 42nd-order model G_{42} that includes the dynamics of the primary and secondary circuits and of all local controllers, except some specific controllers of the secondary circuit that we want to replace; these are named K_{tb} and K_{cd} in Figure 6.16. They control the electrical power and the condenser water level, respectively, and their structures are very simple: K_{tb} is a PI controller and K_{cd} is a second-order, two-input one-output controller that includes an integrator. For the sake of simplicity, K_{tb} and K_{cd} will both be called 'PID' controllers and collectively denoted by K_{pid} in the sequel.

In Figure 6.16, $W_e(t)$ is the electrical power produced by the plant, controlled to follow the demand $W_{ref}(t)$ of the network and directly related to the steam flow in the turbine, which highly depends on the high-pressure turbine inlet control valve aperture $O_{hp}(t)$. $N_{cd}(t)$ is the water level in the condenser and is

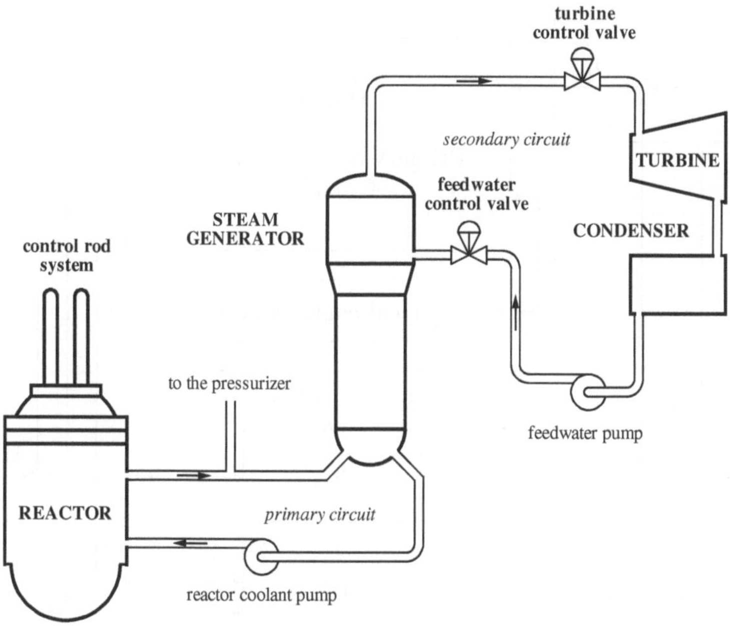

Figure 6.15. PWR plant description

affected by the extraction flow $Q_{ex}(t)$ which, in turn, depends on the extraction control valve aperture $U_{cex}(t)$. N_{sp} is the (constant) setpoint of $N_{cd}(t)$ (0 in the variational linear model we consider). $d_{O hp}(t)$ and $d_{U cex}(t)$ represent disturbances acting on the control inputs (they can also be used as excitation sources for identification), while $c_{O hp}(t)$ and $c_{U cex}(t)$ are the computed values of the two valves apertures. The strong couplings between $W_e(t)$ and $O_{hp}(t)$ on one hand, and between $N_{cd}(t)$ and $U_{cex}(t)$ on the other hand, explain the structure of the present control loops. However, the control performance might be improved by taking the cross-couplings into account.

The extraction valve used for the control of N_{cd}, as well as a constant-speed extraction pump, are located downstream of the condenser. They suck water from the condenser and send it to the feedwater tank located upstream of the feedwater pump. These components of the secondary circuit are not shown in Figure 6.15.

6.6.2 Control Objective and Design

The control objective is to improve the flexibility of the plant in order to make it follow fast medium-range variations in the power demand $W_{ref}(t)$ by acting on the steam flow in the turbine (to which $W_e(t)$ is approximately proportional) using the high-pressure turbine inlet valve (*i.e.*, $O_{hp}(t)$ as control signal) rather

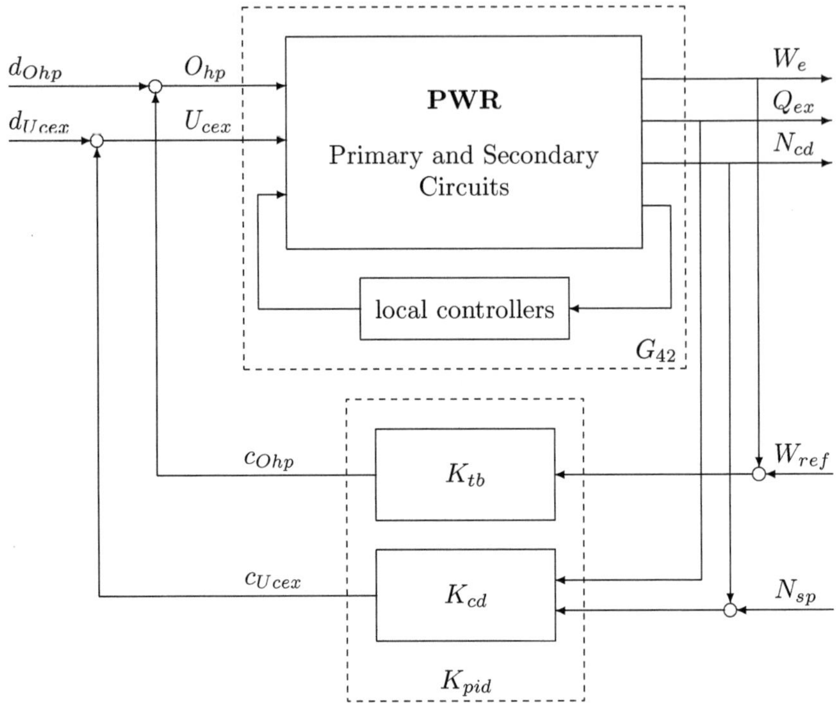

Figure 6.16. Interconnection of G_{42} with the PID controllers

than the control rods of the reactor (*i.e.*, without changing the nuclear thermal power produced in the primary circuit). In order to keep the water level in the steam generator and in the condenser at their respective setpoints, it is necessary to act, respectively, on the feedwater flow and on the extraction water flow. This is made possible by the feedwater tank, which plays the role of a buffer between both parts.

The control of the steam generator water level is not studied here. Our primary concern is the control of the plant load $W_e(t)$ and our secondary concern is the regulation of the condenser water level $N_{cd}(t)$. In order to achieve the required level of flexibility, we shall replace the current PID controllers K_{tb} and K_{cd} by a single multivariable controller able to take the interactions between both subsystems into account. The chosen design method is Linear Quadratic Gaussian (LQG) control using either a low-order approximation $\hat{G}_r$ of the plant model G_{42}, or this model itself. The extraction water flow $Q_{ex}(t)$ is measured and can be used by the state observer of the controller for better state estimation.

The verification of the controller performance will be done with step signals applied at the W_{ref}, d_{Ohp} and d_{Ucex} inputs of the system.

In order to remain consistent throughout the comparative study, the same LQG criterion will be used with each full-order or reduced-order model:

$$J_{LQG}\big(O_{hp}(t), U_{cex}(t)\big) = \int_0^{\infty} \left(\left| F(s) \begin{bmatrix} W_{err}(t) \\ N_{cd}(t) \end{bmatrix} \right|_Q^2 + \left| \begin{bmatrix} O_{hp}(t) \\ U_{cex}(t) \end{bmatrix} \right|_R^2 \right) dt \quad (6.72)$$

where $W_{err}(t) = W_e(t) - W_{ref}(t)$ is the tracking error and $F(s) = \frac{s+20}{s}$ is a filter aimed to ensure a zero static error. This loop-shaping filter is then connected to the corresponding inputs of the designed controller.

All signals are normalised and expressed in engineering units. Since the primary objective is to control $W_e(t)$, more weight is put on the electrical power tracking error than on the condenser water level in J_{LQG}. The following weighting matrices have proved very satisfactory: $Q = \text{diag}(1, 0.01)$, $R = \text{diag}(100, 100)$.

For the design of the Kalman filter, the external signals d_{Ohp}, d_{Ucex} and W_{ref}, which are in the low frequency range, are modelled as independent Gaussian white noises filtered through a low-pass filter $N(s) = \frac{1}{s+0.1}$. Their covariance matrix is unknown, as well as that of the measurement noise, and they can be used as tuning parameters. The following choice has lead to good results: $Q_n = I_{3\times 3}$ for the input disturbances covariance matrix and $R_n = 0.01 I_{3\times 3}$ for the measurement noise covariance matrix (remember that Q_{ex} is measured and used for state estimation, although it is not regulated, which is why R_n is 3×3).

The presence of the filters $F(s)$ and $N(s)$, respectively on the two controlled outputs and on the three exogenous inputs of the (augmented) model used for the design, will yield a controller with order equal to that of the original model ($\hat{G}_r$ or G_{42}) plus 5. Furthermore, $F(s)$ is a loop-shaping filter that has to be explicitly added to the designed controller before it can be used with the original model $\hat{G}_r$ or G_{42} and the total order of the controller will therefore finally be that of $\hat{G}_r$ or G_{42} plus 7.

Because of the loop-shaping filters, this control design method may not be the best regarding the resulting controller order. It is, however, very convenient for illustrating the subsequent use of order reduction techniques.

6.6.3 System Identification

The first approach we consider for the computation of a low-order model for the PWR plant is closed-loop identification. The identification experiments were carried out using the nonlinear simulator as the plant. The identification was carried out in closed loop with excitation signals d_{Ohp} and d_{Ucex} in the frequency range of interest. As seen in the previous chapters, this has the effect of producing a model that is accurate in the frequency range that is important for control design. Our goal was to obtain a reasonably low-order linear model for the plant around the chosen operating point.

A. MIMO state-space description. Consider the system depicted in Figure 6.16. The physical process is described by a discrete-time LTI system around an operating point by the following equations:

$$\begin{bmatrix} x(k+1) \\ y(k) \end{bmatrix} = \begin{bmatrix} A & B \\ C & D \end{bmatrix} \begin{bmatrix} x(k) \\ u(k) \end{bmatrix} \tag{6.73}$$

where $y(k) = \begin{bmatrix} W_e(k) & Q_{ex}(k) & N_{cd}(k) \end{bmatrix}^T$, $u(k) = \begin{bmatrix} O_{hp}(k) & U_{cex}(k) \end{bmatrix}^T$ and $x(k)$ represent respectively the output, the input and the state at sample time kt_s. The sampling period is $t_s = 0.2$ s.

Physical insights and preliminary identification of several SISO and MISO transfer functions were used to provide insight into an appropriate identifiable parametrisation with a low number of parameters, the other entries of A, B, C and D being set to 0 or 1. The electrical power and the water flow are mainly related to the control inputs by third-order systems, while the water level integrates the water flow. Furthermore, O_{hp} affects essentially the electrical power, while U_{cex} affects the water flow. Hence, the system is diagonal dominant with some cross-couplings. These insights led to an identifiable state-space realisation of order 7 with a vector θ of 19 free parameters:

$$\left[\begin{array}{c|c} A(\theta) & B(\theta) \\ \hline C(\theta) & D(\theta) \end{array}\right] = \left[\begin{array}{ccccccc|cc} 0 & 1 & 0 & 0 & 0 & 0 & 0 & b_1 & 0 \\ 0 & 0 & 1 & 0 & 0 & 0 & 0 & b_2 & 0 \\ a_1 & a_2 & a_3 & a_4 & 0 & a_5 & 0 & b_3 & 0 \\ 0 & 0 & 0 & 0 & 1 & 0 & 0 & 0 & b_4 \\ 0 & 0 & 0 & 0 & 0 & 1 & 0 & 0 & b_5 \\ 0 & 0 & a_6 & a_7 & a_8 & a_9 & 0 & 0 & b_6 \\ 0 & 0 & a_{10} & a_{11} & 0 & 0 & 1 & 0 & 0 \\ \hline 1 & 0 & 0 & 0 & 0 & 0 & 0 & d_1 & 0 \\ 0 & 0 & 0 & 1 & 0 & 0 & 0 & 0 & d_2 \\ 0 & 0 & 0 & 0 & 0 & 0 & 1 & 0 & 0 \end{array}\right] \tag{6.74}$$

B. Identification. The output of the model is

$$y(k,\theta) = \Big(C(\theta)\big(zI - A(\theta)\big)^{-1}B(\theta) + D(\theta)\Big)u(k) \tag{6.75}$$

A set of 10000 input/output data samples collected on the closed-loop system made up of the nonlinear simulator with the PID controllers K_{tb} and K_{cd} was used to determine the parameter estimates. A standard quadratic prediction-error criterion was minimised.

A continuous-time model of order 7, $\hat{G}_7^{clid}(s)$, could be obtained by first-order approximation of the identified discrete-time model, *i.e.*, by setting $\dot{x}(t) = (x(k) - x(k-1))/t_s$, *i.e.*, $s = (1 - z^{-1})/t_s$, since the sampling period $t_s = 0.2$ s is much smaller than the fastest natural time constant of the system. $\hat{G}_7^{clid}$ is then validated by checking how well it matches the behaviour of the nonlinear plant when simulated with a new data set (Figure 6.17).

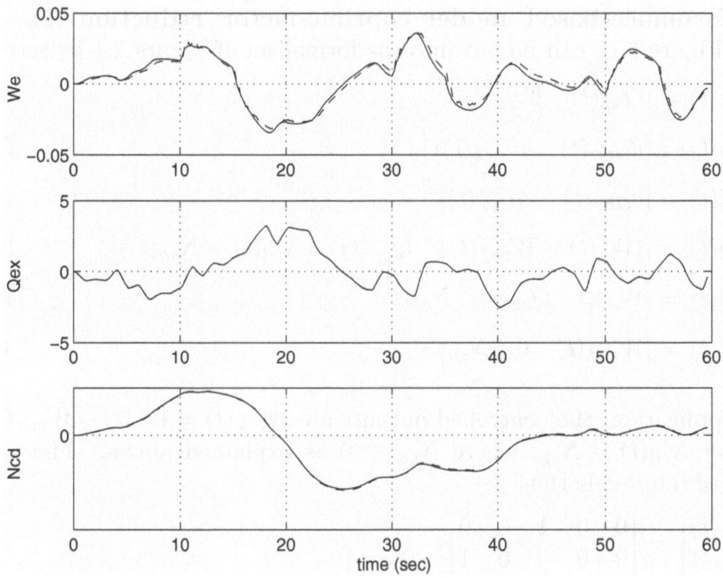

Figure 6.17. Time-domain responses of the nonlinear simulator (—) and $\hat{G}_7^{clid}$ (– –) to validation data

C. Control design. This continuous-time model $\hat{G}_7^{clid}$ was used to compute an LQG controller following the procedure of Subsection 6.6.2. This controller has order 14 and will be called $K_{14}(\hat{G}_7^{clid})$. Its performance will be studied in Subsection 6.6.6.

6.6.4 Model Reduction

Our second approach for obtaining a low-order controller is model reduction applied to the high-order model G_{42} and followed by LQG control design.

A. Open-loop model coprime-factor reduction. Since the nominal model G_{42} is unstable, the reduction procedure was applied to its normalised coprime factors $\left[\tilde{N}_{42} \quad \tilde{M}_{42}\right]$ following the procedure of Subsection 6.4.1:

$$\left[\tilde{N}_r \quad \tilde{M}_r\right] = \text{bt}\left(\left[\tilde{N}_{42} \quad \tilde{M}_{42}\right], r\right) \tag{6.76a}$$

$$\hat{G}_r^{olcf} = \tilde{M}_r^{-1}\tilde{N}_r \tag{6.76b}$$

We computed a 7th-order model $\hat{G}_7^{olcf}$ for the sake of establishing a comparison with the 7th-order model $\hat{G}_7^{clid}$ obtained by identification. It was used to compute a 14th-order LQG controller $K_{14}(\hat{G}_7^{olcf})$ following the procedure of Subsection 6.6.2.

B. Performance-based model coprime-factor reduction. The PWR plant of Figure 6.16 can be put into the formalism of Figure 2.1 by setting

$$u(t) = \begin{bmatrix} O_{hp}(t) & U_{cex}(t) \end{bmatrix}^T \tag{6.77a}$$

$$r_2(t) = \begin{bmatrix} d_{Ohp}(t) & d_{Ucex}(t) \end{bmatrix}^T \tag{6.77b}$$

$$f(t) = \begin{bmatrix} c_{Ohp}(t) & c_{Ucex}(t) \end{bmatrix}^T \tag{6.77c}$$

$$g(t) = \begin{bmatrix} (W_e(t) - W_{ref}(t)) & Q_{ex}(t) & (N_{cd}(t) - N_{sp}(t)) \end{bmatrix}^T \tag{6.77d}$$

$$y(t) = \begin{bmatrix} W_e(t) & Q_{ex}(t) & N_{cd}(t) \end{bmatrix}^T \tag{6.77e}$$

$$r_1(t) = \begin{bmatrix} W_{ref}(t) & 0 & N_{sp} \end{bmatrix}^T \tag{6.77f}$$

In this application, the controlled outputs are $W_{err}(t) = W_e(t) - W_{ref}(t)$ and $N_{err}(t) = N_{cd}(t) - N_{sp}$, where $N_{sp} \equiv 0$ as explained above. The transfer function of interest is thus

$$\begin{bmatrix} W_{err}(t) \\ N_{err}(t) \end{bmatrix} = \underbrace{\begin{bmatrix} 0 & 0 & 1 & 0 & 0 \\ 0 & 0 & 0 & 0 & 1 \end{bmatrix}}_{W_{out}}$$

$$\times \underbrace{\begin{bmatrix} U_{pid} \\ V_{pid} \end{bmatrix} \Phi^{-1} \begin{bmatrix} \tilde{N}_{42} & \tilde{M}_{42} \end{bmatrix}}_{T_o(G_{42}, K_{pid})} \underbrace{\begin{bmatrix} 1 & 0 & 0 & 0 \\ 0 & 1 & 0 & 0 \\ 0 & 0 & 1 & 0 \\ 0 & 0 & 0 & 0 \\ 0 & 0 & 0 & 1 \end{bmatrix}}_{W_{in}} \begin{bmatrix} d_{Ohp}(t) \\ d_{Ucex}(t) \\ W_{ref}(t) \\ N_{sp} \end{bmatrix} \tag{6.78}$$

where $\begin{bmatrix} U_{pid} \\ V_{pid} \end{bmatrix}$ is a right coprime factorisation of K_{pid}. It yields the following reduction criterion[3] :

$$J_{perf}^{(0)}(\hat{G}_r) = \left\| W_l \left(\begin{bmatrix} \tilde{N}_{42} & \tilde{M}_{42} \end{bmatrix} - \begin{bmatrix} \tilde{N}_r & \tilde{M}_r \end{bmatrix} \right) W_r \right\|_\infty \tag{6.79}$$

where $W_r = W_{in}$ and $W_l = W_{out} \begin{bmatrix} U_{pid} \\ V_{pid} \end{bmatrix} \Phi^{-1}$. We had to use $J_{perf}^{(0)}(\hat{G}_r)$, *i.e.*, the zeroth-order approximation of the desired criterion $J_{perf}(\hat{G}_r)$, instead of its

[3]The choice of an $\mathcal{H}_\infty$-based reduction criterion may seem inappropriate, since our control design criterion is $\mathcal{H}_2$-based. We make the following remarks regarding this choice.

- In practice, the choice of an $\mathcal{H}_\infty$-based reduction criterion is immaterial since the reduction is carried out by means of balanced truncation, which does not minimise the $\mathcal{H}_\infty$ norm.
- Our objective is to discuss the potential interest that one can have in taking the closed-loop interconnection into account during model or controller reduction, the reduction method being imposed *a priori*. This is very similar to the problem of computing an uncertainty region that can be used for $\mathcal{H}_\infty$ robust control design using prediction-error identification with an $\mathcal{H}_2$-based criterion, as we do in Chapter 5.
- Most reduction methods are based on $\mathcal{H}_\infty$ robustness criteria.
- $\mathcal{H}_2$ norm approximation is, as we already mentioned, a very difficult problem for which no reliable solution exists.

first-order approximation $J^{(1)}_{perf}(\hat{G}_r)$, in order to avoid numerical issues raised by the high order of the additional weighting in $J^{(1)}_{perf}(\hat{G}_r)$.

Bilateral frequency-weighted balanced truncation was used to produce two low-order models:

$$\begin{bmatrix}\tilde{N}_r & \tilde{M}_r\end{bmatrix} = \text{fwbt}\left(\begin{bmatrix}\tilde{N}_{42} & \tilde{M}_{42}\end{bmatrix}, W_l, W_r, r\right) \tag{6.80a}$$

$$\hat{G}_r^{perf} = \tilde{M}_r^{-1}\tilde{N}_r \tag{6.80b}$$

Firstly, we computed a 7th-order model $\hat{G}_7^{perf}$ for the sake of establishing a comparison with the 7th-order model $\hat{G}_7^{clid}$ obtained by identification. It was used to compute a 14th-order LQG controller $K_{14}(\hat{G}_7^{perf})$ following the procedure of Subsection 6.6.2. Secondly, we computed a 3rd-order model $\hat{G}_3^{perf}$. This is the lowest-order model, computed by this method, that produced a stabilising LQG controller for G_{42}. This controller has order 10 and is denoted by $K_{10}(\hat{G}_3^{perf})$. The performance of these two controllers will be studied in Subsection 6.6.6.

6.6.5 Controller Reduction

A. Design of the optimal high-order controller. The first step in this approach consists in designing the optimal LQG controller for G_{42} following the procedure of Subsection 6.6.2. Therefore, a minimal realisation of G_{42} must be used. Such a realisation has order 37 and the corresponding optimal LQG controller has order 44. It will be denoted by K_{44}. This controller has a faster and less oscillatory response than K_{pid} (see Subsection 6.6.6).

B. Open-loop controller coprime-factor reduction. Since the optimal controller K_{44} is unstable, the reduction procedure was applied to its normalised coprime factors $\begin{bmatrix}U_{44}\\ V_{44}\end{bmatrix}$ following the procedure of Subsection 6.5.1:

$$\begin{bmatrix}U_r\\ V_r\end{bmatrix} = \text{bt}\left(\begin{bmatrix}U_{44}\\ V_{44}\end{bmatrix}, r\right) \tag{6.81a}$$

$$\hat{K}_r^{olcf} = U_r V_r^{-1} \tag{6.81b}$$

Two controllers were computed: a 10th-order controller $\hat{K}_{10}^{olcf}$, for the sake of establishing a comparison with the 10th-order controller $K_{10}(\hat{G}_3^{perf})$ obtained via closed-loop model reduction, and a 7th-order controller $\hat{K}_7^{olcf}$ that will be compared to the 7th-order controller $\hat{K}_7^{perf}$ obtained by performance-based controller reduction: see below. Their performances will be studied in Subsection 6.6.6.

C. Performance-based controller coprime-factor reduction. The same reasoning as in closed-loop model reduction yields the following reduction criterion:

$$J^{(0)}_{perf}(\hat{K}_r) = \left\| W_l \left(\begin{bmatrix} U_{44} \\ V_{44} \end{bmatrix} - \begin{bmatrix} U_r \\ V_r \end{bmatrix} \right) W_r \right\|_\infty \tag{6.82}$$

where $\begin{bmatrix} U_{44} \\ V_{44} \end{bmatrix}$ is a normalised coprime factorisation of K_{44}, $W_l = W_{out}$, and $W_r = \Phi^{-1} \begin{bmatrix} \tilde{N}_{42} & \tilde{M}_{42} \end{bmatrix} W_{in}$. The output weighting filter W_{out} used here is different from that used in closed-loop model reduction: better results were obtained by penalising the degradation of the control signals in $f(t)$ and by filtering $W_{err}(t)$ and $N_{err}(t)$ with a low-pass filter $\frac{1}{s+0.01}$, so as to preserve the low-frequency properties of the controller, *i.e.*, a zero static error:

$$W_l(s) = W_{out}(s) = \begin{bmatrix} 1 & 0 & 0 & 0 & 0 \\ 0 & 1 & 0 & 0 & 0 \\ 0 & 0 & \frac{1}{s+0.01} & 0 & 0 \\ 0 & 0 & 0 & 0 & \frac{1}{s+0.01} \end{bmatrix} \tag{6.83}$$

Two low-order controllers were produced using

$$\begin{bmatrix} U_r \\ V_r \end{bmatrix} = \text{fwbt}\left(\begin{bmatrix} U_{44} \\ V_{44} \end{bmatrix}, W_l, W_r, r \right) \tag{6.84a}$$

$$\hat{K}^{perf}_r = U_r V_r^{-1} \tag{6.84b}$$

These are a 10th-order controller $\hat{K}^{perf}_{10}$, which will be compared with the previously obtained 10th-order controllers $K_{10}(\hat{G}^{perf}_3)$ and $\hat{K}^{olcf}_{10}$, and a 7th-order controller $\hat{K}^{perf}_7$, which is the lowest-order controller obtained by this method that achieves closed-loop stability and acceptable performance with G_{42}. It will be compared with $\hat{K}^{olcf}_7$ computed by open-loop controller reduction. Their performances will be studied in Subsection 6.6.6.

6.6.6 Performance Analysis of the Designed Controllers

A. Controllers computed from low-order models obtained by closed-loop techniques. Here, we compare the performances of the controllers computed from low-order models obtained by means of a closed-loop identification or closed-loop model reduction. The step responses of G_{42} in closed loop with these controllers ($K_{14}(\hat{G}^{clid}_7)$, $K_{14}(\hat{G}^{perf}_7)$ and $K_{10}(\hat{G}^{perf}_3)$) are shown in Figure 6.18. Clearly, they all have a much better performance than the original K_{pid} and their responses are very close to those of the optimal controller K_{44} designed from G_{42}.

B. Controllers computed by closed-loop controller reduction. The two controllers $\hat{K}^{perf}_{10}$ and $\hat{K}^{perf}_7$ obtained by performance-based reduction of K_{44} have performances comparable to that of the latter (see Figure 6.19) and of

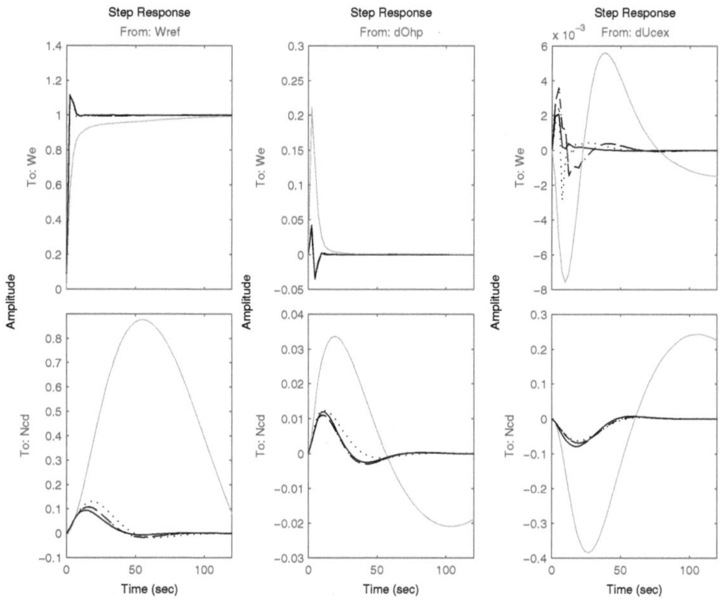

Figure 6.18. Responses of G_{42} to steps on W_{ref}, d_{Ohp} and d_{Ucex} with controllers K_{44} (—), $K_{14}(\hat{G}_7^{clid})$ (−−), $K_{14}(\hat{G}_7^{perf})$ (−·), $K_{10}(\hat{G}_3^{perf})$ (···) and K_{pid} (—)

the controllers obtained *via* closed-loop model identification or reduction (compare Figure 6.19 with Figure 6.18). Closed-loop controller reduction, however, allows one to obtain a satisfactory controller with a lower order (order 7 for $\hat{K}_7^{perf}$) than closed-loop model reduction (order 10 for $K_{10}(\hat{G}_3^{perf})$). The degradation of the response of $W_e(t)$ to disturbances on $d_{Ucex}(t)$ is not an issue, since the amplitude of the response remains in the range of that of the response to $d_{Ohp}(t)$. K_{pid} was over-performing regarding the response of $W_e(t)$ to $d_{Ucex}(t)$, but under-performing regarding its response to $d_{Ohp}(t)$ and regarding the response of $N_{cd}(t)$ to both disturbances.

C. Controllers obtained by open-loop reduction techniques. The three controllers $K_{14}(\hat{G}_7^{olcf})$, $\hat{K}_{10}^{olcf}$ and $\hat{K}_7^{olcf}$ obtained by means of open-loop model or controller reduction have unsatisfactory performances, as shown in Figure 6.20. They all exhibit nonzero static errors, especially $K_{14}(\hat{G}_7^{olcf})$ on $N_{cd}(t)$.

We made further tests showing that a zero static error could be obtained *via* open-loop model reduction only for $r \geq 12$, yielding a controller order greater or equal to 19. Using open-loop controller reduction, a nonzero static error is obtained only for controller order greater or equal to 17 and, using closed-loop stability-based controller reduction (see Subsection 6.5.3), only for controller order greater or equal to 16.

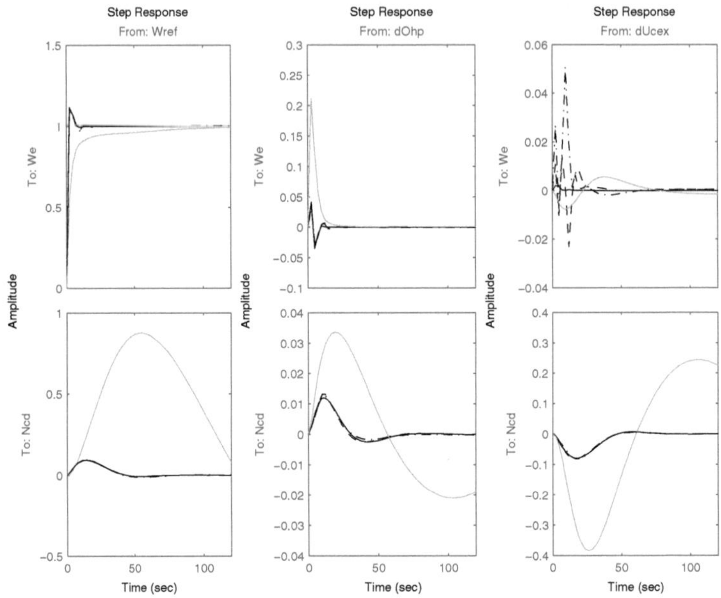

Figure 6.19. Responses of G_{42} to steps on W_{ref}, d_{Ohp} and d_{Ucex} with controllers K_{44} (—), $\hat{K}_{10}^{perf}$ (−−), $\hat{K}_{7}^{perf}$ (−·) and K_{pid} (—)

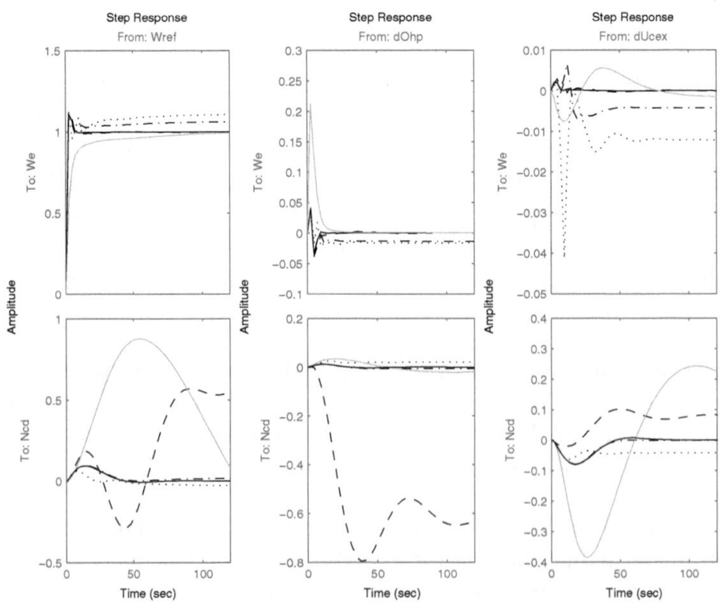

Figure 6.20. Responses of G_{42} to steps on W_{ref}, d_{Ohp} and d_{Ucex} with controllers K_{44} (—), $K_{14}(\hat{G}_{7}^{olcf})$ (−−), K_{10}^{olcf} (−·), K_{7}^{olcf} (···) and K_{pid} (—)

6.7 Classification of the Methods and Concluding Remarks

Following the observations made in Section 6.6 and in the various examples of this chapter, the reduction methods can be classified, as in table Table 6.1, in function of the performances achieved by controllers of a given order (when this order is sufficiently low to make differences appear) or, alternatively, in function of the lowest controller order that can be reached while achieving a pre-specified level of performance. The performance of each controller is evaluated in terms of the discrepancy between its responses and those of the optimal high-order controller.

Table 6.1. Classification of the model and controller reduction methods

Performance (given order)	**Method**	**Controller order** (given performance)
MAX	Performance-based controller coprime-factor reduction Closed-loop Balanced Truncation	MIN
↘	Performance-based model coprime-factor reduction	↗
↘↘	Stability-based controller coprime-factor reduction Unweighted controller coprime-factor reduction	↗↗
MIN	Unweighted model coprime-factor reduction	MAX

The case study of Section 6.6 has also shown that closed-loop identification of a low-order model is a good alternative to model reduction.

As a result, the best approach when one wants to design a low-order controller for a high-order process consists of the following steps:

1. to start from a high-order model of the process. If this model results from some identification step, it is better to carry out the identification in closed loop (see Chapter 3);
2. to design a high-order controller based on this model;
3. to reduce the order of the controller using a performance-based closed-loop method like those presented in Subsection 6.5.2.

Methods based on closed-loop low-order model identification or performance-oriented closed-loop model reduction are very satisfactory too.

As a conclusion to this chapter, the following remarks can be made:

- For a pre-specified level of performance, a given method (model or controller reduction) makes it possible to compute a lower-order controller if it is used in its closed-loop version rather than in open loop. Alternatively, for a given order, the controllers computed by means of closed-loop methods have generally a better performance than those computed by means of open-loop methods.

- Closed-loop identification proves to be a viable alternative to model or controller reduction for the computation of low-order controllers. It constitutes in fact an implicit order reduction method in the $\mathcal{L}_2$ norm that has the added advantage that no high-precision model is necessary to start with (such a model may sometimes be difficult to obtain). Only data collected in closed loop, generally during normal operation of the plant, is required. On the other hand, the absence of a precise model of the plant may pose problems when trying to assess the performance of the controller before its implementation on the real process. As a result, the identified low-order model should always be delivered with an estimation of the committed modelling error in order to do robust control design or controller validation, as explained in Chapter 5.
- The quality of a low-order model obtained by closed-loop identification depends on the controller operating in the loop during data collection. In a similar way, a low-order model obtained by closed-loop model reduction depends on the controller used in the reduction criterion. It is therefore important to use a controller that is sufficiently close to the to-be-designed controller, as stressed in Chapter 3; optimised PID controllers will often suffice. As a result, it is strongly advisable to finely tune the PID controllers one would like to replace before using them for closed-loop identification or model reduction.
- Methods based on closed-loop (respectively open-loop) controller reduction appear to be potentially more powerful than those based on closed-loop (respectively open-loop) model reduction. This result is quite logical since any order reduction step implies a loss of information: it is best to keep all the information as long as possible. In other words, it may be difficult to predict how the error introduced during model reduction will propagate during the subsequent steps of the control design procedure and affect the resulting controller, while in case of controller reduction, on the other hand, the starting point is the optimal high-order controller and the loss of performance is directly related to the approximation error committed during the reduction. Another point in favour of controller reduction is the fact that the use of loop-shaping filters during control design yields controllers with higher order than the design model. It is easier to compensate for this increase in order after the design than before.

6.8 Summary of the Chapter

Starting from the same heuristic observations that motivate the use of closed-loop system identification when the objective is to use the resulting model for control design, a performance-based criterion for model or controller order reduction has been proposed (6.3). It can be used for model order reduction (6.4) as well as for controller order reduction (6.5).

To make the reduction problem tractable, the closed-loop transfer function is first expressed in function of the plant and controller coprime factors; then, a first-order (or zeroth-order) approximation of the criterion can be used to derive frequency weightings that can be used for frequency-weighted plant or controller coprime-factor reduction (6.31), (6.57). A similar approach, for controller reduction, is Closed-loop Balanced Truncation, which is based on an LFT representation of the closed loop rather than on coprime factors. Other reduction methods have also been reviewed.

The performance-based reduction methods have been compared to other methods based either on open-loop transfer function matching or on closed-loop stability preservation. Their superiority has been illustrated by means of numerical examples and a realistic industrial case study.

CHAPTER 7

Some Final Words

7.1 A Unified Framework

In the previous 210 pages, we have proposed a series of modelling tools that may seem somewhat disconnected from each other at first glance. Put together, however, they form a whole procedure for model-based control design consisting of the following steps.

1. **Obtainment of a nominal model.** This can be done by means of various modelling techniques. System identification using data collected on the real plant G_0 is just one possible option among many others (including nonlinear physical modelling followed by a step of linearisation, 'grey-box' modelling using parametrised subsystems where the parameters have a physical meaning, *etc.*). Depending on the chosen method and, in case of prediction-error identification, on the chosen model structure, a full-order (unbiased) or a low-order (biased) model can be obtained. In case of prediction-error identification, it is best to operate in closed loop, provided the current to-be-replaced controller K_{id} is not too different from the to-be-designed controller. Therefore, re-tuning the present PID controllers should not be neglected, even if the ultimate objective is to replace them by some optimal, possibly multivariable, controller; it may also be necessary to iterate. The guidelines of Chapter 4 must be followed in order to avoid nominal closed-loop stability issues caused by some controller singularities.

2. **Model reduction.** This is generally an optional step. If direct closed-loop prediction-error identification is used in the first step, it is indeed possible to identify a reduced-order model with an ideally tuned bias error. However, if a two-stage identification method is used, a step of model reduction will be needed to cancel the quasi-nonminimal modes of the model. Models obtained by first-principles modelling or other techniques may also have a very high order. If the objective is to obtain a model with order lower than that

of the true system, it is necessary to use a closed-loop oriented reduction criterion like $J_{perf}(\hat{G}_r)$ in order to tune the introduced bias error.

3. **Model validation.** After these first two steps, a nominal model G_{nom} is available. If robustness is an issue, it is necessary to have an uncertainty region $\mathcal{D}$ guaranteed to contain the true system (with some probability) at one's disposal. A possible way to build such an uncertainty region is prediction-error identification with an unbiased model structure. If an unbiased model has already been identified in the first step, the covariance matrix of its parameters can be used to build this uncertainty region. Otherwise, a new stage of identification, with an unbiased model structure, must be performed. Using this uncertainty region, it is possible to decide whether G_{nom} is validated. The worst-case ν-gap can then be used to decide if G_{nom} and $\mathcal{D}$ are potentially good for robust control design. Recall that a controller K stabilising G_{nom} is guaranteed to stabilise G_0 as well if $\delta_\nu(G_{nom}, G_0) < b_{G_{nom},K}$. It is therefore important to have $\delta_{WC}(G_{nom}, \mathcal{D}) < \sup_K b_{G_{nom},K}$ and $\delta_{WC}(G_{nom}, \mathcal{D})$ as small as possible to maximise our chance of finding a stabilising controller for G_0.

4. **Control design.** If G_{nom} is validated, it can be used to design a new controller K using any model-based control design technique.

5. **Controller reduction.** If K has a high order, which will typically be the case if it has been designed from an unbiased model by means of optimal control techniques, it may be necessary to reduce its order before implementation. This can be done using a performance-based reduction criterion like those proposed in Chapter 6; *e.g.*, $J_{perf}(\hat{K}_r)$ or $J_{clbt}(\hat{K}_r)$.

6. **Controller validation.** The final step consists of assessing the robustness of the controller regarding closed-loop stability and performance. This can be done using the validation tools $\mu_\phi(M_{\mathcal{D}}^K(e^{j\omega}))$ and $J_{WC}(\mathcal{D}, K, \omega)$ of Chapter 5.

7.2 Missing Links and Perspectives

This design procedure consists of six stages that fit rather well together, each design step (identification, reduction, controller synthesis) being subject to the approval of a directly following validation step. However, there remain some missing links that would make the whole scheme more integrated. For most of them, solutions are currently in the works.

A. Validation with a biased model structure. While a low-order model can be immediatly identified in step 1, it is a pity that one has to identify an unbiased model as well, for the sole purpose of building the uncertainty region

$\mathcal{D}$. Stochastic embedding provides a solution to this problem. The key idea behind prediction-error identification with stochastic embedding assumptions is that the unmodelled dynamics (*i.e.*, the bias error) are considered as the realisation of a zero-mean stochastic process. The whole procedure results in ellipsoids in the Nyquist plane, as is the case with unbiased model structures. See, *e.g.*, (Bombois *et al.*, 2000, 2001*a*, 2003, 2004).

B. Validation tools for multivariable systems. Most validation tools presented in Chapter 5 are only valid for SISO systems. Solutions have however recently been worked out for the MISO case: see (Bombois and Date, 2003*a*, 2003*b*).

C. Control design based on prediction-error uncertainty sets. The validation tools presented in this book allow one to test, after control design, if the designed controller will stabilise and achieve a required level of performance with all systems contained in a prediction-error uncertainty set $\mathcal{D}$. However, these are only verification tools, *i.e.*, $\mathcal{D}$ is not used during control design. Moreover, it is not one of the classical uncertainty descriptions used in mainstream robust control. X. Bombois and collaborators, again, have worked out a robust control design procedure based on such a set: see (Bombois *et al.*, 2002).

D. Optimal identification design with respect to the ν-gap. Since we want uncertainty regions with the smallest possible worst-case ν-gap, it would be interesting to design the identification experiment so as to make it optimal with respect to this objective. A new theory on this subject has recently been proposed in (Hildebrand and Gevers, 2003).

7.3 Model-free Control Design

As all the needed information is contained in the data, one might ask if the intermediate step of modelling is really necessary and whether it would be possible to design an optimal controller directly from the data. It is indeed possible, using a procedure called Iterative Feedback Tuning. Its principle is the following: a quadratic cost function, similar to an LQG or GPC design criterion, must be minimised with respect to the controller parameters. The structure and the order of the controller can be chosen freely and can be adjusted between two iterations if they prove unsatisfactory. In order to minimise the criterion, it is necessary to start with a stabilising controller. The data collected during closed-loop plant operation can then be used to compute an unbiased estimate of the gradient and of the Hessian of the cost function with respect to the controller parameters. It is thus possible to adapt these parameters using an iterative gradient or Gauss-Newton optimisation algorithm, until a (at least local) minimum of the cost function is reached.

The method was first presented in (Hjalmarsson *et al.*, 1994*b*). Since then, it has been successfully applied to various processes ranging from laboratory lightly damped resonant mechanical systems (Hjalmarsson *et al.*, 1995; Codrons *et al.*, 1998; Ceysens and Codrons, 1995, 1997) to industrial chemical processes (Hjalmarsson *et al.*, 1997, 1998). Research is ongoing to improve its efficacy: see, *e.g.*, (Lequin *et al.*, 2003; Sjöberg *et al.*, 2003; Hildebrand *et al.*, 2004*a*, 2004*b*).

References

Anderson, B.D.O., and J.B. Moore (1990). *Optimal Control: Linear Quadratic Methods.* Prentice-Hall. Englewood Cliffs, New Jersey, USA.

Anderson, B.D.O., and Y. Liu (1989). Controller reduction: Concepts and approaches. *IEEE Transactions on Automatic Control* **34**(8), 802–812.

Anderson, B.D.O., X. Bombois, M. Gevers and C. Kulcsár (1998). Caution in iterative modeling and control design. In: *Proceedings of the 1998 IFAC Workshop on Adaptive Systems in Control and Signal Processing.* Glasgow, Scotland. pp. 13–19.

Ansay, P., M. Gevers and V. Wertz (1999). Closed-loop or open-loop models in identification for control. In: *CD-ROM Proceedings of the 5th European Control Conference.* Karlsruhe, Germany. Paper AP3-2 (F544).

Åström, K.J. (1993). Matching criteria for control and identification. In: *Proceedings of the 2nd European Control Conference.* Groningen, The Netherlands. pp. 697–701.

Åström, K.J., and B. Wittenmark (1995). *Adaptive Control.* Series in Electrical Engineering: Control Engineering. 2nd ed. Addison-Wesley.

Åström, K.J., and T. Hägglund (1995). *PID Controllers: Theory, Design, and Tuning.* 2nd ed. Instrument Society of America. USA.

Bernstein, D.S., and D.C. Hyland (1984). The optimal projection equations for fixed-order dynamic compensation. *IEEE Transactions on Automatic Control* **29**, 1034–1037.

Bitmead, R.R., M. Gevers and V. Wertz (1990). *Adaptive Optimal Control – The Thinking Man's GPC.* Series in systems and control engineering. Prentice-Hall. Englewood Cliffs, New Jersey, USA.

Blondel, V., M. Gevers and R.R. Bitmead (1997). When is a model good for control design? In: *Proceedings of the 36th IEEE Conference on Decision and Control.* San Diego, California, USA. pp. 1283–1288.

Bombois, X. (2000). Connecting Prediction Error Identification and Robust Control Analysis: A New Framework. PhD thesis. Université Catholique de Louvain. Louvain-la-Neuve, Belgium.

Bombois, X., and P. Date (2003*a*). Connecting PE identification and robust control theory: the multiple-input single-output case. Part I: Uncertainty region validation. In: *CD-ROM Proceedings of the 13th IFAC System Identification Symposium (SYSID 2003).* Rotterdam, The Netherlands. Paper WeA01-02.

Bombois, X., and P. Date (2003*b*). Connecting PE identification and robust control theory: the multiple-input single-output case. Part II: Controller validation. In: *CD-ROM Proceedings of the 13th IFAC System Identification Symposium (SYSID 2003).* Rotterdam, The Netherlands. Paper WeA01-03.

Bombois, X., B.D.O Anderson and M. Gevers (2001*a*). Frequency domain image of a set of linearly parametrized transfer functions. In: *CD-ROM Proceedings of the 6th European Control Conference*. Porto, Portugal. pp. 1416–1421. paper WeA07-1.

Bombois, X., B.D.O Anderson and M. Gevers (2003). Quantification of frequency domain error bounds with guaranteed confidence level in prediction error identification. submitted to Systems and Control Letters.

Bombois, X., B.D.O Anderson and M. Gevers (2004). Frequency domain uncertainty sets with guaranteed probability level in prediction error identification. In: *16th International Symposium on Mathematical Theory of Networks and Systems*. Leuven, Belgium.

Bombois, X., G. Scorletti, B.D.O. Anderson, M. Gevers and P.M.J. Van den Hof (2002). A new robust control design procedure based on a PE identification uncertainty set. In: *Proceedings of the 15th IFAC World Congress*. Barcelona, Spain. pp. 2842–2847.

Bombois, X., M. Gevers and G. Scorletti (2000). Controller validation for stability and for performance based on a frequency domain uncertainty region designed with stochastic embedding. In: *CD-ROM Proceedings of the 39th IEEE Conference on Decision and Control*. Sydney, Australia. Paper TuM06-5.

Bombois, X., M. Gevers, G. Scorletti and B.D.O. Anderson (2001*b*). Robustness analysis tools for an uncertainty set obtained by prediction error identification. *Automatica* **37**(10), 1629–1636.

Boulet, B., and B.A. Francis (1998). Consistency of open-loop experimental frequency-response data with coprime factors plant models. *IEEE Transactions on Automatic Control* **43**(12), 1680–1691.

Boyd, S.P., and J. Doyle (1987). Comparison of peak and RMS gains for discrete time systems. *Systems and Control Letters* **9**, 1–6.

Boyd, S.P., L. El Ghaoui, L. Feron and V. Balakrishnan (1994). *Linear Matrix Inequalities in System and Control Theory*. SIAM Studies in Applied Mathematics. SIAM. Philadelphia, Pennsylvania, USA.

Campi, M.C., A. Lecchini and S.M. Savaresi (2000). Virtual reference feedback tuning (VRFT): A new direct approach to the design of feedback controllers. In: *Proceedings of the 39th IEEE Conference on Decision and Control*. Sydney, Australia. pp. 623–628.

Campi, M.C., A. Lecchini and S.M. Savaresi (2002). Controller validation: A probabilistic approach based on prediction error identification. Submitted for publication.

Ceton, C., P.M.R. Wortelboer and O.H. Bosgra (1993). Frequency-weighted closed-loop balanced truncation. In: *Proceedings of the 2nd European Control Conference*. Groningen, The Netherlands. pp. 697–701.

Ceysens, B., and B. Codrons (1995). Synthèse itérative de contrôleurs sans identification: Étude et validation expérimentale. Master's thesis. Université Catholique de Louvain, Faculté des Sciences Appliquées. Louvain-la-Neuve, Belgium. Original French version.

Ceysens, B., and B. Codrons (1997). Iterative identificationless control design. *Journal A* **38**(1), 26–30.

Chen, J. (1997). Frequency-domain tests for validation of linear fractional uncertainty models. *IEEE Transactions on Automatic Control* **42**(6), 748–760.

Codrons, B. (2000). Experiment Design Issues in Modelling for Control. PhD thesis. Université Catholique de Louvain. Louvain-la-Neuve, Belgium.

Codrons, B., B.D.O. Anderson and M. Gevers (2002). Closed-loop identification with an unstable or nonminimum phase controller. *Automatica* **38**(12), 2127–2137.

Codrons, B., F. De Bruyne, M. Gevers and M. De Wan (1998). Iterative feedback tuning of a nonlinear controller for an inverted pendulum with a flexible transmission. In: *Proceedings of the 1998 IEEE International Conference on Control Applications*. Trieste, Italy. pp. 1281–1285.

De Bruyne, F. (1996). Aspects on System Identification for Robust Process Control. PhD thesis. Université Catholique de Louvain. Louvain-la-Neuve, Belgium.

De Bruyne, F., and L. Kammer (1999). Iterative Feedback Tuning with guaranteed stability. In: *Proceedings of the 1999 American Control Conference.* San Diego, California, USA. pp. 3317–3321.

De Bruyne, F., B.D.O. Anderson and N. Linard (1998). The Hansen scheme revisited. In: *Proceedings of the 37th IEEE Conference on Decision and Control.* Tampa, Florida, USA. pp. 706–711.

de Vries, D.K., and P.M.J. Van den Hof (1995). Quantification of uncertainty in transfer function estimation: A mixed probabilistic–worst-case approach. *Automatica* **31**, 543–558.

Dennis, J.E., and R.B. Schnabel (1983). *Numerical Methods for Unconstrained Optimization and Nonlinear Equations.* Prentice-Hall.

Desborough, L.D., and T.J. Harris (1992). Performance assessment measures for univariate feedback control. *Canadian Journal of Chemical Engineering* **70**, 1186.

DeVries, W.R., and S.M. Wu (1978). Evaluation of process control effectiveness and diagnosis of variation in paper basis weight via multivariate time series analysis. *IEEE Transactions on Automatic Control* **23**(4), 702.

Enns, D. (1984*a*). Model Reduction for Control System Design. PhD thesis. Stanford University. Stanford, California, USA.

Enns, D. (1984*b*). Model reduction with balanced realizations: an error bound and a frequency weighted generalization. In: *Proceedings of the 23rd IEEE Conference on Decision and Control.* Las Vegas, Nevada, USA. pp. 127–132.

Forssell, U. (1999). Closed-Loop Identification: Methods, Theory, and Applications. PhD thesis. Linköping University. Linköping, Sweden.

Forssell, U., and L. Ljung (1999). Closed-loop identification revisited. *Automatica* **35**, 1215–1241.

Francis, B.A. (1987). In: *A Course in H_∞ Control Theory.* Vol. 88 of *Lecture Notes in Control and Information Sciences.* Springer-Verlag. London, UK.

Gangsaas, D., K.R. Bruce, J.D. Blight and U.-L. Ly (1986). Application of modern synthesis to aircraft control: Three case studies. *IEEE Transactions on Automatic Control* **31**, 995–1104.

Georgiou, T.T., and M.C. Smith (1990). Optimal robustness in the gap metric. *IEEE Transactions on Automatic Control* **35**, 673–686.

Gevers, M. (1991). Connecting identification and robust control: a new challenge. In: *Proceedings of the IFAC/IFORS Symposium on Identification and Parameter Estimation.* Budapest, Hungary. pp. 1–10.

Gevers, M. (1993). Towards a joint design of identification and control? In: *Essays on Control: Perspectives in the Theory and its Applications* (H.L. Trentelman and J.C. Willems, Eds.). pp. 111–151. Birkhauser. New York, USA.

Gevers, M., and L. Ljung (1986). Optimal experiment designs with respect to the intended model application. *Automatica* **22**, 543–554.

Gevers, M., R.R. Bitmead and V. Blondel (1997). Unstable ones in understood algebraic problems of modelling for control design. *Mathematical Modelling of Systems* **3**(1), 59–76.

Gevers, M., X. Bombois, B. Codrons, F. De Bruyne and G. Scorletti (1999). The role of experimental conditions in model validation for control. In: *Robustness in Identification and Control* (A. Garulli, A. Tesi and A. Vicino, Eds.). Vol. 245 of *Lecture Notes in Control and Information Sciences.* pp. 72–86. Springer-Verlag. London, UK.

Gevers, M., X. Bombois, B. Codrons, G. Scorletti and B.D.O. Anderson (2003*a*). Model validation for control and controller validation in a prediction error identification framework – Part I: Theory. *Automatica* **39**(3), 403–415.

Gevers, M., X. Bombois, B. Codrons, G. Scorletti and B.D.O. Anderson (2003*b*). Model validation for control and controller validation in a prediction error identification framework – Part II: Applications. *Automatica* **39**(3), 417–427.

Giarré, L., and M. Milanese (1997). Model quality evaluation in H_2 identification. *IEEE Transactions on Automatic Control* **42**(5), 691–698.

Giarré, L., M. Milanese and M. Taragna (1997). H_∞ identification and model quality evaluation. *IEEE Transactions on Automatic Control* **42**(2), 188–199.

Glad, T., and L. Ljung (2000). *Control Theory: Multivariable and Nonlinear Methods.* Taylor & Francis. New York, USA.

Glover, K. (1984). All optimal Hankel-norm approximations of linear multivariable systems and their L_∞ error bounds. *International Journal of Control* **39**, 1115–1193.

Glover, K., and J.C. Doyle (1988). State-space formulae for all stabilizing controllers that satisfy an H_∞ norm bound and relations to risk sensitivity. *Systems and Control Letters* **11**, 167–172.

Goddard, P.J., and K. Glover (1993). Controller reduction: Weights for stability and performance preservation. In: *Proceedings of the 32nd IEEE Conference on Decision and Control.* San Antonio, Texas, USA. pp. 2903–2908.

Goddard, P.J., and K. Glover (1994). Performance-preserving frequency weighted controller approximation: a coprime factorisation approach. In: *Proceedings of the 33rd IEEE Conference on Decision and Control.* Lake Buena Vista, Florida, USA. pp. 2720–2725.

Goodwin, G.C., M. Gevers and B. Ninness (1992). Quantifying the error in estimated transfer functions with application to model order selection. *IEEE Transactions on Automatic Control* **37**, 913–928.

Goodwin, G.C., S.F. Graebe and M.E. Salgado (2001). *Control System Design.* Prentice-Hall. Upper Saddle River, New Jersey, USA.

Gu, G. (1995). Model reduction with relative/multiplicative error bounds and relation to controller reduction. *IEEE Transactions on Automatic Control* **40**, 1478–1485.

Gu, G., and P.P. Khargonekar (1992). A class of algorithms for identification in H_∞. *Automatica* **28**, 299–312.

Hakvoort, R.G., and P.M.J. Van den Hof (1997). Identification of probabilistic system uncertainty regions by explicit evaluation of bias and variance errors. *IEEE Transactions on Automatic Control* **42**(11), 1516–1528.

Hansen, F.R. (1989). A Fractional Representation Approach to Closed Loop System Identification and Experiment Design. PhD thesis. Stanford University. Stanford, California, USA.

Harris, T.J., C.T. Seppala and L.D. Desborough (1999). A review of performance monitoring and assessment techniques for univariate and multivariate control systems. *Journal of Process Control* **9**(1), 1–17.

Helmicki, A.J., C.A. Jacobson and C.N. Nett (1991). Control oriented system identification: A worst-case/deterministic approach in H_∞. *IEEE Transactions on Automatic Control* **36**, 1163–1176.

Hildebrand, R., A. Lecchini, G. Solari and M. Gevers (2004*a*). Asymptotic accuracy of Iterative Feedback Tuning. *IEEE Transactions on Automatic Control.* Accepted for publication.

Hildebrand, R., A. Lecchini, G. Solari and M. Gevers (2004*b*). Optimal prefiltering in Iterative Feedback Tuning. *IEEE Transactions on Automatic Control.* Accepted for publication.

Hildebrand, R., and M. Gevers (2003). Identification for control: Optimal input design with respect to a worst-case ν-gap cost function. *SIAM Journal on Control and Optimization* **41**(5), 1586–1608.

Hjalmarsson, H. (1994). Aspects on Incomplete Modeling in System Identification. PhD thesis. Department of Electrical Engineering, Linköping University. Linköping, Sweden.

Hjalmarsson, H., and L. Ljung (1992). Estimating model variance in the case of undermodeling. *IEEE Transactions on Automatic Control* **37**, 1004–1008.

Hjalmarsson, H., M. Gevers and F. De Bruyne (1996). For model-based control design, closed-loop identification gives better performance. *Automatica* **32**, 1659–1673.

Hjalmarsson, H., M. Gevers and O. Lequin (1997). Iterative Feedback Tuning: Theory and applications in chemical process control. *Journal A* **38**(1), 16–25.

Hjalmarsson, H., M. Gevers, F. De Bruyne and J. Leblond (1994*a*). Identification for control: Closing the loop gives more accurate controllers. In: *Proceedings of the 33rd IEEE Conference on Decision and Control*. Orlando, Florida, USA. pp. 4150–4155.

Hjalmarsson, H., M. Gevers, S. Gunnarsson and O. Lequin (1998). Iterative Feedback Tuning : Theory and applications. *IEEE Control Systems Magazine* **18**(4), 26–41.

Hjalmarsson, H., S. Gunnarsson and M. Gevers (1994*b*). A convergent iterative restricted complexity control design scheme. In: *Proceedings of the 33rd IEEE Conference on Decision and Control*. Orlando, Florida, USA. pp. 1735–1740.

Hjalmarsson, H., S. Gunnarsson and M. Gevers (1995). Model-free tuning of a robust regulator for a flexible transmission system. *European Journal of Control* **1**(2), 148–156.

Huang, B., and S.L. Shah (1999). *Performance Assessment of Control Loops: Theory and Applications*. Advances in Industrial Control series. Springer-Verlag. London, UK.

Ingason, H.T., and G.R. Jonsson (1998). Control of the silicon ratio in ferrosilicon production. *Control Engineering Practice* **6**, 1015–1020.

Kim, S.W., B.D.O. Anderson and A.G. Madievski (1995). Error bounds for transfer function order reduction using frequency weighted balanced truncation. *Systems and Control Letters* **24**, 183–192.

Kosut, R.L. (1995). Uncertainty model unfalsification: A system identification paradigm compatible with robust control design. In: *Proceedings of the 34th IEEE Conference on Decision and Control*. New Orleans, LA, USA.

Kozub, D.J. (1997). Controller performance monitoring and diagnosis: Experiences and challenges. Vol. 93 of *AIChE Symposium Series*. p. 83.

Kozub, D.J., and C.E. Garcia (1993). Monitoring and diagnosis of automated controllers in the chemical process industries. In: *Proceedings of the AIChE Annual Meeting*. St. Louis.

Landau, I.D., A. Karimi, A. Voda and D. Rey (1995*a*). Robust digital control of flexible transmissions using the pole placement/sensitivity function shaping method. *European Journal of Control* **1**(2), 122–133.

Landau, I.D., D. Rey, A. Karimi, A. Voda and A. Franco (1995*b*). A flexible transmission system as a benchmark for robust digital control. *European Journal of Control* **1**(2), 77–96.

Lee, W.S., B.D.O. Anderson, R.L. Kosut and I.M.Y. Mareels (1993). A new approach to adaptive robust control. *International Journal of Adaptive Control and Signal Processing* **7**, 183–211.

Lequin, O., M. Gevers, M. Mossberg, E. Bosmans and L. Triest (2003). Iterative Feedback Tuning of PID parameters: comparison with classical tuning rules. *Control Engineering Practice* **11**(9), 1023–1033.

Liu, K., and R.E. Skelton (1990). Closed-loop identification and iterative controller design. In: *Proceedings of the 29th IEEE Conference on Decision and Control*. Honolulu, Hawaii. pp. 482–487.

Liu, Y., B.D.O. Anderson and U. Ly (1990). Coprime factorization controller reduction with Bezout identity induced frequency weighting. *Automatica* **26**(2), 233–249.

Ljung, L. (1978). Convergence analysis of parametric identification methods. *IEEE Transactions on Automatic Control* **23**, 770–783.

Ljung, L. (1985). Asymptotic variance expressions for identified black-box transfer function models. *IEEE Transactions on Automatic Control* **30**(9), 834–844.

Ljung, L. (1997). Identification, model validation and control. Plenary lecture, 36th IEEE Conference on Decision and Control, San Diego, California, USA.

Ljung, L. (1998). Identification for control – what is there to learn? In: *Workshop on Learning, Control and Hybrid Systems*. Bangalore, India.

Ljung, L. (1999). *System Identification: Theory for the User*. 2nd ed. Prentice-Hall. Upper Saddle River, New Jersey, USA.

Ljung, L. (2000). Model error modeling and control design. In: *CD-ROM Proceedings of the 12th IFAC Symposium on System Identification (SYSID 2000)*. Santa Barbara, California, USA. Paper WeAM1-3.

Ljung, L., and L. Guo (1997). The role of model validation for assessing the size of the unmodelled dynamics. *IEEE Transactions on Automatic Control* **42**(9), 1230–1239.

Maciejowski, J.M. (2002). *Predictive Control with Constraints*. Pearson Education. Harlow, Essex, England, UK.

Mäkilä, P.M., and J.R. Partington (1999). On robustness in system identification. *Automatica* **35**(5), 907–916.

Mäkilä, P.M., J.R. Partington and T.K. Gustafsson (1995). Worst-case control-relevant identification. *Automatica* **31**(12), 1799–1819.

McFarlane, D., and K. Glover (1990). Robust controller design using normalized coprime factor plant descriptions. Vol. 138 of *Lecture Notes in Control and Information Sciences*. Springer-Verlag. Berlin, Germany.

Meyer, D.G. (1988). A fractional approach to model reduction. In: *Proceedings of the 1988 American Control Conference*. pp. 1041–1047.

Meyer, D.G. (1990). Fractional balanced reduction: Model reduction by fractional representation. *IEEE Transactions on Automatic Control* **35**, 1341–1345.

Milanese, M., and M. Taragna (1999). SM identification of model sets for robust control design. In: *Robustness in Identification and Control* (A. Garulli, A. Tesi and A. Vicino, Eds.). Vol. 245 of *Lecture Notes in Control and Information Sciences*. pp. 17–34. Springer-Verlag. London, UK.

Moore, B.C. (1981). Principal component analysis in linear systems: Controllability, observability, and model reduction. *IEEE Transactions on Automatic Control* **26**(1), 17–32.

Ninness, B., H. Hjalmarsson and F. Gustafsson (1999). The fundemental role of general orthonormal bases in system identification. *IEEE Transactions on Automatic Control* **44**(7), 1384–1406.

Nordin, M., and P.O. Gutman (1995). Digital QFT design for the benchmark problem. *European Journal of Control* **1**(2), 97–103.

Obinata, G., and B.D.O. Anderson (2000). *Model Reduction for Control System Design*. Communication and Control Engineering series. Springer-Verlag. London, UK.

Poolla, K., P.P. Khargonekar, A. Tikku, J. Krause and K. Nagpal (1994). A time-domain approach to model validation. *IEEE Transactions on Automatic Control* **39**, 951–959.

Safonov, M.G., and R.Y. Chiang (1988). Model reduction for robust control: a schur relative error method. *International Journal on Adaptive Control and Signal Processing* **2**, 259–272.

Safonov, M.G., and R.Y. Chiang (1989). A Schur method for balanced-truncation model reduction. *IEEE Transactions on Automatic Control* **43**, 729–733.

Safonov, M.G., and T.-C. Tsao (1997). The unfalsified control concept and learning. *IEEE Transactions on Automatic Control* **42**(6), 843–847.

Schelfhout, G. (1996). Model Reduction for Control Design. PhD thesis. Katholieke Universiteit Leuven. Leuven, Belgium.

Schrama, R.J.P. (1992*a*). Accurate identification for control: the necessity of an iterative scheme. *IEEE Transactions on Automatic Control* **37**, 991–994.

Schrama, R.J.P. (1992*b*). Approximate Identification and Control Design. PhD thesis. Delft University of Technology. Delft, The Netherlands.

Sjöberg, J., F. De Bruyne, M. Agarwal, B.D.O. Anderson, M. Gevers, F.J. Kraus and N. Linard (2003). Iterative controller optimization for nonlinear systems. *Control Engineering Practice* **11**(9), 1079–1086.

Skelton, R.E. (1989). Model error concepts in control design. *International Journal of Control* **49**(5), 1725–1753.

Skogestad, S., and I. Postlethwaite (1996). *Multivariable Feedback Control: Analysis and Design*. John Wiley & sons. Chichester, England, UK.

Smith, R.S., and M. Dahleh (1994). *The modelling of uncertainty in control systems*. Vol. 192 of *Lecture Notes in Control and Information Sciences*. Springer-Verlag. Berlin, Germany.

Söderström, T., and P. Stoica (1989). *System Identification*. Prentice-Hall International. Hemel Hempstead, Hertfordshire, UK.

Stanfelj, N., T.E. Marlin and J.F. MacGregor (1993). Monitoring and diagnosing process control performance: the single-loop case. *Ind. Eng. Chem. Res.* **32**, 301.

Tjärnström, F., and L. Ljung (1999). L_2 model reduction and variance reduction. Technical report LiTH-ISY-R-2158, Department of Electrical Engineering, Linköping University, Linköping, Sweden.

Van den Hof, P.M.J., and R.J.P. Schrama (1995). Identification and control – closed-loop issues. *Automatica* **31**, 1751–1770.

Varga, A. (1998). Computation of normalized coprime factorizations of rational matrices. *Systems and Control Letters* **33**, 37–45.

Venkatesh, S.R., and M.A. Dahleh (1997). Identification in the presence of classes of unmodeled dynamics and noise. *IEEE Transactions on Automatic Control* **42**(12), 1620–1635.

Vidyasagar, M. (1984). The graph metric for unstable plants and robustness estimates for feedback stability. *IEEE Transactions on Automatic Control* **29**, 403–417.

Vidyasagar, M. (1985). *Control System Synthesis: A Factorization Approach*. MIT Press. Cambridge, Massachusetts, USA.

Vidyasagar, M. (1988). Normalized coprime factorizations for non-strictly proper systems. *IEEE Transactions on Automatic Control* **33**(3), 300–301.

Vinnicombe, G. (1993*a*). Frequency domain uncertainty and the graph topology. *IEEE Transactions on Automatic Control* **38**, 1371–1383.

Vinnicombe, G. (1993*b*). Measuring the Robustness of Feedback Systems. PhD thesis. Cambridge University. Cambridge, UK.

Vinnicombe, G. (2000). *Uncertainty and Feedback – H_∞ loop-shaping and the ν-gap metric*. Imperial College Press. London, UK.

Wortelboer, P.M.R. (1994). Frequency-Weighted Balanced Reduction of Closed-Loop Mechanical Servo-Systems: Theory and Tools. PhD thesis. Technische Universiteit Delft. Delft, The Netherlands.

Wortelboer, P.M.R., and O.H. Bosgra (1994). Frequency weighted closed-loop order reduction in the control design configuration.

Zang, Z., R.R. Bitmead and M. Gevers (1991). H_2 iterative model refinement and controller enhancement. In: *Proceedings of the 30th IEEE Conference on Decision and Control*. Brighton, UK. pp. 279–284.

Zang, Z., R.R. Bitmead and M. Gevers (1995). Iterative weighted least squares identification and weighted LQG control design. *Automatica* **31**, 1577–1594.

Zhou, K. (1995). A comparative study of H_∞ controller reduction methods. In: *Proceedings of the American Control Conference*. Seattle, Washington, USA. pp. 4015–4019.

Zhou, K., and J. Chen (1995). Performance bounds for coprime factor controller reduction. *Systems and Control Letters* **26**, 119–127.

Zhou, K., and J. Doyle (1998). *Essentials of Robust Control*. Prentice-Hall. Upper Saddle River, New Jersey, USA.

Zhu, Y. (2001). *Multivariable System Identification for Process Control*. Elsevier Science. Oxford, UK.

Zhu, Y.C. (1989). Black-box identification of MIMO transfer functions: Asymptotic properties of prediction error models. *International Journal of Adaptive Control and Signal Processing* **3**, 357–373.

Index

D

E

L

M

N

O

P

Q

R

MIX
Papier aus verantwortungsvollen Quellen
Paper from responsible sources
FSC® C105338

If you have any concerns about our products,
you can contact us on
ProductSafety@springernature.com

In case Publisher is established outside the EU,
the EU authorized representative is:
Springer Nature Customer Service Center GmbH
Europaplatz 3, 69115 Heidelberg, Germany

Printed by Libri Plureos GmbH
in Hamburg, Germany